AF876497

ENVIRONMENTAL POLLUTION: NEW RESEARCH

ENVIRONMENTAL POLLUTION: NEW RESEARCH

RACHEL H. PLATTENBERG
EDITOR

Nova Science Publishers, Inc.
New York

Copyright © 2007 by Nova Science Publishers, Inc.

All rights reserved. No part of this book may be reproduced, stored in a retrieval system or transmitted in any form or by any means: electronic, electrostatic, magnetic, tape, mechanical photocopying, recording or otherwise without the written permission of the Publisher.

For permission to use material from this book please contact us:
Telephone 631-231-7269; Fax 631-231-8175
Web Site: http://www.novapublishers.com

NOTICE TO THE READER

The Publisher has taken reasonable care in the preparation of this book, but makes no expressed or implied warranty of any kind and assumes no responsibility for any errors or omissions. No liability is assumed for incidental or consequential damages in connection with or arising out of information contained in this book. The Publisher shall not be liable for any special, consequential, or exemplary damages resulting, in whole or in part, from the readers' use of, or reliance upon, this material.

Independent verification should be sought for any data, advice or recommendations contained in this book. In addition, no responsibility is assumed by the publisher for any injury and/or damage to persons or property arising from any methods, products, instructions, ideas or otherwise contained in this publication.

This publication is designed to provide accurate and authoritative information with regard to the subject matter cover herein. It is sold with the clear understanding that the Publisher is not engaged in rendering legal or any other professional services. If legal, medical or any other expert assistance is required, the services of a competent person should be sought. FROM A DECLARATION OF PARTICIPANTS JOINTLY ADOPTED BY A COMMITTEE OF THE AMERICAN BAR ASSOCIATION AND A COMMITTEE OF PUBLISHERS.

Library of Congress Cataloging-in-Publication Data

Environmental pollution : new research / Rachel H. Plattenberg, editor.
p. cm.
Includes index.
ISBN 13 978-1-60021-285-7
ISBN 10 1-60021-285-9
1. Pollution--Research. I. Plattenberg, Rachel H.
TD178.5.E58 2006
628.5--dc22 2006018340

Published by Nova Science Publishers, Inc. ✣New York

CONTENTS

PREFACE

Pollution is the release of chemical, physical, biological or radioactive contaminants to the environment. Principal forms of pollution include: air pollution, the release of chemicals and particulates into the atmosphere. Common examples include carbon monoxide, sulfur dioxide, chlorofluorocarbons (CFCs), and nitrogen oxides produced by industry and motor vehicles. Ozone and smog are created as nitrogen oxides and hydrocarbons react to sunlight. Water pollution affects oceans and inland bodies of water. Examples include organic and inorganic chemicals, heavy metals, petrochemicals, chloroform, and bacteria. Water pollution may also occur in the form of thermal pollution and the depletion of dissolved oxygen. Soil contamination often occurs when chemicals are released by spill or underground storage tank leakage. Contaminants include hydrocarbons, heavy metals, MTBE, herbicides, pesticides and chlorinated hydrocarbons. Often occurs with water pollution, thanks to surface runoff and groundwater. Radioactive contamination was added in the wake of 20th-century discoveries in atomic physics. Noise pollution encompasses roadway noise, aircraft noise, industrial noise as well as high-intensity sonar. Light pollution, includes light trespass, over-illumination and astronomical interference. Visual pollution, which can refer to the presence of overhead power lines, highway billboards, scarred landforms (as from strip mining), open storage of junk or municipal solid waste. The nature, distribution and ecological effects of all types and forms of pollutants in air, soil and water are the subject of this new book from Nova Science.

The group of enzymes called B-esterases comprises the hydrolases acetylcholinesterase (AChE), butyrylcholinesterase (BChE) and carboxylestererase (CbE), among other enzymes. For decades, inhibition of AChE in the nervous system or blood has been the most sensitive and rapid biomarker for assessing exposure to, and effect of, organophosphorus (OP) and carbamate (CB) pesticides. Comparison of enzyme activity levels between exposed and non-exposed organisms has been sufficient for conclusion-making related to pesticide impact. Occasionally, chemical analysis of OP or CB residues in abiotic or biotic samples has assisted AChE measurements to give a clear evidence of pesticide contamination. However, other features of B-esterases in relation to pesticide intoxication (blood ChE and CbEs) and field monitoring (e.g., use of oximes) has not been investigated in detail. Chapter 1 summarizes recent experimental and field data on the use of B-esterases as biomarkers of pesticide contamination in a broad group of organisms. Particular concern is focused on plasma BChE and CbE activities. Advantages and limitations of the use of pralidoxime as complementary index of OP exposure are discussed, and a set of considerations are summarized when this methodology is used for assessing wildlife exposure to OP pesticides. The potential use of B-esterases in field biomonitoring and standardized toxicity tests for making ecological risk assessments examined and recommendations are given for obtaining a solid line-of-evidence about pesticide exposure and effect. Although BChE and CbE activities are not directly linked to the acute toxicity of OP and CB agrochemicals, they could contribute to predict the sensitivity of the species to these pesticides: a biochemical endpoint (index of pesticide tolerance) linked to an effect at individual level (identification of species at risk).

Earthworms are widely used to assess the bioavailability and deleterious effects of metals in contaminated soils. Chapter 2 discusses some of the direct and indirect methods that have recently been used to determine the ligand-binding speciation and the extent of the toxicologically-important bio-reactive metal and metalloid fractions in the cells of earthworms as receptor organisms. It proceeds to describe suites of cell-based biomarkers (e.g. morphometrics; neutral red retention time and immuno-competence assays), and their variants, that are more-or-less widely used for reporting chemically-evoked stress in earthworms under laboratory, semi-field, and field exposure conditions. NMR-based metabonomic profiling of the tissues of stressed earthworms is reviewed, and the potential of infra-red microspectroscopy for determining the biochemical profiles of specific cells in different functional states is highlighted. The way that the rapidly expanding genomic database is beginning to inform and propel research designed to further understanding of the fundamental mechanisms underpinning metal trafficking and toxicosis at the level of cells, partly through the provision of immuno-histochemical and in situ hybridization 'tools', is discussed. The chapter concludes with the prediction that the future will see the adoption of high-throughput cellular, as well as molecular-genetic, biomarker techniques in earthworm ecotoxicology, possibly with parallel metal fractionation measurements done as a desirable component of experimental design.

Lead (Pb) contamination at shooting range sites is under increasingly environmental concern. Chapter 3 investigated Pb bullet transformation, Pb contamination, and phosphate-induced Pb immobilization in shooting range soils. Soil samples were collected from three shooting ranges, referred to as SR1, SR2, and SR3 ranges. Mineralogical analysis by powder X-ray diffraction (XRD) and scanning electron microscopy-energy dispersive X-ray spectroscopy (SEM-EDS) revealed that spent Pb bullets was transformed into soluble cerussite ($PbCO_3$) or hydrocerussite ($Pb_3(CO_3)_2(OH)_2$) in both SR1 and SR2 soils, while in the SR3 soil Pb was transformed into insoluble hydroxypyromorphite ($Pb_5(PO_4)_3OH$) due to the presence of high level of soil phosphorus (P). The Pb transformation is proposed to follow a four-stage process: Oxidation, carbonation, dissolution, and precipitation. As a result of Pb transformation, the three soils were heavily contaminated with Pb, with Pb concentrations ranging from 12 to 57 g kg^{-1}. In SR1 and SR2 soils, as much as 68 % and 26 % of Pb was extracted by Toxicity Characteristic Leaching Procedure (TCLP), whereas SR3 soil leached only 5.9 % of Pb by TCLP due to the formation of hydroxypyromorphite. Amendment with phosphate rock (PR), phosphoric acid (PA), or monocalcium phosphate (MCP) to SR1 and SR2 soils significantly reduced TCLP-Pb concentrations. The treatments of PR+PA or MCP reduced TCLP-Pb to be below USEPA for non-hazadous regulatory limit of 5 mg L^{-1} at P:Pb molar ratios ≥ 3.0 after 28-d reaction in both two soils. Lead immobilization in the phosphate-treated SR1 and SR2 soils resulted mainly from the formation of insoluble hydroxypyromorphite at the expense of soluble Pb carbonates. Modeling showed that Pb leachability in the SR3 soil was controlled by hydroxypyromorphite, while cerussite determined Pb leaching in the SR1 and SR2 soils. Upon application of PR+PA or MCP to the

SR1 and SR2 soils, Pb activity was controlled by the formed hydroxypyromorphite. Our study indicated that shooting ranges are significantly contaminated with Pb due to the transformation (weathering) of Pb bullets. The end products of the transformation depend on soil chemical conditions and environment conditions. Application of phosphate is an effective alternative for stabilizing Pb to minimize adverse impacts of Pb to the environment. Among the amendments studied, the treatment combining phosphate rock and phosphoric acid is more cost-effective in immobilizing Pb in shooting range soils and poses a lower risk of secondary P leaching.

Increasing the amount of C in soils may be one method to reduce the concentration of CO_2 in the atmosphere. We measured organic C stored in southern Idaho soils having long term cropping histories that supported native sagebrush vegetation (NSB), irrigated moldboard plowed crops (IMP), irrigated conservation -chisel- tilled crops (ICT) and irrigated pasture systems (IP). The CO_2 emitted as a result of fertilizer production, farm operations and CO_2 lost via dissolved carbonate in irrigation water, over a 30 year period, was included. Net organic C in ecosystems decreased in the order IP>ICT>NSB>IMP. In Chapter 4, if NSB were converted to IMP, 0.15 g C m^{-2} would be emitted to the atmosphere, but if converted to IP 3.56 g C m^{-2} could be sequestered. If IMP land were converted to ICT, 0.95 g C m^{-2} could be sequestered in soil and if converted to IP 3.71 g C m^{-2} could be sequestered. There are $2.6x10^8$ ha of land worldwide presently irrigated. If irrigated agriculture were expanded 10% and the same amount of rainfed land were converted back to native grassland, an increase of $3.4x10^9$ Mg C (5.9% of the total C emitted in the next 30 yr) could potentially be sequestered. The total projected release of CO_2 is $5.7x10^{10}$ Mg C worldwide during the next 30 years. Converting rainfed agriculture back to native vegetation while modestly increasing areas in irrigated agriculture could have a significant impact on CO_2 atmospheric concentrations while maintaining or increasing food production.

Sediments provide habitat for biota, substrate for aquatic plants and hunting ground for predators. They are sinks for nutrients but at the same time they can interact with the many organic and inorganic pollutants that are released to the aquatic systems. As a consequence, sediments are able to concentrate a wide range of chemicals at levels that largely exceed those present in the water column posing a risk to the whole aquatic ecosystem. However, the complex and variable composition of the natural particles makes it very difficult to predict the potential bioaccumulation and harmful effects to biota. It is widely accepted that the bioavailability of sediment-bound contaminants depends on three main factors: the composition and characteristics of the natural particles, the physical and chemical properties of the contaminants, and the behavior and physiological characteristics of the organisms. All these aspects will be reviewed in Chapter 5.

As concerns for sediment quality criteria increase, several approaches have been proposed in literature to understand and clarify the problem, such as the use of several types of formulated control sediments and also the use of artificial particles. For this chapter it was investigated the usefulness of a negatively charged artificial particle (with defined surface properties), in the presence and absence of commercially available humic acids, as an experimental model for natural sediments. Acute bioassays (48 h) were conducted to determine the bioaccumulation of several organic contaminants by the freshwater fingernail clam *Sphaerium corneum* exposed in systems containing either water only, a sample of a natural sediment, or the artificial particles with and without humic acids. Radiolabelled molecules were used to analyze nmol L^{-1} concentrations, octanol-water partition coefficients

(Kow) were applied to investigate the extent of uptake, particle-water partition coefficients (Kd) were calculated to describe aqueous and particulate sources, and dissociation constants were applied to distinguish the contributions of ionized and undissociated forms of the molecules. The results are analyzed to identify the major factors involved in bioaccumulation studies and they show that the negatively charged particles coated with humic acids may be considered as a very good experimental model for the natural particles. The significance of this work is in applying novel features to a better understanding of the role of sediment in the uptake and accumulation of organic contaminants by benthic organisms.

As presented in Chapter 6, aznalcollar pyrite mine is located on the bank of the middle to upper stretch of the Guadiamar River (SW Spain). On 25 April 1998, the walls of the two contiguous ponds containing the ore-processing residues from of this pyrite mine broke open, spilling some 45 x 10^5 m^3 of polluted water and toxic tailings, affecting roughly 46 km^2. Most of the Cu, Zn, and Cd penetrated the soils in the solution phase of the spill (polluted water); meanwhile, the As, Pb and Tl penetrated mostly as part of the solid phase (tailings). Subsequently, drying and aeration of the tailings deposited over the soils rapidly oxidized sulphides to sulphates, lowered the pH, and increased the solubility of the pollutants. These soluble elements infiltrated the soils with the rainwater, swiftly augmenting the soil pollution. Consequently, a rapid removal of the tailings was recommended. In an extensive clean-up operation (May to December 1998) the tailings and 10 cm of the upper soil were removed using heavy machinery, which resulted in a deficient clean-up, as part of the tailings remained mixed with the soil. The oxidation of the leftover tailings lowered the pH and solubilized the contaminants of the cleaned soils. In non-carbonate soils, the pH fell to 3.0, the elements solubilized, rising to the soil surface and forming a crust that lowered the infiltration rate and thus encouraged the spread of the pollutants dissolved in the liquid phase of the runoff water. In carbonate soils, the acidity was largely neutralized ($pH \geq 6.5$), causing the pollutants to precipitate without forming a surface crust and reducing the spread of pollutants by the liquid phase of the runoff water. Nevertheless, the soil-erosion rate increased, intensifying the spread of the pollutants in the solid phase of the runoff. In both soil types, though for different reasons, 15 months after spill, the pollutants had not penetrated deeper than 10 cm. This implies that the speed of the clean-up was not as critical as a thorough removal of the spill without leaving behind remains of tailings mixed with soils. In addition, a random systematic sampling and analysis of the soils at 3 depths (0-10 cm, 10-30 cm, and 30-50 cm) at the end of 1998 (just after of the clean-up), enabled an estimate of the contamination level of the soils before and after the spill. The soils were contaminated prior to the accident, which raised the concentration in the first 10 cm of 60% of the soils. In 24% of the soils, the contaminant infiltrated to 10-30 cm in depth and only in the 12% reached 30-50 cm in depth. Because of the heavy pollution of the soils after tailings were removed, different remediation measures were taken (new cleaning of the most polluted soils, liming with sugar-refinery scum, addition of organic matter plus iron-rich clayey materials, and tilling the upper 20-25 cm of the soils). The analysis of the soils after each cleaning and remediation operation revealed that, although the measures were effective in reducing the concentration of contaminating elements in the soils, 6 years after the spill, 65% of the soils analysed continue to exceed the level of 50 mg $kg^{-1}$of As and 30% double this concentration.

Over the course of many years different methods for the extraction of solid phase environmental samples have been developed. Traditionally these methods have been concerned with maximising analyte recovery with a view to determining the total contaminant concentration. However, in light of the body of knowledge relating to compound availability and ageing these extraction methods are unsuitable in determining the fraction that may be bioavailable. It is evaluation of available contaminant concentrations that are most significant with respect to meaningful contaminated land assessment under present day risk based frameworks of legislation. However, to date no review has been made of extraction methodologies of solid phase environmental samples despite this shift in paradigm. Chapter 7 appraises both traditional exhaustive techniques and more contemporary non-exhaustive techniques against this back drop of evolving perspectives in contaminated land assessment and legislation. The review highlights the variability in techniques considered to be exhaustive thereby questioning their value in meaningful assessment of contaminated land. In complement the review draws comparison between non-exhaustive extraction techniques and biological endpoints, specifically compound biodegradability, that supports the use of certain non-exhaustive extraction methods for inference of risk and estimation of bioremediation potential.

The Strait of Gibraltar and the Alborán Sea are the only connection between the Atlantic Ocean and the Mediterranean Sea. Intense shipping activities occur in the area, including massive transport of chemicals and oil. Thus, it is relevant to have a dispersion model that can be used in an emergency situation after an accident, to help the decision-making process. Such dispersion models require a deep knowledge of the physical oceanography of the region of interest, with detailed simulations of tides and residual (average) circulation. In Chapter 8, a particle-tracking dispersion model that can be used to simulate the dispersion of oil, radioactive and chemical substances in the system Strait of Gibraltar-Alborán Sea is described. Tides are simulated using a barotropic model and for the average circulation a reduced-gravity model is applied. This model is able to reproduce the main features of the Alborán circulation (the well known western Alborán gyre). The dispersion model is run off-line, using previously computed tidal and residual currents. The contamination patch is simulated by a number of particles whose individual paths are computed, diffusion being modelled using a Monte Carlo method. Contaminant concentrations may be obtained from the density of particles. Specific processes for each contaminant have to be included (for instance radioactive decay or buoyancy and biodegradation of oil) and they are discussed in detail. Several examples of dispersion computations for several contaminants and under different wind and circulation conditions are presented. It is interesting to note that, under certain conditions, contaminants are trapped in the gyre.

In: Environmental Pollution: New Research
Editor: Rachel H. Plattenberg, pp. 1-45
ISBN: 1-60021-285-9
© 2007 Nova Science Publishers, Inc.

Chapter 1

ECOTOXICOLOGICAL PERSPECTIVES OF B-ESTERASES IN THE ASSESSMENT OF PESTICIDE CONTAMINATION

Juan C. Sanchez-Hernandez
Laboratory of Ecotoxicology, University of Castilla-La Mancha,
Avda. Carlos III s/n, 45071, Toledo, Spain

Abstract

The group of enzymes called B-esterases comprises the hydrolases acetylcholinesterase (AChE), butyrylcholinesterase (BChE) and carboxylestererase (CbE), among other enzymes. For decades, inhibition of AChE in the nervous system or blood has been the most sensitive and rapid biomarker for assessing exposure to, and effect of, organophosphorus (OP) and carbamate (CB) pesticides. Comparison of enzyme activity levels between exposed and non-exposed organisms has been sufficient for conclusion-making related to pesticide impact. Occasionally, chemical analysis of OP or CB residues in abiotic or biotic samples has assisted AChE measurements to give a clear evidence of pesticide contamination. However, other features of B-esterases in relation to pesticide intoxication (blood ChE and CbEs) and field monitoring (e.g., use of oximes) have not been investigated in detail. This chapter summarizes recent experimental and field data on the use of B-esterases as biomarkers of pesticide contamination in a broad group of organisms. Particular concern is focused on plasma BChE and CbE activities. Advantages and limitations of the use of pralidoxime as a complementary index of OP exposure are discussed, and a set of considerations are summarized when this methodology is used for assessing wildlife exposure to OP pesticides. The potential use of B-esterases in field biomonitoring and standardized toxicity tests for making ecological risk assessments are examined and recommendations are given for obtaining a solid line-of-evidence about pesticide exposure and effect. Although BChE and CbE activities are not directly linked to the acute toxicity of OP and CB agrochemicals, they could contribute to predict the sensitivity of the species to these pesticides: a biochemical endpoint (index of pesticide tolerance) linked to an effect at the individual level (identification of species at risk).

Introduction

Inhibition of acetylcholinesterase (AChE) activity by organophosphorus (OP) and carbamate (CB) pesticides has been one of the most universal phenomenons demonstrated in toxicology. This enzyme is responsible for the normal functioning of the synapses of the central and autonomic nervous systems, as well as at the neuromuscular junctions. Acetylcholinesterases belong to a group of hydrolases classified by Aldridge [1] as B-esterases, which are inhibited by OPs and CBs, so called anti-cholinesterase (anti-ChE) pesticides. For decades, the AChE inhibition has been the principal biological response measured in organisms exposed to anti-ChE pesticides. These agrochemicals display a relatively low persistence in the environment and within organisms, therefore their detection by chemical analysis can resultin an underestimated measure of bioavailability. The measurement of AChE inhibition has become an alternative biochemical analysis for detecting the presence of anti-ChE pesticides, and to assess their bioavailability and toxicity.

One of the challenges in the ecotoxicology of B-esterases has been the establishment of a relationship between the degree of brain AChE inhibition and mortality. Thus, Ludke et al. [2] suggested that an inhibition of brain AChE exceeding 20 % in birds indicated exposure, whereas an inhibition greater than 50 % is indicative of mortality due to OPs. Likewise, Hall and Clark [3] reported similar limits of exposure and effects of OP pesticides in reptiles. They found that a 40 % of brain AChE inhibition is a level of pesticide exposure, while an inhibition degree higher than 50 % lead to death. In mammals, inhibition of brain AChE exceeding 50 % suggest sublethal exposure and mortality is associated to an inhibition degree higher than 80 % [4]. Fulton and Key [5] proposed on the basis of available data in the literature that fish AChE inhibition higher than 70 % is related to mortality due to OP pesticides. However, such a relationship was not possible to establish for invertebrates because of high variability of responses [5].

Butyrylcholinesterase (BChE) and carboxylesterase (CbE) activities are other two common B-esterases that currently have gained increasing concern in field monitoring of pesticide pollution. Although they are not directly related to toxicity of anti-ChE pesticides, the non-destructive nature of its analysis and their high sensitivity to anti-ChE pesticides have led to their popularity as biomarkers of pesticide exposure. Furthermore, these B-esterases have a noteworthy role in pesticide detoxification. Carbamate insecticides are primarily hydrolyzed by CbEs, whereas circulating anti-ChE pesticides can bind to blood BChE and CbEs. Despite their importance in pesticide detoxification, there is a paucity of ecotoxicological studies concerning this perspective of CbE and BChE activities. We could ask the following questions: what would be the contribution of measurements of BChE or CbE activity levels in the relationship “mortality and brain AChE inhibition” in an ERA process?, or would it be convenient to consider a relationship “survival and B-esterase activity levels” as alternative approach in the exposure and effect assessments?.

This contribution does not attempt to give an exhaustive compilation of the available data on B-esterase responses to pesticide exposure in a comparative approach, but its purpose is going over the main aspects related to practical and ecotoxicological concerns of BChE and CbEs. Based on experimental evidences, it is suggested that the measurement of these B-esterases should be included in monitoring programs for assessing pesticide impact on biota. They could contribute to identify, at the regional scale, the species at risk or the most tolerant species to pesticide contamination. Finally, future lines of research are suggested to increase the understanding of B-esterases in pesticide tolerance in non-target organisms and their implications in the assessment of wildlife exposure to anti-ChE agrochemicals.

Functions and Properties of B-esterases

The International Union of Biochemistry applies the term 'esterases' to hydrolases that act on the ester link. They are divided, in turn, into subclasses according to the type of ester bond that hydrolyze (e.g., carboxylester, phosphoric ester, thiolic ester, sulphuric ester). On the other hand, Prof. Aldridge [1] classified the esterases in function of their interaction with OP compounds. Thus, A-esterases are able to hydrolyze OPs, B-esterases are inhibited by them, and C-esterases do not exert any interaction with these chemicals. Currently, this classification continues to be used by most of the researchers. In mammals, there are two types of B-esterases: cholinesterases (ChE) and carboxylcstcrases (CbE). The former comprises acetylcholinesterases (AChE) and butyrylcholinesterases (BChE, EC 3.1.1.8), whereas mammalian CbEs include a number of isoenzymes of similar physicochemical and immunological characteristics and with a low substrate specificity [6]. Summarized next are those main functions and properties of ecotoxicological concern described for B-esterases.

Acetylcholinesterase (EC 3.1.1.7). This is a key element in the normal functioning of cholinergic synapses of the nervous system and neuromuscular junction. The neurotransmitter acetylcholine released at the presynaptic terminal is efficiently hydrolyzed by AChE thereby avoiding a persistent stimulation of cholinergic receptors on the postsynaptic cell. In addition to this classical function, other biological functions have been attributed to AChE such as the proliferation and differentiation of nerve cells [7], or the development and maintenance of retina [8].

Acute toxicity of OP and CB pesticides is primary linked to inhibition of AChE activity in the nervous tissue, which can be described in a two-steps process [9]. Organophosphates react with the serine hydroxyl moiety at the active site of the enzyme to give an enzyme-inhibitor complex (Fig. 1A). The moiety between the P atom and the X group is then weakened, and the leaving group releases. This phosphorylated stage of the enzyme is highly stable and, depending on the groups attached to the P atom, the inhibition can be irreversible. If dealkylation or the removal of one alkyl group (R or R') from the phosphoryl moiety takes place, then the enzyme activity cannot be reversed. This stage is known as aging, and the enzyme activity cannot be reactivated even under treatment with reactivating substances (e.g., oximes). The mechanism of AChE inhibition by CBs is similar to that described by OPs (Fig. 1B), but the carbamylated enzyme is very unstable and the enzyme activity is spontaneously reversed in the presence of nucleophilic molecules such as H_2O [10]. This acute toxicity mechanism of anti-ChE pesticides has led to the popularity of AChE as a biomarker of pesticide exposure and effect. Many reviews and monographs have discussed the good and well-known qualities of AChE activity as a biomarker of pesticide contamination in a broad range of organisms [5, 11-13].

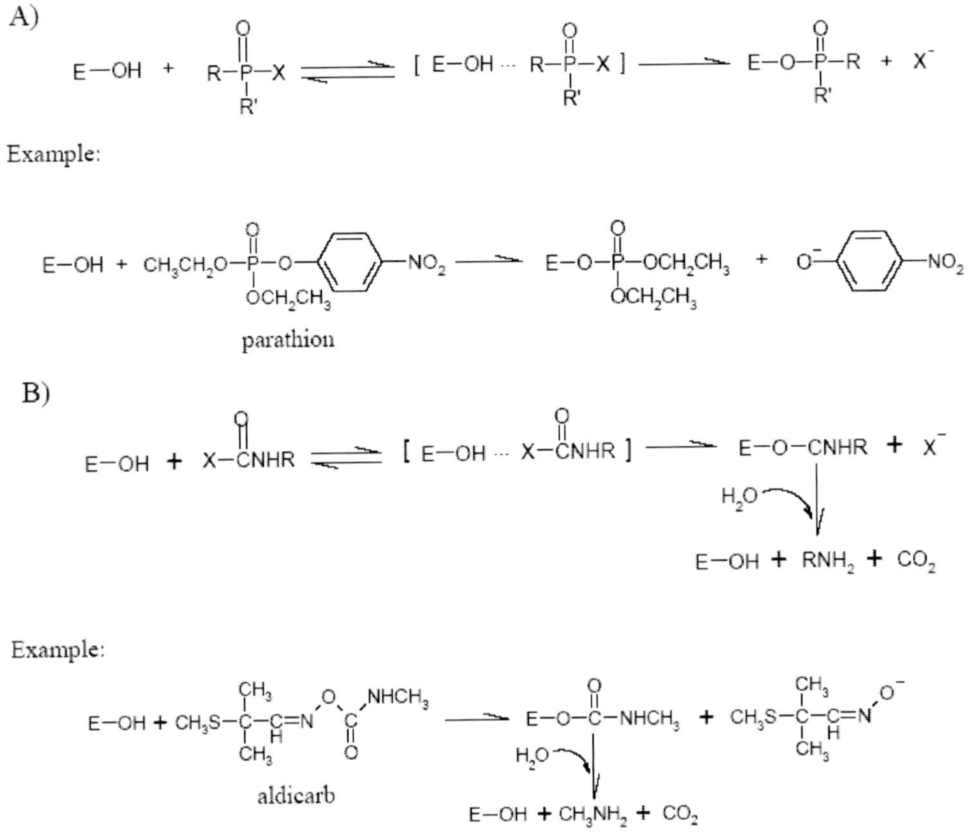

Figure 1. Proposed mechanism of acetylcholinesterase (E-OH) inhibition by organophosphorus (A) and carbamate (B) pesticides. Adapted from Fukuto [9].

Butyrylcholinesterase (EC 3.1.1.8). Butyrylcholinesterase, frequently called non-specific esterase or pseudocholinesterase, hydrolyzes a number of cholinesters such as butyrylcholine, acetylcholine, propionylcholine or benzoylcholine. This enzyme is found throughout the animal kingdom, and it is particularly abundant in blood of many vertebrate species. Inhibition of BChE activity by OP and CB compounds takes place in the same way like for AChE (Fig. 1). From an enzimological viewpoint, BChE activity has been distinguished from AChE activity attending mainly to their substrate preferences and specific inhibitors (Table 1). Historically, no physiological function has been attributed to BChE, being a matter of continue discrepancy [14,15]. However, BChE seems to be involved in a vast variety of biological functions. For example, there is solid evidence supporting a participation of BChE activity, together to AChE, in the neuronal proliferation and differentiation [7]. Human serum BChE activity is able to inactive deoctanoylation of ghrelin, a peptide that stimulates the growth hormone release from the pituitary gland [16]. In the central nervous system, BChE activity can hydrolyze the neurotransmitter acetylcholine when AChE activity is inhibited or absent [17]. Recently, it has been examined the available data supporting a role of ChEs in the tumorogenesis [18]. Their anomalous expression seems to account for cancer growth and development. Furthermore, variations in BChE, and AChE, activity levels are highly dependent on cancer type. However, the absence of physiological abnormalities in individuals with an inactivating mutation in the BChE gene, among other experimental evidences, has led to consider the mammalian BChE as a vestigial enzyme [14].

Table 1. Enzimological properties for distinguishing mammalian acetylcholinesterase (AChE) from butyrylcholinesterase (BChE) activities.

Cholinesterase	Substrate specificity	Common selective inhibitors	Others
AChE	Hydrolyze preferentially AcSCh and does not split BuSCh. But Ac□MeSCh is believed to be a specific substrate for AChE.	BW284C51 and GD-42	Inhibition of AChE activity occurs at high substrate concentrations. No ions are essential for activity, although Ca^{+2} and Mg^{+2} can enhance the AChE activity.
BChE	Hydrolyze BuSCh at a higher rate than other cholinesters. It does not hydrolyse Ac□MeSCh.	Iso-OMPA*, DFP, and ethopropazine.	It is not inhibited at high substrate concentrations. No ions are required for activity.

AcSCh = acetylthiocholine; Ac□MeSCh = acetyl-□-(methyl)thiocholine; BuSCh = butyrylthiocholine; Iso-OMPA = tetraisopropyl pyrophosphoramide; DFP = diisopropylfluorophosphate; BW284C51 = 1,5-bis(4-allyldimethyl-ammonimphenyl)pentan-3-one dibromide.

* Also known as *NN'*-diisopropylphosphorodiamidic anhydride, which abbreviation is DPDA [27].

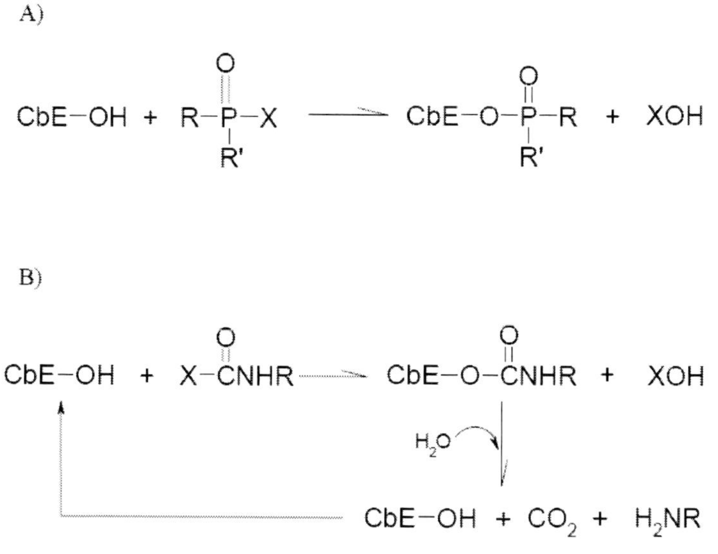

Figure 2. Proposed mechanism of carboxylesterase (CbE-OH) inhibition by organophosphorus (A) and carbamate (B) pesticides. Carbamylated CbE is spontaneously reactivated in the presence of water, and the carbamic acid formed is rapidly decomposes in CO_2 and an amine; whereas phosphorylated CbE is irreversibly inhibited. In both type of interactions an alcohol (XOH) is released. Adapted from Sogorb and Vilanova [10].

The purified human BChE activity is a potent bioscavenger against OP poisoning in mammals [19]. This property could explain, partially, the tolerance of certain non-target organisms to OP intoxication [20]. In a laboratory experiment, Sanchez et al. [21] observed that inhibition of brain AChE activity of lizards treated with parathion took place when serum BChE activity was strongly inhibited. A 'buffer' capacity was attributed to lizard serum BChE activity, which leads to minimize the OP toxic effects at the brain AChE. Yet, there is not available data to date that demonstrate a significant role of blood BChE activity in the tolerance to OP pesticides in non-target vertebrates.

Carboxylesterases (EC 3.1.1.1). This group of B-esterases – so called aliesterases – is able to hydrolyze esters of carboxylic acids [10]. Carboxylesterases participate in the detoxification of pesticides by two ways: i) hydrolysis of ester bonds (CBs, pyrethroids and those OPs that contain this type of bond) and ii) binding of OPs and CBs to CbE (inhibition). Carboxylesterases have a serine residue in their active site which is inhibited by OP and CB pesticides in the same way than AChE (Fig. 2); inhibition by CBs (e.g., eserine) occurs at high concentrations. Carbamylated CbE is spontaneously reactivated, whereas the phosphorylation of CbE implies its irreversible inhibition. Although, Jokanovic et al. [22], and later Maxwell and Brecht [23] found a rapid recovery of plasma and liver CbE activity of rats treated with war agents (e.g., soman, sarin, VX) and OP pesticides. However, the reactivation rates seem related to the size of the OP and the pH of the reaction medium [23]. Inhibition of CbEs by anti-ChE pesticides is also considered a detoxification mechanism because of decrease of circulating pesticide and consequently reduction of the impact on brain AChE activity (target for acute toxicity). Carboxylesterases take part, together with other enzymatic systems, in the development of insect tolerance to OPs. In relation to this assumption, a number of studies have documented that inhibition of CbEs by OPs results in an increase of toxic effects by other anti-ChE pesticides such as CBs [24]. Like BChE activity, the physiological function of CbEs remains still uncertain, although they are involved in the metabolism of drugs, lipids and steroids in mammals [6,25].

Ecotoxicological Studies Involving B-esterases as Biomarkers

A. Aquatic organisms

Cholinesterases. These B-esterases have been extensively measured in aquatic organisms for monitoring the impact of OP and CB pesticides. Research on ChE responses of fish and aquatic invertebrates to anti-ChE pesticides has been constantly reviewed [5,12,26]. Most of the recent studies on ChEs of aquatic organisms have been prompted to characterize kinetic properties and tissue distribution of ChEs (Table 1). The use of specific substrates and inhibitors typical for distinguishing mammalian AChE from BChE activities, have enabled to perform this type of enzimological studies. In general, eserine (formulated as eserine hemisulfate) is used to confirm the presence of cholinesterases, because this carbamate is a specific inhibitor of cholinesterase. The substrate acetyl-□-(methyl)thiocholine (Ac□MeSCh) is specific for AChE activity, while the substrate butyrylthiocholine (BuSCh) is believed to be selective to BChE activity. On the other hand, the use of specific inhibitors has also enabled to distinguish AChE from BChE [27]. Thus, AChE activity is inhibited in the presence of 1,5-bis(4-allyldimethyl-ammonimphenyl)pentan-3-one dibromide (BW284C51), whereas BChE activity is specifically inhibited by tetraisopropyl pyrophosphoramide (iso-OMPA). With some reservations, most of researchers have used these selective substrates and inhibitors to characterize ChE activity in aquatic organisms. However, a single ChE of these organisms often show enzimological properties of both mammalian ChEs (Table 2). In fish, several authors have found atypical responses of BChE to classic substrates and inhibitors (Table 2). For example, Monteiro et al. [28] found that BChE activity was inhibited by BW284C51, a specific inhibitor of AChE activity. Sturm et al. [29] also reported an inhibition of muscle BChE activity of several fish by the specific inhibitor of AChE activity. This result also was observed in the muscle BChE activity of *Oreochromis niloticus* [30], and in the fish *Dicentrarchus labrax* [31]. The methodological approach by which of non-mammalian ChEs is distinguished is thereby questioned. For example, Bocquené et al. [32] described in different tissues of the common oyster *Crassostrea gigas* an insensitive AChE to OP and CB inhibitory effects, coexisting with a sensitive AChE. Furthermore, atypical behaviours of both oyster ChEs were observed against treatment with substrate excess or specific inhibitors. It is concluded from this study that typical classification of mammalian ChEs into AChE and BChE should not be applied for invertebrate ChEs.

In invertebrates, ChE activity has been detected and characterized for a wide range of organs or tissues (e.g., gills, digestive gland, and muscle) and even fluids (haemolymp). In general, extensive variation in ChE biochemistry exists among tissues and species (Table 2), and it would be unwise to make generalizations about ChE typology. This justify that enzimological characterization of ChEs should be a previous step to their use in biological surveys of pesticide pollution. Otherwise, there exists a high risk for making errors in the interpretation of ChE activity levels, especially when unspecific substrates are used for ChE determinations (e.g., acetylthiocholine) or when both type of ChEs (AChE and BChE) occur in the same tissue. The interpretation errors can increase still whether measurements of ChE activity is made in the whole organisms and ChEs show different sensitivity to OP/CB pesticides.

Although the tissue distribution and enzimological properties of ChEs is an important step for using them as index of OP/CB exposure, another not less important aspect is to investigate the sensitivity of ChEs to these pesticides. For example, Chuiko [33] found that serum BChE activity of six cyprinid species was more sensitive to *in vitro* inhibition by dichlorvos ($pI_{50} \sim 9$) than brain AChE activity (pI_{50} = 5–6). Wogram et al. [34] found that liver and muscle BChE activity of *Gasterosteus aculeatus* exposed to 1 μg L^{-1} of parathion for 1 h was significantly inhibited, while brain AChE activity was not affected in all analysed tissue. The muscle AChE activity of *Catostomus occidentalis* resulted more sensitive to diazinon exposure (5, 10 and 50 μg L^{-1} treatments for 96 h) than brain AChE activity [35]. These studies lead inevitably to reflect in the possibility of a protective effect of BChE activity against OP intoxication; a property well documented for plasma BChE activity of laboratory mammals (see section *B-esterases as pesticide scavengers*).

Table 2. Enzymological characterization of cholinesterases according to preferential substrates for hydrolysis and sensitivity to inhibitors.

Species	Tissue or organ	Specific substrates	Specific ChE inhibitors	Main remarks on ChE characterization	Ref.
Molluscs					
Ostrea edulis and *Mytilus galloprovincialis*	Gills	AcSCh, AcβMeSCh, BuSCh, PrSCh	Eserine, Iso-OMPA, BW284C51	– ChE decreased at high substrate concentrations (> 1 mM), and it was drastically inhibited by eserine (IC_{50}s < 10^{-7} M).	(143)
Crassostrea rhyzophorae	Gills	AcSCh	Eserine, iso-OMPA	– ChE activity measured in the S9 fraction of oyster gills showed inhibition at high substrate concentration (> 3.5 mM), was not inhibited by iso-OMPA and it displayed inhibition by eserine (IC_{50} = 0.91 x 10^{-6} M). – It was suggested that gill ChE of *C. rhyzophorae* is an AChE.	(144)
Mytilus edulis	Gill, posterior adductor muscle, foot and digestive gland	AcSCh, AcβMeSCh, BuSCh, PrSCh	Iso-OMPA, BW284C51	– High levels of ChE activity was observed in the mitochondrial fraction (7.000 X g for 10 min) of foot and in the microsomal fraction (105.000 X g for 100 min) of gill. Measurement of ChE activity in the microsomal fraction of gills is suggested for field monitoring of OP contamination. – ChE activity in mussels tissues appear to be an AChE because its preferential substrates AcSCh and Ac□MeSCh, no apparent inhibition by iso-OMPA (10 and 100 x 10^{-6} M), and inhibition by BW284C51 (10 and 100 x 10^{-6} M).	(145)

Table 2. Continued

Species	Tissue or organ	Specific substrates	Specific ChE inhibitors	Main remarks on ChE characterization	Ref.
M. galloprovinciales and *Corbicula fluminea*	Whole body, gills, digestive gland, muscle and mantle	AcSCh, BuSCh, PrSCh	–	– AcSCh (2 mM) was more efficiently hydrolyzed by ChE activity of *M galloprovincialis*, whereas ChE of *C. fluminea* hydrolyzed preferentially the substrate PrSCh (5 mM). – ChE inhibition by the substrate was not observed (up to 20 mM of AcSCh or PrSCh). – Maximum ChE activity was found in the gills of *M. galloprovinciales*, whereas the highest ChE activity of *C. fluminea* was measured in the mantle.	(146)
M. edulis	Haemolymph, whole body	AcSCh, BuSCh, PrSCh	Eserine, iso-OMPA, DFP	– AcSCh was the preferred substrate for haemolymph ChE activity, although hydrolysis of BuSCh was also observed. Conversely, BuSCh and PrSCh were preferentially breakdown by ChE from tissue homogenate. – Haemolymph ChE showed properties of vertebrate AChE: it hydrolyzed preferentially AcSCh, high substrate concentrations (> 6 mM) inhibited ChE activity, and it was completely inhibited by eserine (IC_{50} < 0.01 mM) and not by iso-OMPA. – ChE from tissue homogenate preferred BuSCh and PrSCh as substrates, and it was inhibited (77 % inhibition) by iso-OMPA (3×10^{-3} M).	(40)

Table 2. Continued

Species	Tissue or organ	Specific substrates	Specific ChE inhibitors	Main remarks on ChE characterization	Ref.
Crustaceans					
Palaemon serratus	Eyes	AcSCh, AcβMeSCh, BuSCh, PrSCh	Eserine, iso-OMPA	– Eyes of the common prawn posses only a ChE (k_i = 1.3 x 10^{-6} M^{-1} min^{-1} for eserine) with enzymatic properties of mammalian AChE (inhibition at high substrate concentration, no inhibition by iso-OMPA, inactive to BuSCh, and high hydrolysis rate with AcSCh).	(147)
Tigriopus brevicornis	Whole organism	AcSCh, BuSCh, PrSCh	Eserine, iso-OMPA	– *T. brevicornis* posses one only ChE showing properties of mammalian AChE: it hydrolysed AcSCh at a higher rate than BuSCh and PrSCh; its activity decreased at high AcSCh concentrations (< 7 mM); it was inhibited by eserine (k_i = 2.5 x 10^{-6} M^{-1} min^{-1}) but not by iso-OMPA (*ki* = 6.1 x 10^{-1} M^{-1} min^{-1}).	(148)
Eurytemora affinis	Whole organism	AcSCh, BuSCh, PrSCh	Eserine, iso-OMPA	– One ChE activity was purified and characterized in this copepod species. It showed properties of typical mammalian AChE activity: it hydrolysed AcSCh at a higher rate than BuSCh and PrSCh; its activity notably decreased at high substrate concentrations (< 10 mM); it was inhibited by eserine (k_i = 1.7 x 10^{-6} M^{-1} min^{-1}) but not by iso-OMPA (*ki* = 8.3 x 10^{-1} M^{-1} min^{-1}).	(149)

Table 2. Continued

Species	Tissue or organ	Specific substrates	Specific ChE inhibitors	Main remarks on ChE characterization	Ref.
Palaemonetes pugio	Whole organism	AcSCh, AcβMeSCh, BuSCh, PrSCh	Eserine, Iso-OMPA, BW284C51	– ChE activity displayed the main properties that characterize the AChE activity: i) high rate of AcSCh hydrolysis, minimum hydrolysis of BuSCh, ii) strong inhibition by eserine (10^{-5} M) and BW284C51 (10^{-6} M) using the substrates AcSCh, AcβMeSCh, and PrSCh, iii) no inhibition by iso-OMPA (10^{-4} M), and iv) hydrolysis of AcβMeSCh.	(150)
Artemia salina and *A. parthenogenetica*	Whole organism	AcSCh, BuSCh, PrSCh	Eserine, Iso-OMPA, BW284C51	– The substrate PrSCh was preferred by ChEs of *A. parthenogenetica* and *A. salina*, the latter at substrate concentrations were higher than 1.28 mM. – No significant ChE inhibition by excess of substrate was observed. – Eserine inhibited ChE activity from both *Artemia* species at concentration higher than 6.25 x 10^{-6} M. – Iso-OMPA concentrations higher than 0.5 x 10^{-3} M inhibited ChE activity of *A. parthenogenetica*, but did not cause inhibition of *A. salina* ChE activity. Likewise, BW284C51 inhibited ChE activity of *A. salina* (> 25 x 10^{-6} M) and *A. parthenogenetica* (> 800 x 10^{-6} M).	(151)

Table 2. Continued

Species	Tissue or organ	Specific substrates	Specific ChE inhibitors	Main remarks on ChE characterization	Ref.
Equinoid					
Paracentrotus lividus	Ambulacral podia	AcSCh, BuSCh, PrSCh	Eserine, Iso-OMPA, BW284C51	– The ChE characterized in the ambulacral podia of this organism showed features of both mammalian ChEs: AcSCh and PrSCh were the preferential substrate for ChE activity, while BuSCh was poorly hydrolyzed. There was a ChE inhibition at high substrate concentration (2.6 mM), and furthermore it was completely inhibited by 50 x 10^{-6} M of BW284C51 (91.9 % of inhibition) and by 8 x 10^{-3} M iso-OMPA (73.05 %).	(152)
Fish					
Pomatoschistus microps	Head	AcSCh, AcβMeSCh, BuSCh, PrSCh	Eserine, Iso-OMPA, BW284C51	– ChE was inhibited by eserine. – Presence of both ChEs, but AChE contributed to 95 % of total ChE activity. – BChE activity was inhibited by BW284C51 (a specific inhibitor of AChE activity).	(28)
Limanda limanda *Platychthys flesus* *Serranus cabrilla*	Brain Muscle	AcSCh, AcβMeSCh, BuSCh, PrSCh	Eserine, Iso-OMPA, BW284C51	– Total ChE activity in brain was exclusively AChE (inhibition by BW284C51 and resistant to iso-OMPA). – Presence of both ChEs in muscle tissue, but BChE showed atypical properties (hydrolysis of AcβMeSCh, inhibition by high substrate concentrations and BW284C51).	(29)

Table 2. Continued

Species	Tissue or organ	Specific substrates	Specific ChE inhibitors	Main remarks on ChE characterization	Ref.
Oreochromis niloticus	Brain Muscle Live	AcSCh, BuSCh, PrSCh	Eserine, Iso-OMPA, BW284C51	– ChE was inhibited by eserine (2 x 10^{-6} M). – AChE was present in brain. An atypical BChE was suggested in muscle and liver because its activity was decreased in the presence of BW284C51.	(30)
Dicentrarchus labrax	Brain Muscle	AcSCh, BuSCh, PrSCh	Eserine, Iso-OMPA, BW284C51	– ChEs of both tissues hydrolyzed preferentially AcSCh and they were inhibited by eserine (10^{-6} M). Brain ChE activity was attributed to AChE, while BChE activity seemed to be present in muscle tissue, although it showed an inhibition to BW284C51 (a selective inhibitor of mammalian AChE).	(31)
Fish belonging to: *Coregonidae* (1 specie), *Cyprinidae* (7 species), *Esocidae* (1 specie) and *Percidae* (2 species).	Brain Serum	AcSCh, BuSCh, PrSCh	No selective inhibitors were used	– Serum AChE activity was found in *Esocidae*, *Percidae* and in the common carp (*Cyprinidae*). – Serum BChE activity was observed only in cyprinids except the common carp. Serum BChE activity was much more sensitive to dichlorvos inhibition ($pI_{50} \sim 9$) than brain AChE activity ($pI_{50} \sim 5$).	(33)
Danio rerio	Whole embryos	AcSCh, AcβMeSCh, BuSCh, PrSCh	Eserine, Iso-OMPA, BW284C51	– ChE activity was attributed to AChE, however no inhibition of enzyme activity was found at high AcSCh concentration (30 mM). – Presence of non-specific esterases able to hydrolyse the substrates and are not inhibited by eserine, iso-OMPA or BW284C51 (50 x 10^{-6} M).	(42)

Table 2. Continued

Species	Tissue or organ	Specific substrates	Specific ChE inhibitors	Main remarks on ChE characterization	Ref.
Catostomus occidantalis	Brain Muscle	AcSCh	Iso-OMPA, BW284C51	– AChE activity was the only ChE presents in the brain and muscle. – Muscle AChE appeared to be more sensitive to inhibition by the OP diazinon than brain AChE activity.	(35)
Gasterosteus aculeatus	Brain Muscle	AcSCh, AcβMeSCh, BuSCh, PrSCh	Eserine, Iso-OMPA	– AChE activity is present in brain and it showed some typical properties of mammalian AChE: it cleaved AcSCh but not BusCh, and it was not affected by iso-OMPA – ChE activities in muscle were attributed to both AChE and BChE. However, BChE activity showed atypical properties in comparison to mammalian BChE: it also was able to hydrolyse the substrate AcβMeSCh (selective for AChE).	(153)
G. aculeatus	Brain Muscle Liver Gills	AcSCh, BuSCh	Iso-OMPA	– AChE was not detected in liver, and BChE activity was not found in brain. – BChE was inhibited by parathion exposure, while AChE activity was not affected in all tissues.	(34)
G. aculeatus	Muscle	AcSCh, AcβMeSCh, BuSCh, PrSCh	Iso-OMPA, BW284C51	– It is suggested the presence of AChE and an atypical BChE activity in the muscle of this species. However, some atypical ChE properties were observed: i) the specific AChE inhibitor BW284C51 inhibited the ChE activity using the three substrates, ii) the specific BChE inhibitor iso-OMPA inhibited completely the ChE activity on BuSCh but also on AcSCh and AcβMeSCh (25 %).	(154)

Table 2. Continued

Species	Tissue or organ	Specific substrates	Specific ChE inhibitors	Main remarks on ChE characterization	Ref.
Gambusia yucatana	Head Muscle	AcSCh, BuSCh, PrSCh	Eserine, Iso-OMPA, BW284C51	– The predominant ChE in the head and muscle of mosquitofish is AChE. It did not vary in the presence of iso-OMPA; it was inhibited at high concentration of AcSCh (20.48 mM); and it was drastically inhibited (95.5 %) by 10^{-4} M of BW284C51.	(155)
Terrestrial invertebrates					
Locusta migratoria manilensis	Head, Abdomen, Hindleg femora	AcSCh, BuSCh	Eserine, BW284C51, ethopropazine	– ChE activity was dominant in head. – Purified ChE activity from head resulted to be an AChE. It was completely inhibited by 10-6 M of eserine or BW284C51, but not by ethopropazine (a selective BChE inhibitor). AChE was also depressed at high substrate concentrations.	(156)
Aonidiella aurantii	Whole body	AcSCh, BuSCh, PrSCh	Iso-OMPA, BW284C51	– ChE activity in this insect species was believed to be AChE because it showed properties of typical mammalian AChE: i) it was inhibited at substrate concentrations higher than 10 mM (AcSCh) or 20 mM (PrSCh), ii) ChE activity was inhibited by BW284C51 ($K_i = 5.3 \times 10^{-7}$ M), but not by iso.OMPA. – PrSCh was the preferential substrate for AChE activity.	(98)

Table 2. Continued

Species	Tissue or organ	Specific substrates	Specific ChE inhibitors	Main remarks on ChE characterization	Ref.
Reptiles					
Gallotia galloti	Brain Serum	AcSCh, BuSCh, PrSCh	Iso-OMPA	– Both AChE and BChE activities were present in serum, and BChE represented the 83 % of the total ChE activity. – Serum AChE activity was not inhibited by excess of AcSCh (50 mM). – Solely AChE activity is present in brain.	(99)
Pogona vitticeps	Plasma	AcSCh	Iso-OMPA	– Plasma BChE activity represented the 75 % of total ChE activity.	(61)
Birds					
Nycticorax nyticorax, Bubulcus ibis, Egretta thula, E. caerulea and *Plegadis falcinellus*	Serum	AcSCh	Iso-OMPA	– AChE activity in the four heron species represented from 44 to 54 % of the total ChE activity. – Only the 5 % of the total serum ChE activity in the glossy ibis (*P. falcinellus*) was attributed to AChE.	(50)
Cardinalis cardinalis	Plasma	AcSCh	Iso-OMPA	– AChE activity represented about 10 % of the total ChE activity.	(52)

AcSCh = acetylthiocholine; Ac□MeSCh = acetyl-β-(methyl)thiocholine; BuSCh = butyrylthiocholine; PrSCh = propionylthiocholine; iso-OMPA = tetraisopropyl pyrophosphoramide; BW284C51 = 1,5-bis(4-allyldimethyl-ammonimphenyl)pentan-3-one dibromide; DFP = diisopropylfluorophosphate; pI_{50} = is the negative logarithm of the molar concentration of pesticide causing 50 % inhibition.

References: Valbonesi et al. [143], Monserrat et al. [144], Brown et al. [145], Mora et al. [146], Galloway et al. [40], Frasco et al. [147], Forget and Bocquené, [148], Forget et al. [149], Key and Fulton [150], Varò et al. [151], Cunha et al. [152], Monteiro et al. [28], Sturm et al. [29], Rodríguez-Fuentes and Gold-Bouchot [30], Varò et al. [31], Chuiko [33], Küster [42], Whitehead et al. [35], Sturm et al. [153], Wogram et al. [34], Sturm et al. [154], Osten et al. [155], Yerushalmi and Cohen [98], Sanchez-Hernandez and Moreno Sanchez [99], Bain et al. [61], Parsons et al. [50], Maul and Farris [52], Ma et al. [156].

Carboxylesterases. These enzymes are responsible for the development of insect resistance to insecticides [36]. An over-production of CbEs increases the detoxification capacity by two mechanisms: hydrolysis and inhibition. In comparison to ChEs, CbEs have had a scarce interest in ecotoxicology despite to their role in the detoxification of pesticides. In a descriptive study, CbE activity was measured in serum and liver of rainbow trout [37]. The higher CbE activity levels were found in the serum respect to gill or liver; and juveniles had higher CbE activity in comparison with adults or embryos. Comparatively, the CbE activity measured in the fish liver were similar to those from rat liver. On the basis of these results, authors suggested that the high CbE activity found in rainbow trout can significantly have influence on the bioaccumulation and toxicity of pesticides.

An observation generalized for invertebrates is that CbE activity is more sensitive to OP inhibition than AChE activity. For example, the cladoceran *D. magna* presents a higher CbE activity and a greater sensitivity to inhibition by OPs than AChE activity. This could be determinant in the pesticide tolerance of this organism [38]. In fact, when CbE activity is inhibited in the presence of triphenyl phosphate (TPP) or 2-(*O*-cresyl)-4H-1,3,2-benzodioxaphosphorin-2-oxide (CDPB), the mortality of *D. magna* significantly increases in response to OP or CB exposure [38]. The same result was reported by Escartín and Porte [39] in the mussel *Mytilus galloprovincialis*. *In vitro* experiments showed that the CbE activity measured in the gills and digestive gland of this mollusc was more sensitive to fenitrothion, and its oxon metabolite, than AChE activity. In other mussel species (*M. edulis*), Galloway et al. [40] also observed that CbE activity was slightly more sensitive to paraoxon (IC_{50} = 0.4 x 10^{-3} M) and chlorpyrifos (IC_{50} = 10.1 x 10^{-3} M) than AChE activity (IC_{50} = 1.8 x 10^{-3} M for paraoxon and 15.0 x 10^{-3} M for chlorpyrifos). In a biomonitoring study, Fourcy et al. [41] also found that CbE activity of the polychaeta *Nereis diversicolor* sampled in an area treated with the pesticide Abate® 500E (containing temephos as active ingredient) was more depressed than the AChE activity 24 and 72 h after treatment. This higher sensitivity of CbE activity to OPs also occurs in fish. Wogram et al. [34] studied the sensitivity of AChE, BChE and CbE in various tissues of the three-spined stickleback (*Gasterosteus aculeatus*) to paraoxon. *In vitro* inhibition of liver CbE activity was more pronounced (IC_{50} = 5.6 x 10^{-7} M) than brain (IC_{50} = 7.5 x 10^{-6}) or muscle (IC_{50} = 9.4 x 10^{-6} M) AChE inhibition. *In vivo* experiments also support a higher sensitivity of CbE activity to OP exposure. For example, a higher percent of inhibition, although not statistical significant, was found in the CbE activity respect to AChE activity of whole embryo homogenates of zebrafish (*Danio rerio*) after 48 h exposure with 0.4 μM of paraoxon-methyl [42].

Surprisingly, there exist a limited number of studies aimed to characterize the CbE activity. Topics such as tissue distribution of CbEs, optimal conditions for CbE measurement, natural variation of its activity, impact of environmental and biological variables on CbE activity, time of recovery after OP/CB inhibition or ability of oximes to reactivate the phosphorylated CbE would be areas of study for future research. This knowledge is necessary to define the qualities of CbEs as biomarkers of pesticide exposure. Some researchers have reported significant data on CbE properties of ecotoxicological concern. For example, Galloway et al. [40] performed a characterization of CbE activity towards phenylthioacetate in *M. edulis*, suggesting an optimum substrate concentration for kinetic determinations of 800 μM. In the study by Barron et al. [37], CbE activity was measured in a broad range of fish tissues. Kinetic parameters (K_m and V_{max}) and optimum temperature for the enzyme assay were established. CbE activity was determined using 4-nitrophenyl acetate, which is considered a model substrate. They reported that CbE activity in serum and liver homogenates displayed a Michaelis-Menten kinetic, and the CbE activity was higher in serum than in liver or gill. Barata et al. [38] documented that CbE activity of *D. magna* recovered its activity after exposure to OP pesticides within 96 h, showing a shorter recovery time when CbE was inhibited by the carbamate carbofuran (< 12 h). In the study by Küster [42], CbE activity was measured during the development of the zebrafish embryo, and it was found that the enzyme activity remained unchanged during the fish development. In addition, CbE activity was much higher than AChE or BChE activities in the early development stages. This finding encourages future investigations on the buffer capacity – binding to circulating OPs – of this B-esterase against anti-ChE pesticides, like occur in pest species.

B. Terrestrial organisms

Cholinesterases. Birds and mammals have been the pioneer groups in ecotoxicological studies involving ChE inhibition as biomarker of OP/CB exposure [12,13,43]. In the case of birds, several authors have tried to establish a threshold for brain AChE inhibition by which sublethal exposure can be distinguished from mortality due to OP intoxication [2,4]. The level of brain ChE inhibition has been also related to abnormal behavioural responses in birds [44-46]. The recovery rate of plasma or brain ChE activities after inhibition by anti-ChE pesticides has been another matter of ecotoxicological concern. In general, a logarithmic pattern seems to explain the rapid initial recovery of enzyme activity attributed to a combination of spontaneous reactivation and new synthesis of the ChE, followed by a slower recovery rate to normal enzyme activity mainly due to de novo synthesis [47,48]. Enzimological characterization of brain and blood ChEs have been also investigated in some terrestrial organisms (Table 2). Most of the total ChE activity in the blood of birds and reptiles is attributable to BChE activity. This B-esterase is an attractive methodology because of certain advantageous features such as high sensitivity to OPs and CBs, and especially because it can be measured easily by non-destructive methods. This is a desirable feature in the case of endangered or protected species or when the density of natural population is low at the study area. In addition, a number of biomonitoring studies have made clear that chemical reactivation of avian blood ChE using oximes (e.g., 2-PAM) is a good complementary approach to assess wildlife exposure to OP pesticides [49-52].

The substantial body of literature describing the effects of OP/CB pesticides on bird ChEs contrasts with the paucity of available information on herpetofauna. Fortunately, the need for increasing the knowledge on toxic effects of pollutants on amphibians and reptiles stressed by some authors [53-56], have resulted in a surge of investigations about adverse effects to these vertebrates from anti-ChE pesticides [57-59]. Regarding ChEs, Clark et al. [60] performed a field monitoring to assess the adverse impact on four reptile species from pesticide applications. Among other chemical and biological parameters, ChE activity levels were measured in the blood of reptiles. Curiously, an inverse correlation between blood ChE activity and the body mass was found in both male and female adults of the turtle *Trachemys scripta* collected from the agricultural area only. Anti-ChE pesticides could be implicated in this relationship, although clear evidences of such a link are not provided in that study. In a laboratory investigation, Bain et al. [61] determined the plasma ChE activity of the Australian agamid *Pogona vitticeps* orally dosed with fenitrothion. One of the most interesting results was the extremely slow recovery of plasma ChE activity after OP inhibition. They observed that plasma ChE had not yet reversed to its normal values after three weeks of fenitrothion treatment. This plasma ChE feature was previously reported by Sanchez et al. [21] in the lizard *Gallotia galloti* orally dosed with parathion. In that study, serum BChE activity took more than 24 days to full recovery. Additional investigations should be performed to confirm whether this slow recovery of blood ChE activity is a common property of this group of vertebrates. In the case of birds, for instance, blood ChE recovers its activity after OP inhibition very quickly (in term of a few hours or days), and its potential use for field OP monitoring purposes is therefore fairly limited. Measurement of plasma BChE activity levels in the lizard *G. galloti* is an excellent approach for assessing field exposure to anti-ChE pesticides. In areas of intensive pesticide applications (Canary Islands, Spain), Sanchez-Hernandez et al. [62] found a high proportion of lizards having plasma BChE activity inhibited by CB and OP pesticides (Fig. 3). Chemical reactivation of BChE activity using 2-PAM and spontaneous reactivation after water dilution of the plasma corroborated the inhibition of the BChE activity by OP and CB pesticides, respectively. However, the implications at higher level of biological organization of this intense and long-term exposure to anti-ChE agrochemicals remain obscure. Curiously, population density of this lizard species in agricultural areas is higher than in its natural environment. Whether plasma BChE activity and other detoxification systems or strategies are implied in the lizard tolerance to pesticides is a hypothesis to be tested.

Soil invertebrates have occupied a notable place in the toxicity testing of new substances to be discharged into the environment or to assess the potential toxicity of a contaminated soil. Toxicity test protocols have been developed and standardized for a wide range of representative organisms of the soil community such as earthworms, isopods, nematodes, enchytraeids, collembolans, arachnids or gastropods [63,64]. Soil is the primary reservoir for many agrochemicals and in spite of this, soil invertebrates are had little attention in the study of the B-esterases [64-67]. By contrary, the role of these enzymes in the molecular mechanism of resistance to insecticides has led to an intensive investigation in pests such as some Lepidoptera [68], acarids [69] or mosquitoes [70].

An increase in the knowledge of B-esterases in non-target invertebrates is necessary to use them as biomarkers of pesticide contamination, and to know at what extend these enzymes can contribute to a likely resistance to insecticides like in pest species. A few studies have examined, however, invertebrate AChE response to OP pesticides in the laboratory. The AChE activity in the foot and visceral of the snail *Helix aspersa* was investigated after a continuous exposure to dimethoate-contaminated food [71]. A strong AChE inhibition (≥ 75 %) was found in the snails exposed to food contaminated with 250 $\mu g\ g^{-1}$ after 28 d, which was more drastic (≥ 97 %) at higher OP concentrations. The authors suggested that the decrease in the growth of the exposed animals could be due to inhibition of AChE activity. Earthworms are considered excellent bioindicators of environmental change in agroecosystems [72]. These organisms have been historically used for assessing soil pollution by heavy metals. Indeed, they have experienced a spectacular development in the understanding of biomarkers related to metal exposure such as metallothioneins or lysosomal fragility [73,74]. Conversely, a reduced number of studies have measured earthworm AChE responses in relation to OP or CB contamination. For example, a substantial inhibition of AChE activity was found in *Eisenia fetida* after 2, 7 and 14 d of exposure to contaminated artificial soils with carbaryl [75], and in juveniles of *Aporrectodea caliginosa* after 24 h of exposure to sublethal concentrations of diazinon or chlorpyrifos [76]. Using the OP chlorpyrifos and in the same earthworm species, Rao et al. [77] reported an AChE inhibition of 60 % after 12 h of exposure, which increased up to 91 % 48 h later. They also observed toxic effects from chlorpyrifos exposure in the form of morphological damage such as rupture of the cuticle or fragmentation of the body. Toxic effects of anti-ChE pesticides on the earthworm reproduction system have been also described in *E. fetida*. Sorour and Larink [78] showed a gradual damage on the male reproduction system (abnormal cytophores and malformed spermatides) in individuals exposed for a week to sublethal concentrations (8.3–112 mg kg^{-1}) of benomyl. Similarly, Espinoza-Navarro and Bustos-Obregon [79] reported alterations in the male reproduction system of *E. fetida* after exposure to malathion (80–600 mg kg^{-1}). Besides a loss of body weight up to 50 % in the treated animals respect to the non-

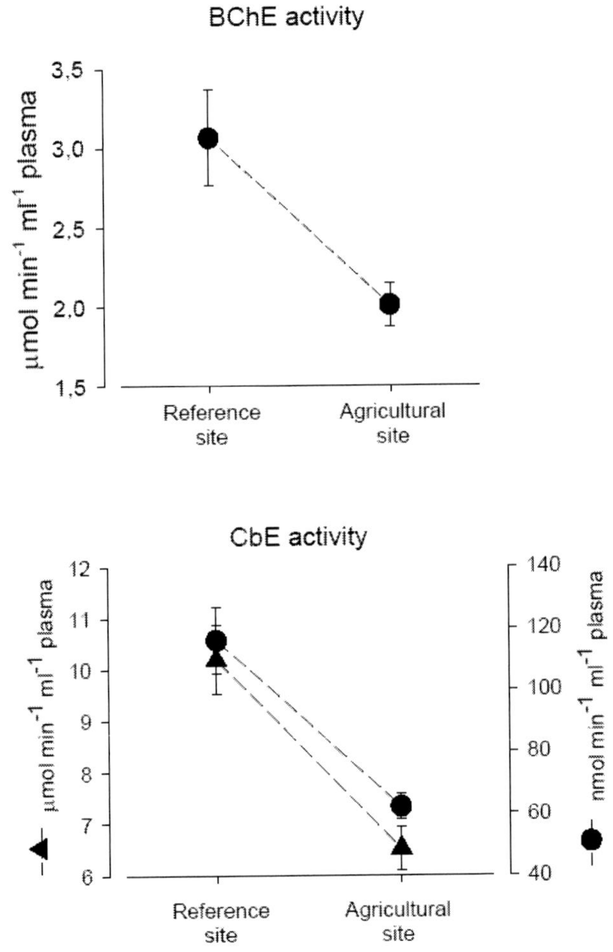

Figure 3. Significant differences (Wilks λ = 0.63, $F_{4,245}$ = 35.5, $P < 0.001$) of plasma butyrylcholinesterase (BChE) and carboxylesterase (CbE) activities in the lizard *Gallotia galloti palmae* collected from reference and agricultural sites. Plotted values are means and vertical bars denote 0.95 % confidence intervals. BChE activity data were taken from Sanchez-Hernandez et al. [62]; data of CbE activities are unpublished. CbE activity toward α-naphthyl acetate (triangle symbols) was measured according to Walker [140], while CbE activity using *p*-nitrophenyl valerate as substrate (circle symbols) was determined following the method by Carr & Chambers [141].

exposed earthworms, they also found vacuolization of spermatheca and fragmentation of DNA in a high proportion of spermatogonia. Changes in growth and fecundity was also related to ChE inhibition in the earthworm *A. caliginosa* exposed for 28 days to contaminated soils by diazinon and chlorpyrifos [80].

One of the most important features of a good biomarker is the stability of its response to pollutants, especially when it is used in the field. In the case of phosphorylated ChEs, a slow recovery rate enables to detect the OP impact over a longer period of time after OP applications. This is a desirable feature for assessing anti-ChE exposure because of most of these pesticides show a low persistence in the environment [81]. Panda and Sahu [82] explored the time to full recovery of AChE activity in the tropical earthworm *Drawida willsi* after exposure to malathion and carbofuran. They found an extremely slow recovery of AChE activity of *D. willsi* (45–75 days), which was not related to the pesticide concentration in soil. However, in this study earthworms were continuously exposed to the contaminated soils and therefore it is difficult to make conclusions about the real time to full recovery of AChE. Ideally, the animals would be transferred to clean soil (free of pesticides) when ChE activity reaches a certain inhibition degree. This approach would be more environmentally realistic than keep the earthworms continuously in the contaminated soils for a long time, especially when it is known that earthworms tend to avoid contaminated soils. However, this extremely slow recovery of earthworm ChEs has been also demonstrated by Dell'Omo et al. [83], whose exposed the earthworm *Eisinia* spp. to contaminated soils with dimethoate (150–200 mg kg^{-1}). After 48 h of OP exposure, earthworms were transferred to clean soil and the ChE activity remained inhibited (35 % respect to controls) for 40 days.

Carboxylesterases. Like for aquatic organisms, little information is available on the use of CbE activity for ecotoxicological purposes. These B-esterases haven been extensively studied in mammals probably due to their role in the drug metabolism and protection against nervous gases intoxication [6,23]. In non-mammal vertebrates, levels of CbE activity seem to be related to either diet or size of the bird [11]. The highest CbE activities are associated to smallest birds or the most diverse diets (omnivorous or herbivores). The role of CbE in the detoxification of lipophilic esters in the diet could account for the high level of CbE activity. Sanchez et al. [84] found a higher serum CbE activity (using α-naphthyl acetate as substrate) in the lizard *G. galloti* (11.0 ± 2.8 μmol min^{-1} ml^{-1} serum; mean ± standard deviation) than in the bird species *Sylvia melanocephala* (9.87 ± 2.8), *Serinus canaria* (2.36 ± 0.5), *Parus caeruleus* (6.97 ± 2.6) and *Erithacus rubecula* (1.96 ± 0.2). These data, although insufficient to make generalizations, could represent an evidence of the assumption by Thompson and Walker [11]. *Gallotia galloti* is an omnivorous lizard and this could explain its relatively high serum CbE activity levels respect to birds. Recently, Roy et al. [85] determined CbE activity (using α-naphthyl acetate as substrate) in 729 European raptors belonging to 20 species. In all analyzed species the plasma CbE activities was very low (< 2 μmol min^{-1} ml^{-1} plasma) or even non existent in some species. The comparison of bird CbE activity measured by Roy et al. [85] in raptors species with those measured by Sanchez et al. [84] in passerines; give further evidence on the inverse relationship between CbE activity levels and size.

Avian plasma CbE activity has not as popularity as ChE activity for monitoring anti-ChE pesticide effects [84,86,87]. In these vertebrates, CbE recover its activity rapidly after inhibition by OPs [11]. In addition, an increase of plasma CbE activity has been reported in starlings treated with low doses of demeton-S-methyl [88]. Similar results were obtained by

Sanchez et al. [21] in lizards orally dosed with low doses of parathion (0.5 and 2.5 mg kg^{-1}). In mammals, plasma CbE is likely synthesized in liver and then secreted into the bloodstream [6]. A possible explanation why blood CbE activity increases at low dose of OP could be the high induction of the enzyme in liver. Figure 4 shows a relationship between serum and liver CbE activities observed in lizards dosed with 0.5, 2.5, 5 and 7.5 mg kg^{-1} of parathion [21]. At the lowest dose of the OP, serum CbE activity did not vary (0.5 mg kg^{-1}) or slightly decrease (2.5 mg kg^{-1}) respect to the control group (black symbol in Fig. 4), however liver CbE activity increases three- and two-fold. At the higher doses of the OP, serum CbE activity decreased drastically, whereas liver CbE activity decrease progressively with OP dose increase; its activity was significantly inhibited respect to the control group at the highest OP dose only (7.5 mg kg^{-1}). The highly significant correlation found between serum and liver CbE activities in the treated lizards leads to suppose two phenomenons. First, the liver is probably the main source of serum CbE activity in this species, like for mammals [6]. Second, serum and liver CbE activities participate in the detoxification of OPs. The over-expression and OP sequestration capacity attributed to CbEs in pest species (mechanism of insecticide resistance) can also occur in lizards expose to relatively low doses of OP pesticides.

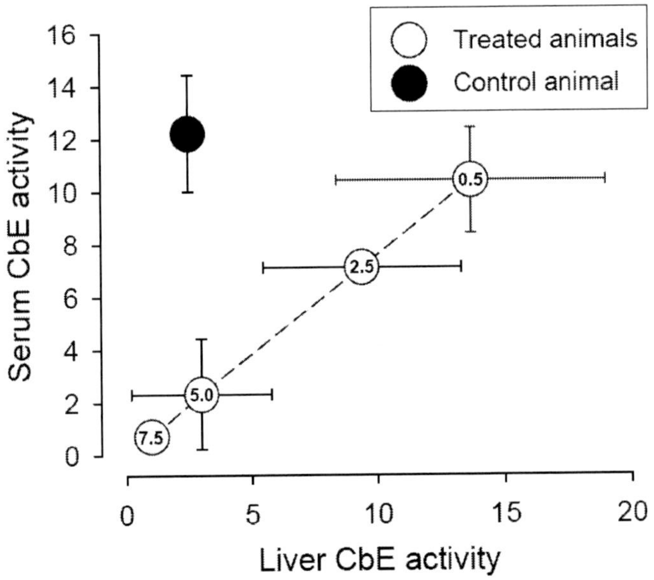

Figure 4. Relationship between liver and serum carboxylesterase (CbE) activities in lizards (*Gallotia galloti*) after 6 h of an acute treatment with the organophosphorus parathion. Plotted values are means and bars denote standard errors. Data taken from Sanchez et al. [21]. Numbers inside white symbols indicate the parathion dose in mg a.i. kg^{-1}. Serum CbE activity is expressed as µmol min^{-1} ml^{-1} serum, while liver CbE activity is expressed as mmol µin^{-1} mg^{-1} of microsomal protein.

Beside blood BChE activity, the measurement of CbE activity can be useful as complementary index of OP exposure in field situations. For example, in the field study by Sanchez-Hernandez et al. [62], together to inhibition of plasma BChE activity, a high proportion of lizards collected from agricultural areas had plasma CbE depressed (Fig. 3). In addition, a good correlation between plasma BChE activity levels and CbE activity was observed in these lizards (Fig. 5). By contrary, such a correlation was not observed in the lizards from the reference site, although this finding should be taken with reservations because the number of lizards collected from the reference areas (N = 52) was much lower than that captured in the agricultural areas (N = 198).

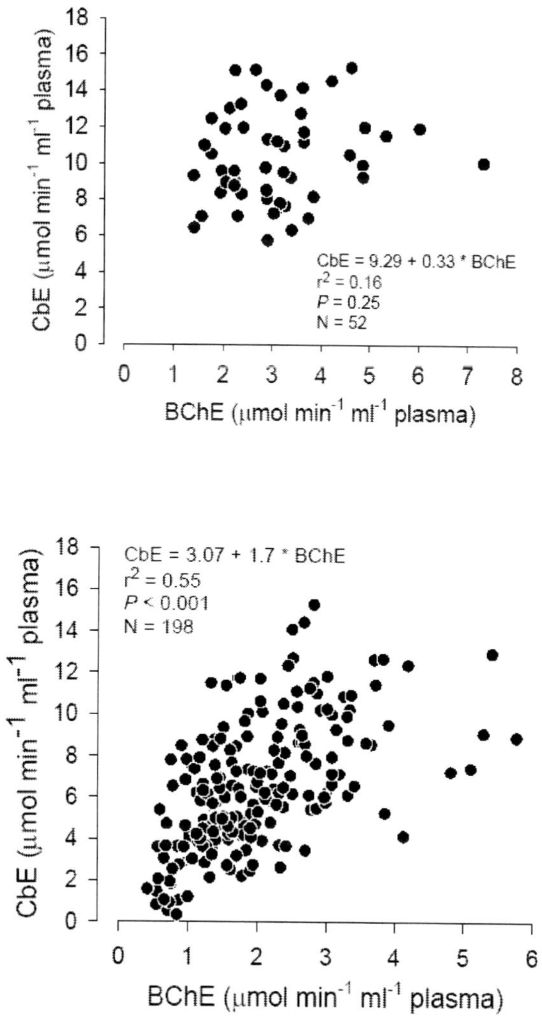

Figure 5. Relationship between plasma butyrylcholinesterase (BChE) and carboxylesterase (CbE) activities in lizards (*Gallotia galloti palmae*) collected from an agricultural and from a reference area. BChE activity data were taken from Sanchez-Hernandez et al. [62]; while data of CbE activities are unpublished. CbE activity was determined according to Walker [140] using □-naphthyl acetate as substrate.

B-esterases in Ecological Risk Assessment

Toxicity testing is an essential element in the ERA scheme. They are necessary for the exposure and effect assessment of new chemicals before release into the environment, or to assess the ecological impact of a new aqueous or atmospheric emission source (predictive

ERA). Toxicity assays are also used in a retrospective approach of ERA to assess the historical contamination with possible ongoing ecological consequences. Survival, growth and reproduction are the main traits measured in the toxicity tests. However, they are assessed under laboratory conditions and extrapolation to field situation present serious limitations. Scientific literature offers some examples of how environmental variables influence on pesticide toxicity leading to a different conclusion respect to that from toxicity test results. For example, Relyea and Mills [89] demonstrated that toxicity of carbaryl increased when *Hyla versicolor* tadpoles were exposed simultaneously to predatory cues. Using the same carbamate pesticide, Boone and James [59] found that the density of larval amphibians and the pond hydroperiod (constant or drying situations) in a mesocosm study were determinant factors in the carbaryl toxicity. In the past, it has been demonstrated that OP toxicity is dependent on the ambient temperature [90]. These interactions between environmental fluctuating variables and pollutants, leading to potentiation or antagonist effects on biota, is one of the new challenges of modern ecotoxicology [91], and they should be therefore considered as key variables in the toxicity tests.

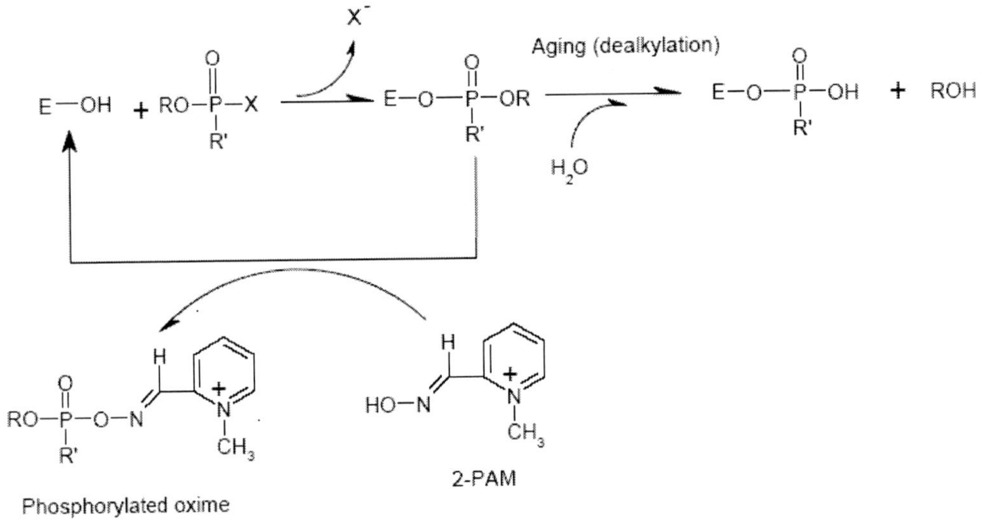

Figure 6. Phosphorylation of cholinesterases (E-OH) by organophosphorus pesticides. The phosphorylated ChE can be either chemically reactivated in the presence of oximes (e.g., 2-PAM) or suffer a dealkylation (aging of enzyme); a non-enzymatic reaction in which one alkyl side chain of the phosphoryl moiety is removed and a hydroxyl group holds its place.

Recently, biomarkers have gained ecotoxicological meaning when they form part of the ecological weight-of-evidence (WOE) framework [92]. Appropriate biomarkers may be applied in standardized bioassays in order to provide evidence of the cause-effect relationship between pollutants and toxic effects in the individuals. In aquatic toxicity testing, the biomarker approach has brought about promising results. For example, Neuparth et al. [93] included certain biomarkers (MT induction, DNA strand breakage and lipid peroxidation) in a standardized sediment toxicity test to assess toxic effects at multiple biological organization levels. They concluded that the use of biomarkers in these ecotoxicity tests can help to distinguish biological responses to contaminant exposure from those originated from physicochemical variables of the sediment. B-esterases can be used in a similar way because they give valuable information about the bioavailability (inhibition of non-vital ChEs and/or CbEs) and the bioactive fraction (inhibition of AChE in the nervous tissue) of pollutants. For example, Booth and O'Halloran [80] measured ChE activity in juvenile earthworms (*A. caliginosa*) exposed to OP pesticides following the guidelines of standardized toxicity testing. Cholinesterase inhibition was a more sensitive indicator for low pesticide exposure level than classical toxicity endpoints, and ChE response was suggested as a suitable index for predicting adverse effects on earthworm reproduction or growth. In a complementary study, ChE inhibition of *A. caliginosa* observed in the laboratory was not reproduced in mesocosm and field scenarios of pesticide exposure Booth et al. [76]. This is a clear example of difficulties in the extrapolation of toxicity test results to field situation. An understanding of the impact of fluctuating environmental variables on biomarker responses is required to establish concentration(dose)-response relationship, if any, under multiple stressors other than pollutants.

The pollutant related responses of biomarkers can have implications at higher levels of biological organization when they are directly related to the three factors that define the population fitness, i.e., survival, growth and reproduction. In this context, biomarkers such as CbE and BChE activities that can be involved in the organism tolerance to pesticides should be considered in biological monitoring because they can give information about features (tolerance to pesticides) of organism's survival. However, most of the field biomonitoring involving B-esterases evaluate levels of these enzyme activities under different exposure approaches (e.g., caged animals or comparison of natural populations with different pesticide exposure history) to assess pesticide exposure. In previous sections, we have shown that enzimological characterization and tissue distribution of ChEs are two issues of current concern in the ecotoxicology of ChEs. However, other practical and ecotoxicological aspects should be equally considered. In the following sections, it is examined three ecotoxicological features of B-esterases which need still increase their knowledge: 1) the chemical reactivation of inhibited B-esterases using oximes as a complementary – or definitive – method of assessing OP intoxication, 2) the role of B-esterases (scavengers) in the tolerance of wildlife to anti-ChE pesticides, and 3) the need of testing consequences from B-esterase inhibition at higher levels of biological organization, i.e., perturbation of behavioural responses leading to ecologically relevant consequences.

C. Chemical reactivation using oximes

Oximes are nucleophilic reagents clinically used in the therapy for reduction the acute toxic effects from organophosphate compounds. These reactivators are able to attack the phosphoryl group bound to the active site of the ChE (or CbE), resulting in the recovery of enzyme activity to its normal values (Fig. 6). One of the main limitations of using ChE for assessing wildlife OP exposure is the high enzyme variability among individuals making difficult the identification of pesticide exposed individuals. The use of oximes as a complementary indicator of pesticide exposure becomes an attractive methodology. Pyridine-2-aldoxime methochloride (2-PAM) or pralidoxime is the most often oxime used in the biomonitoring of OP contamination. Generally, the sample suspected to have the ChE inhibited by an OP is incubated in the presence of 2-PAM. The ChE activity is measured before and after the addition of the oxime. If an increase in the enzyme activity is detected after 2-PAM incubation, then it is interpreted that the enzyme was phosphorylated. Therefore, each sample act as its control and non-exposed individuals are not necessary in order to compare enzyme activity levels. However, this methodology presents several limitations and often reactivation of phosphorylated ChE is not always achieved (Fig. 6). *First*, the reactivation rate mainly depends on the type of OP bound to the active site of the ChE. Maxwell et al. [94] observed that the type of OP that inhibits the rat plasma CbE activity influenced the ability of the oxime diacetylmonoxime (DAM) to recover completely the enzyme activity. The alkoxy-containing phosphonates were efficiently reactivated in the presence of DAM, while aryloxy-containing OPs caused the CbE aging, and the enzyme activity could not be reversed. *Second*, the time elapsed since OP exposure is another important factor affecting the ability of the oxime to reactivate the phosphorylated ChE. If the ChE remains long time phosphorylated, it can occur that one alkyl group is released from the OP-AChE complex (Fig. 6). This dealkylation leads to a permanent inhibition of the enzyme known as aging. This is the main argument discussed by many authors when treatment with 2-PAM, or other types of oximes, is not enough to full recover of ChE activity. *Third*, the 2-PAM concentration in the incubation media is a critical factor to achieve a satisfactory reactivation of ChE activity. For example, the AChE activity of mussels (*M. galloprovincialis*) exposed for 48 h to fenitrothion was not reactivated in the presence of 5, 10 or 20 mM of 2-PAM [39]. This ineffectiveness of 2-PAM to recover the mussel AChE activity could be explained by the 2-PAM concentration. Monserrat and Bianchini [95] reported an inhibitory effect by 2-PAM on the purified electric eel AChE (IC_{50} = 2.0 x 10^{-4} M), on the brain AChE of the fish *Odontesthes argentinensis* (IC_{50} = 2.5 x 10^{-4} M) and on the thoracic ganglia AChE of the crab *Callinectis sapidus* (IC_{50} = 8.2 x 10^{-4} M). The authors suggested that concentrations of 2-PAM higher than 3.0 x 10^{-4} M should be avoid because of enzyme inhibition and non-enzymatic hydrolysis of the substrate interfering in the ChE kinetic assay. An inhibition of BChE activity was also found when serum of starlings previously exposed to demeton-*S*-methyl was incubated in the presence of 12.5 x 10^{-2} M of 2-PAM [86]. In other study [96], 2-PAM resulted toxic (LC_{50} = 0.14 μmol L^{-1}) to the 72-h old larvae of *Artemia salina*. In addition, they found that acute toxicity by the carbamate carbaryl was enhanced in the presence of 2-PAM. It was suggested that oximes exert an allosteric action on ChE in the presence of carbaryl; this interaction results thereby in an enhanced inhibition degree. This potentiation effect between the carbaryl and the oxime has been also reported by others [97,98]. The effect of 2-PAM concentration in the reactivation rate of phosphorylated ChE was also reported by Sanchez-Hernandez and Sanchez [99]. They found that 10^{-4} M of 2-PAM resulted to be a more effective concentration for reactivating serum BChE activity of lizards than 10^{-2} M. More recently, an *in vivo* experiment performed with the zebra mussel *Dreissena polymorpha*, Binelli et al. [100] demonstrated that the addition of 2-PAM into the test media was able to reactive the phosphorylated AChE. In that study, mussels were exposed to three 2-PAM concentrations (0.1, 1 and 10 μg L^{-1}) after an initial exposure regime to 100 ng L^{-1} of chlorpyrifos-oxon for 48 h. Although a full recovery of the AChE activity was achieved by treatment with 0.1 and 1 □g L^{-1} of 2-PAM, the highest oxime concentration did not recover the AChE activity. All these limitations in the use of 2-PAM forces to characterise and optimise the 2-PAM reactivation protocol in each particular situation (species, type of ChE, type of OP when it is known, among other factors).

Despite these practical considerations, chemical reactivation of ChE using oximes is an alternative approach to chemical analysis of pesticide residues in abiotic or biotic matrices.

Chemical analysis of residues represents an indirect measurement of bioavailability of pesticides, whereas the reactivation of ChE using oxime gives a direct measurement (qualitative) of the active fraction of pesticide. In the last ten years, several biomonitoring studies have employed 2-PAM as a complementary index of bird exposure to anti-ChE pesticides. McInnes et al. [49] evaluated the field exposure the impact on nestlings belonging to three species (*Agelaius phoeniceus, Passer domesticus* and *Molothrus ater*) from OP pesticides by measuring the plasma ChE activity levels, reactivation of blood ChEs using 2-PAM and chemical analysis of chlorpyrifos residues (suspected to be applied in the sampling area) in gastrointestinal tissues and contents, and carcass. They reported significant increase of the plasma AChE activity after 2-PAM treatment in the 31 % of the nestlings (N = 52 samples) sampled from the agriculture areas. Furthermore, residues of chlorpyrifos were detected in several samples of carcass, gastrointestinal tissues and contents, supporting the exposure of nestlings to this OP. In a similar field study, breeding colonies of wading birds (*Nycticorax nyticorax, Bubulcus ibis, Egretta thula, E. caerulea* and *Plegadis falcinellus*) were monitored during six years for exposure to anti-ChE pesticides [50]. Positive reactivations (from 5.1 to 76.9 % of the assayed samples) of serum ChE activity were detected after incubation with 2-PAM. This specific exposure index significantly correlated with the land agricultural usage, suggesting that exposure to OP pesticides was the primary cause for depressing of serum ChE activity. More recently, Maul and Farris [52,101] monitored pesticide impact on bird populations through, among other indicators, 2-PAM reactivation of plasma ChEs. In a same study area but at different periods of sampling, they recorded a 5.6 % (N = 51 samples) of plasma ChE reactivations in eight passerines species [101]; and an 8.7 % (N = 128) of positive ChE reactivations in plasma samples of adults of *Cardinalis cardinalis* [52]. 2-PAM reactivation of avian plasma ChE was also used by Iko et al. [51] to assess exposure of the mountain plover (*Charadrius montanus*) to anti-ChE pesticides. However, no positive reactivation was recorded in birds collected from two sampling areas (reference and agricultural). The authors suggested that recent agrochemicals application did not occur in the sampling period, which was corroborated by the chemical analysis of anti-ChE pesticide residues in foodwash samples from both sampling areas.

In a field study, 2-PAM reactivation approach was used to assess exposure of lizards to OPs. Positive reactivation of BChE activity was observed in the 24 % of the assayed serum samples, which increased the BChE activity up to 60 % [102]. In another related study [62], it was recorded BChE reactivation after 2-PAM treatment in the 11 % of the assayed samples, however they found that most of the samples (N = 134) showed a spontaneous increase in the BChE activity after dilution of the plasma with deionised water (an indicative of ChE inhibition by carbamate pesticides). In the field study by Parsons et al. [50], it also was found spontaneous reactivation of plasma ChE of birds (*N. nytocorax*) when samples were water diluted, indicating exposure to CB pesticides. Cholinesterases of non-target invertebrates have been also explored for chemical reactivation using pralidoxime, although respect to vertebrates, the number of studies is less and they have been performed under laboratory conditions. For example, the muscle AChE of the red crayfish *P. clarkii* exposed to 100 μg L^{-1} of fenitrothion for 12 h recovered fully its activity after incubation (1 h at 25°C) with 10 mM of 2-PAM [103].

The reactivation capacity of oximes has been also examined with CbEs. Thus, phosphorylated CbE activity of the earthworm *Lumbricus terrestris* recovers completely its activity after treatment (15-min incubation time) with 10^{-4} M of 2-PAM [104]. OP-inhibited

CbE activity of the rat plasma is also reversed in the presence of different types of oximes, although the best one was the diacetylmonoxime [94].

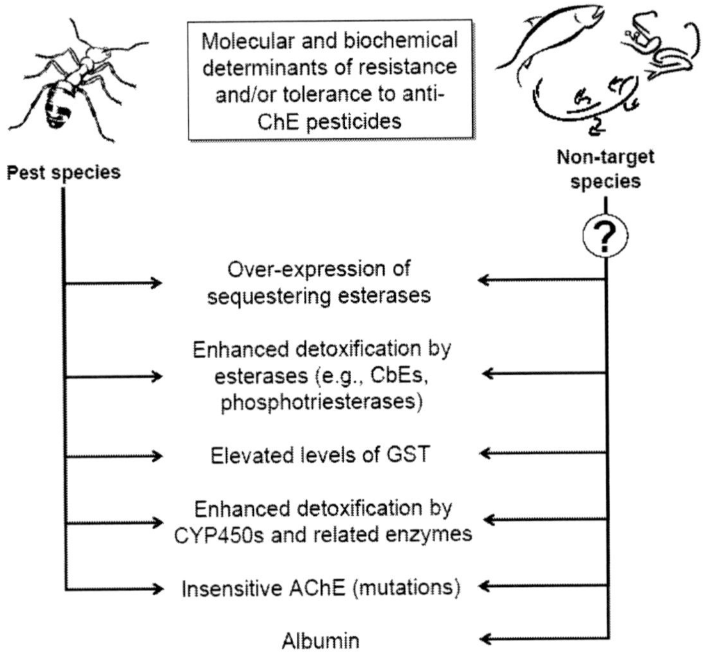

Figure 7. Main biochemical and molecular mechanisms involved in the insect resistance to anti-cholinesterase insecticides (A). Similar mechanisms, and others (e.g., albumin), could operate in non-target organisms to develop tolerate to the acute toxic action (inhibition of brain AChE activity) from OPs and CBs. See text for examples.

D. Role of B-esterases as pesticide scavengers

Some B-esterases (e.g., blood BChE and CbEs) have the ability of binding to circulating OP compounds, decreasing thereby the toxic effects on the molecular target (e.g., brain AChE activity). Human serum BChE activity has been demonstrated to be an efficient OP scavenger in several mammal species [19,105, 106]. Similarly, rat plasma CbE has revealed as a suitable OP bioscavenger because of its high reactivity to several OP compounds. Moreover, phosphorylated CbE does not suffer aging and spontaneous reactivation; these are ideal features for a good OP bioscavenger [23].

The figure 7 shows the main biochemical-based mechanisms contributing to insect resistance to insecticides [107-109]. Similarly, there exists evidence that the same biochemical systems, and other (e.g., albumin), respond similarly against pesticide exposure in non-target organisms. However, the implications of these biochemical systems on pesticide tolerance in non-target species have not been yet studied in detail. In insects, three enzymatic groups are involved in the biochemical mechanism of resistance to insecticides (Fig. 7): esterases, glutathione-*S*-transferase (GST) and cytochrome P450-dependent monooxygenases (CYP450s). Esterase-based resistance appears to be due mainly to an overproduction of esterases (e.g., CbEs), which are able to bind circulating insecticides [109], or an increase in the detoxification (hydrolysis) activity [108,110]. In an ecotoxicological context, enhance of detoxification capacity by esterases should be investigated in non-target species, especially when they are used as indicator or sentinels species in pesticide impacted areas [36]. Several studies give some evidence of such an esterase induction. In birds, Thompson et al. [86] and Lari et al. [111] found an increase of serum CbE activity at low doses of OP insecticides. Induction of liver CbE activity of lizards seems also occurs at low OP doses [21].

Glutathione-*S*-transferasas are other biochemical system involved in insect resistance to insecticides (Fig. 7). An induction of this enzyme activity has been reported for non-target organisms as well. Three different GSTs (pi-, sigma-, and omega-GST) were induced in the digestive glands of oysters (*Crassostrea gigas*) after two weeks of pesticide exposure [112]. Likewise, a significant induction of GST activity in the earthworm *A. caliginosa* was found after 14 days of exposure to diazinon (56 % of GST induction respects to controls) and chlorpyrifos (148 %) [80].

Comparatively, it could be supposed that vertebrates can display pesticide tolerance by increase of the enzymatic detoxification systems, or that differences in pesticide tolerance among species could be due, among other factors, to either presence or high levels of esterases involved in pesticide detoxification. Some studies postulate that certain blood proteins such as albumin as well as blood esterases (BChE, AChE and CbEs) can play a significant role in the sequestration of circulating OPs and CBs (Fig. 7). This 'buffer' effect of blood BChE and CbE activities has been observed in lizards exposed to parathion [21]. In that study, it was described a non-linear correlation ($r^2 > 0.85$, $P < 0.05$) between brain AChE activity and serum B-esterase activities after 6 and 24 h of OP treatment (Fig. 8). These relationships showed a clear discontinuity, and two stages were distinguished: 1) a first in which the brain AChE activity remained unchanged, while serum BChE and CbE (not shown in fig. 8) activities progressively decreased in relation to the parathion doses; 2) and a second stage in which a drastic inhibition of brain AChE activity took place (with a different regression slope) corresponding with the highest dose of parathion. The breakpoints of the non-linear correlations corresponded with the maximum inhibition levels of serum B-esterases, which would be the saturation point of the buffer capacity. Unfortunately, this OP-sequestration effect by blood B-esterases has not been investigated in other non-target vertebrates (e.g., fish or birds) to give solid evidence for this phenomenon. Bain et al. [61] investigated the toxic effects on the lizard *P. vitticeps* after fenitrothion treatment. Although plasma BChE activity was inhibited after 24 h of OP dosing, physiological (body temperature and standard metabolic rate) or behavioural (feeding rate) endpoints were not affected by fenitrothion exposure. The authors suggested that the ability of plasma BChE to sequester circulating OP could explain the absence of toxic effects.

Several studies have given clear evidence for a detoxification capacity of anti-ChE pesticides by albumin. This protein acts on anti-ChE pesticides by two ways: binding circulating pesticide or hydrolysis. For example, mice serum albumin is able to bind circulating OPs, acting as a bioscavenger [113]. Villanova and coworkers noticed that rabbit [114] and human [115] serum albumins are able to hydrolyse carbaryl. The authors have characterized this carbarylase activity by albumins in terms of metal ions dependence or inhibitors (e.g., fatty acids). Their studies have reported a hydrolytic activity from albumin toward this CB; a role attributed mainly to CbEs. Their findings have significant ecotoxicological implications because albumin is a universal protein in vertebrates; it is the most abundant protein in the bloodstream; and it is present in extravascular tissues such as the liver. Future investigations should be addressed to quantify the contribution of the albumin in

CB hydrolysis in comparison with CbEs, for instance. Moreover, this carbarylase activity of albumins needs to be examined in other non-target vertebrate species. These findings encourage future investigations to look at the role of albumins in pesticide detoxification and tolerance.

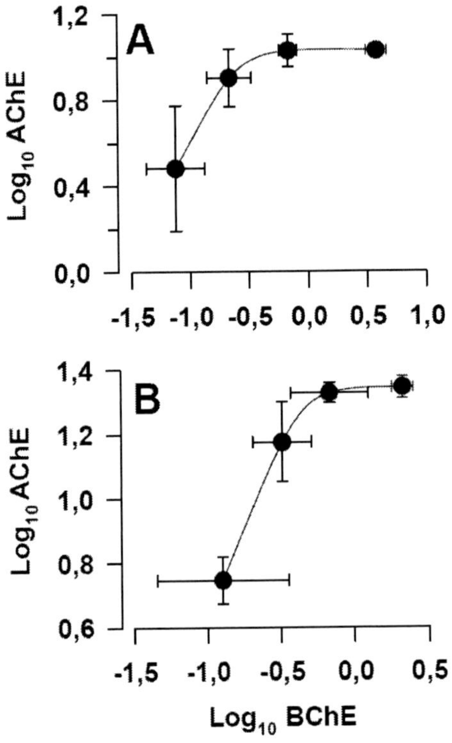

Figure 8. Correlations between serum BChE and brain AChE activities of *Gallotia galloti* after 6 h (graph A) and 24 h (graph B) of an acute treatment of parathion. Taken from Sanchez-Hernandez and Walker [142].

E. Linking ChE responses to behavioral perturbation

Behaviour of an organism is the final integrated result of a diversity of biochemical and physiological processes, which interact with the abiotic and biotic components of the environment. For example, Figure 9 shows the theoretical cascade of biological responses leading to impairment of behavior in earthworms exposed to pollutants. In this case, abnormal behavior of earthworms could derive in adverse effects at population level or even to alter certain properties of soil such as organic matter content. Although behavior is controlled by the motor output of the nervous system, the sensory, hormonal and metabolic systems, as well as the genetic attributes are also involved in the behavior performance. In turn, they represent the primary target systems of many pollutants.

Behavioral responses are included in the definition of the term biomarker by several authors [116,117]. Nevertheless they have had low consideration in the ecotoxicological research in comparison to molecular biomarkers. One reason could be the difficulty in quantify the behavior changes, particularly under field conditions. Despite this, behavioral abnormal responses to pollution are becoming a matter of increasing concern in ecotoxicology, and in several textbooks [e.g., 118], it is referred even as 'behavior toxicology'. The most common behavior responses to judge pollutant exposure are related to the three endpoints that define the species fitness, i.e., growth, survival and reproduction [118]. Behaviour responses such as avoidance, migration, locomotor activity, foraging, learning, predation, competition, reproductive behavior or social interactions are the most common behaviour endpoints in Ecotoxicology.

The relationship between behavior changes and inhibition of AChE activity has been studies in a wide range of organisms. The ability to retract the mantle fringes and close the valves in mussels was altered when animals were exposed to acephate [119] or dichlorvos [120]. The sheltered behavior of juvenile lobsters (*Homarus americanus*) was also perturbed under continuous exposure to azamethiphos [121]. The insecticide dimethoate (0.35 □g cm^{-2} for 36 h) caused significant changes in the locomotor activity of the collembolan *Folsomia candida* [122]. Capowiez et al. [123] examined the response of two common biochemical biomarkers (AChE and GST) and the burrowing behavior of two earthworm species exposed to the chloronicotinyl insecticide imidacropid. The burrowing behavior was a more sensitive endpoint than biochemical biomarkers which did not respond to the insecticide. In the case of vertebrates, a considerable number of laboratory and field studies have examined the relationship of behavioral disturbances of birds to AChE inhibition [13,44-46]. Sublethal OP exposure causing around 50 % enzyme inhibition seems to be sufficient to lead to significant changes in a wide range of behavior responses (e.g., foraging, flying, resting posture, singing, etc.). Similarly, a substantial volume of literature describing perturbation or disruption of physiological systems directly involved in fish behavior has been recently reviewed [124]. An example of the direct link between inhibition of AChE activity and behavioral perturbation is shown in the investigations by Beauvais and coworkers. A good correlation was found between inhibition of brain AChE activity and decreasing swimming speed in larval rainbow trout exposed (24- and 96-h exposures) to carbaryl [125]. In other related study, larval rainbow trout were exposed to malathion (20 and 40 μg L^{-1}) and diazinon (250, 500 and 1,000 μg L^{-1}), and significant correlations were obtained between AChE inhibition and swimming speed or distance [126].

Comparatively, investigations on behavior perturbations related to ChE inhibition in herpetofauna are quite limited [127,128]. For example, Bain et al. [61] examined the adverse effects of sublethal doses (2.0 and 20 mg kg^{-1}) of fenitrothion on the thermal preferences and prey-capture ability of the Australian lizard *P. vitticeps*. Although plasma ChE remained inhibited (22 % respect to a control group) for a long period of time (21 d) after the higher OP dosage, no impact on behavioral responses was observed. A reduction in swimming speed of neonates of two snake species (*Seminatrix pygaea* and *Nerodia rhombifer*) has been also observed after exposure to 5.0 mg L^{-1} of carbaryl [57]. Although, ChE inhibition was not measured in that study, disruption of swimming performance could be related to AChE inhibition.

The abnormal behavior responses related to pollution can lead to serious consequences at population level. Caquet and Lagadic [129] suggested the following sequence of events leading to alterations in benthic communities of aquatic insects caused by OP and CB pesticides: functional disruption of nervous system (AChE inhibition), behavioral changes (hyperactivity), increasing of drift, and finally alteration in the community (changes in species diversity and richness). Despite this example, behavioral impairments caused by pesticides exposure are underutilized in the ERA. As previously mentioned, some of the main limitations could be the difficulties in recording the behavior responses in the field, the standardization of behavior endpoints in toxicity testing or the extrapolation of behavioral effects from laboratory to field [118]. However, efforts are being performed to make a quantitative analysis of behavior based on the image processing [130, 131], although these sophisticated and relatively high cost recording systems probably discourage their use in monitoring pollution effects especially when legal requirements and environmental management decisions are required.

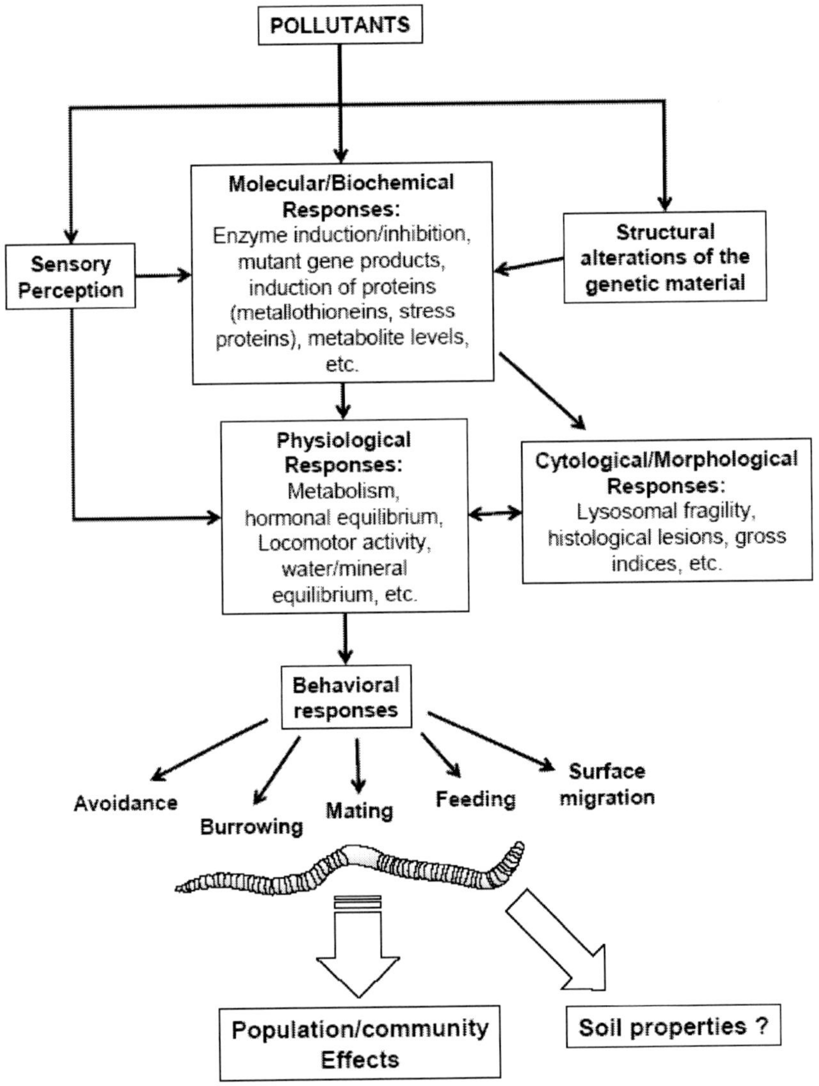

Figure 9. Conceptual scheme of biological responses of earthworms to pollutants. Pollutant-induced behavioral changes are the result from the integration of many biochemical and physiological systems altered by pollutants. Abnormal behavioral responses can have adverse consequences at population or community levels and, perhaps, in the ecological and physicochemical properties of soil because of their contribution to improve soil structure and function. Adapted from Sanchez-Hernandez [67].

Standardized behavioral bioassays are becoming a common approach in the toxicity tests. A standardized behavioral endpoint is the ability to avoid the contaminated medium. Thus, behavioural avoid response of earthworms has been used for developing a rapid, low cost and sensitive toxicity test [e.g., 132]. The most common design to carry out the test is a two-chamber system. This is a rectangular container which is divided in two equal compartments by a removable plastic separator. A control soil is placed in a compartment whereas the contaminated soil is placed in the other. Earthworms are then released in the middle of the rectangular container after remove the plastic split. Elapsed a period of exposure, the plastic separator is inserted again in the middle of the rectangular container and individuals are counted in each soil. The avoidance response is judged as positive when a percent of live earthworms higher than 80 % is found in the compartment containing the reference soil. This simple test can have an ecologically relevant meaning, because this escape behavior could lead to alter the community integrity of the soil or changes in physicochemical properties of soil. It is well known that earthworms contribute significantly to improve the soil structure and function, and they are important member of the agroecosystems community [72,133]. A considerable number of studies have demonstrated that the avoidance response of earthworms occurs often at low levels of metal concentration, at which survival and reproduction, for instance, are not affected [134-137]. However, this avoidance behavior test has been scarcely used for assessing pesticide-contaminated soils. One of the most attractive attribute of this test is that it could be used under field conditions without the need of sophisticated apparatus for behavior recording. Earthworms could be released in a mesocosm system located in the study area, and avoidance behavior responses could be monitored in relation to pesticide applications. Figure 10 represents a tentative of such an *in situ* exposure bioassay to monitor pesticide effects. This toxicity approach would enable not only examine the toxic effects of pesticide on earthworm, but also to investigate the contribution of fluctuating environmental variables on chemical toxicity. Avoidance behavior test is based on the ability of earthworm to detect toxic substance by its chemoreceptors located in the prostomium and anterior segment. This ability of chemical detection by earthworms makes the voidance behavior test a simple measure of repellence instead of a measure of toxicity as pointed out by Capowiez and Bérard [131]. However, the meaning of the avoidance behavior test can be modified whether earthworms are released in the chamber containing the contaminated soil and after a fixed period of time, the separator is removed enabling that earthworm can move toward the clean soil. Under this approach avoidance ability, locomotor activity and ChE inhibition can be evaluated all together and quantitative relationships can be established in relation to pesticide exposure; an important aspect not considered in the standardized avoidance behavior test. A more ecological realism is given in the relationships between biochemical and behavioral biomarkers when this approach is performed in the natural environment (Fig. 10).

Changes in behavioral responses should be assisted by measuring of biochemical and physiological biomarkers in order to give solid evidences of the chemical or chemicals implicated in the behavior alterations. In the case of anti-ChE pesticides this link seems relatively easy to establish because AChE is specifically inhibited by this type of agrochemicals. Although other contaminants (e.g., certain heavy metals or detergents; [12]) can also inhibit this enzyme, their effective concentrations are quite high and probably no environmentally realistic. According to the concept of a hierarchical cascade of biological responses to pollution occurring at different levels of biological organization (Fig. 9), behavioral responses could be the key biomarkers for making predictive assessments of pollution at population or community levels. When quantitative relationships can be established between ChE inhibition, behavior alterations and detrimental effects at population level, then we might consider ChE as a true ecotoxicological biomarker and not an exposure biomarker only.

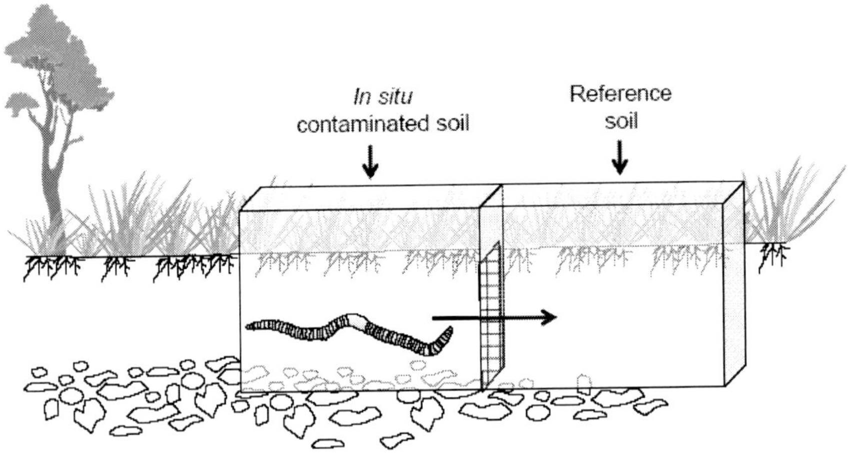

Advantages
- Homogeneous earthworms (e.g., weight, sexual maturation) can be used to minimize variations of B-esterase responses due to biological variables.
- Influence of fluctuating environmental variables on B-esterases can be assessed in the control (non-exposed) group.
- Impact of environmental variables on B-esterase responses to pesticides can be investigated in a real scenario.
- The avoidance behaviour response in relation to exposure degree can be controlled by using the separator.
- Realistic exposure conditions.

Limitations
- Vandalism and damage by adverse atmospheric conditions.
- Time consuming and relatively high effort (e.g., cage construction and deployment).
- Effects on earthworm behaviour from the deployment system.
- Use of a suitable reference soil. It should have similar physicochemical properties than *in situ* contaminated soil.

Figure 10. A tentative design for assessing *in situ* exposure of earthworms to pesticides. Responses of ChE inhibition can be studied in relation to the avoidance behaviour response under the impact of fluctuating environmental variables. See text for explanation. Furthermore, with the aid of a separator placed between the two test chambers it is possible to control the earthworm exposure in the compartment containing the polluted soil. The consequences of AChE inhibition on the ability of earthworm to escape from the polluted soil can be assessed under natural environmental conditions. Adapted from Sanchez-Hernandez [67].

Perspectives of B-esterases as Ecotoxicological Biomarkers

In 2003, an editorial entitled “Relevant environmental science” appeared in the scientific journal Environmental Toxicology and Chemistry [138]. In this article, a set of recommendations, grouped into four basic rules, are suggested to improve the relevance of the environmental studies. I would underlie two of them:

Rule 1: Ensure that phenomena being demonstrated are environmentally relevant.
Rule 2: Do not continue to demonstrate phenomena that are well established.

Owing of ChE activity is ubiquitous in animals and acute toxicity by CB and OP pesticides is the inhibition of this enzyme, the interaction between these two elements is expected to be the same in the animal kingdom. Even, this interaction has led to ChEs to be one of the principal components of biosensors for toxicity monitoring [139]. However, we continue to demonstrate this well documented phenomenon using model anti-ChE pesticides (e.g., paraoxon, dichlorvos or carbaryl). Whether ChE activity is used for field monitoring of pesticide contamination, we should test other ChE properties other than its sensitivity to anti-ChE pesticides. For instance, time to full recovery of enzyme after inhibition, natural variation of the enzyme activity, impact of fluctuating environmental variables on enzyme responses to pesticides, or advantages and limitations of reactivation technique employing oximes need still further understanding. Based in the subjects discussed in this chapter, some lines of future research are recommended to gain knowledge on the ecotoxicology of B-esterases.

Chemical reactivation of phosphorylated ChE using oximes is a complementary – or definitive in many cases – methodology for identifying field exposure to OP pesticides. It is useful when a reference group is not available or is not well represented, or when it is necessary to know the type of pesticide involved in the ChE inhibition. However, several methodological aspects should be considered when we use this approach. Optimum concentration of oxime must be determined to get a successful reactivation rate of the phosphorylated enzyme and, in the same time, avoid ChE inhibition by the oxime. Other practical aspects to be considered in future investigations using 2-PAM could be: i) to check if the oxime is able to reactivate the phosphorylated ChE through *in vitro* experiments, and ii) the ability of 2-PAM reactivation in relation to the time elapsed since OP exposure. This period of efficacy in the enzyme reactivation may be also determined through *in vitro* and *in vivo* experiments.

Behavior responses should be considered in future research because of their implications at population levels. Behavioral avoidance response of earthworms to pesticide contaminated soils is an attractive model for examining such a relationship. The introduction of biomarkers such as B-esterases in these experimental protocols could be an opportunity to investigate the relationship between B-esterase responses and behavior perturbation.

Tolerance/Resistance to pesticides is primarily attributed to A- and B-esterases. The latter contribute to decrease the impact of anti-ChE pesticides at the molecular target level (brain AChE activity) binding to circulating pesticides. Although this relationship between levels or over-expression of B-esterases and pesticide resistance seems to be clear for pest species, few data support a similar relationship that occurs in non-target organisms. However, to what extent B-esterases participate in the tolerance of non-target species to insecticides remains to be clarified. This could be a significant ecotoxicological endpoint to identify the species at risk or the most tolerant species to pesticide contamination. Field studies are still necessary to investigate the potential of B-esterases (blood BChE and CbE activities) as biomarker of pesticide exposure, especially under the approach of pesticide tolerance. What is the role of these B-esterases in a pesticide-polluted environment other than to be sensitive, rapid and relatively low cost indicators of pesticide contamination?

Natural variations of B-esterase activities as well as their response to environmental fluctuating variables need still to be investigated in detail. In some cases such as CbE activity the knowledge about normal variations of enzyme activity is practically inexistent.

Acknowledgments

Part of the material in this contribution was funded by the University of Castilla-La Mancha (Spain) and by the Consejería de Educación y Ciencia (Junta de Castilla-La Mancha, Spain).

References

[1] Aldridge, W.N. (1953). Two types of esterases (A and B) hydrolysing p-nitrophenyl acetate, propionate and butyrate, and a method for their determination. *Biochem. J.*, **53**, 110–117.
[2] Ludke, J.L., Hill, E.F. & Dieter, M.P. (1975). Cholinesterase (ChE) response and related mortality among birds fed ChE inhibitors. *Arch. Environ. Contam. Toxicol.* **3**, 1–21.
[3] Hall, R.J. & Clark D.R. Jr. (1982). Responses of the iguanid lizard Anolis carolinensis to four organophosphorus pesticides. *Environ. Pollut.*, **28**, 45–52.
[4] Walker, C.H. (1998). The use of biomarkers to measure the interactive effects of chemicals. *Ecotoxicol. Environ. Saf.* **40**, 65–70.
[5] Fulton, M.H. & Key, P.B. (2001). Acetylcholinesterase inhibition in estuarine fish and invertebrates as an indicator of organophosphorus insecticide exposure and effects. *Environ. Toxicol. Chem.*, **20**, 37-45.
[6] Satoh, T. & Hosokawa, M. (1998). The mammalian carboxylesterases: from molecules to functions. *Annu. Rev. Pharmacol. Toxicol.*, **38**, 257–288.
[7] Brimijoin, S. (2005). Can cholinesterase inhibitors affect neural development? *Environ. Toxicol. Pharmacol.*, **19**, 429–432.
[8] Layer, P.G., Allebrandt, K., Andermann, P., Bodur, E., Boopathy, R., Bytyqi, A.H. & Paraoanu, L.E. (2005). On the multifunctionality of cholinesterases. *Chemico-Biological Interactions*, **157**, 37–41.
[9] Fukuto, T.R. (1990). Mechanism of action of organophosphorus and carbamate insecticides. *Environ. Health. Perspect.*, **87**,245–254.
[10] Sogorb, M.A. & Vilanova, E. (2002). Enzymes involved in the detoxification of organophosphorus, carbamate and pyrethroid insecticides through hydrolysis. *Toxicol. Letters*, **128**, 215–228.
[11] Thompson, H.M. & Walker, C.H. (1994). Blood esterases as indicators of exposure to organophosphorus and carbamate insecticides. In: Fossi, M.C. & Leonzio, C. (eds.), Non-destructive Biomarkers in Vertebrates. Lewis Publisher, Chelsea, MI, U.S.A., pp. 37-62.
[12] Sanchez-Hernandez, J.C. (2001). Wildlife exposure to organophosphorus insecticides. *Rev. Environ. Contam. Toxicol.*, **172**, 21–63.
[13] Hill, E.F. (2003). Wildlife toxicology of organophosphorus and carbamate pesticide. In: Hoffman, D.J., Rattner, B.A., Burton, G.A., Cairns, J.J. (eds), *Handbook of*

Ecotoxicology, 2nd Edition. Lewis Publishers, Boca Raton, Florida, U.S.A., pp. 281–312.

[14] Cooper, J.R. (1994). Unsolved problems in the cholinergic nervous system. *J. Neurochem.*, **63**, 395–399.

[15] Small, D.H. (1995). A function for butyrylcholinesterase? *J. Neurochem.*, **64**, 466–467.

[16] De Vriese, C., Gregoire, F., Lema-Kisoka, R., Waelbroeck, M., Robberecht, P. & Delporte, C. (2004). Ghrelin degradation by serum and tissue homogenates: identification of the cleavage sites. *Endocrinology*, **145**, 4997–5005.

[17] Li, B., Stribley, J.A., Ticu, A., Xie, W., Schopfer, L.M., Hammond, P., Brimijoin, S., Hinrichs, S.H. & Lockridge, O. (2000). Abundant tissue butyrylcholinesterase and its possible function in the acetylcholinesterase knockout mouse. *J. Neurochem.*, **75**, 1320–1331.

[18] Vidal, C.J. (2005). Expression of cholinesterases in brain and non-brain tumours. *Chemico-Biological Interactions*, **157**-158, 227–232.

[19] Allon, N., Raveh, L., Gilat, E., Cohen, E., Grunwald, J. & Ashani, Y. (1998). Prophylaxis against soman inhalation toxicity in guinea pigs by pre-treatment alone with human serum butyrylcholinesterases. *Toxicol. Sci.*, **43**, 121–128.

[20] Salles, J.B., Cunha Bastos, V.L.F., Silva Filho, M.V., Machado, O.L.T., Salles, C.M.C., Giovanni de Simone, S. & Cunha Bastos, J. (2005). A novel butyrylcholinesterase from serum of *Leporinus macrocephalus*, a neotropical fish. Biochimie, in press.

[21] Sanchez, J.C., Fossi, M.C. & Focardi, S. (1997). Serum "B" esterases as a non-destructive biomarker in the lizard *Gallotia galloti* experimentally treated with parathion. *Environ. Toxicol. Chem.*, **16**, 1954–1961.

[22] Jokanovic, M., Kosanovi, M. & Maksimovic, M. (1996). Interaction of organophosphorus compounds with carboxylesterases in the rat. *Arch. Toxicol.*, **70**, 444–450.

[23] Maxwell, D.M. & Brecht, K.M. (2001). Carboxylesterases: specificity and spontaneous reactivation of an endogenous scavenger for organophosphorus compounds. *J. Appl. Toxicol.*, **21**, S103–S107.

[24] Jokanovic, M. (2001). Biotransformation of organophosphorus compounds. *Toxicology*, **166**, 139–160.

[25] Satoh, T. (2005). Toxicological implications of esterases – from molecular structures to functions. *Toxicol. Appl. Pharmacol.*, **207**, S11 –S18.

[26] Kozlovskaya, V.I., Mayer, F.L., Menzikova, O.V. & Chuyko, G.M. (1993). Cholinesterase of aquatic animals. *Rev. Environ. Contam. Toxicol.* **132**, 117–141.

[27] Austin, L. & Berry, W.K. (1953). Two selective inhibitors of cholinesterase. *Biochem. J.*, **54**, 695–700.

[28] Monteiro, M., Quintaneiro, C., Morgado, F., Soares, A.M.V.M. & Guilhermino, L. (2005). Characterization of the cholinesterases present in head tissues of the estuarine fish *Pomatoschistus microps*: Application to biomonitoring. *Ecotoxicol. Environ. Saf.*, **62**, 341–347.

[29] Sturm, A., da Silva da Assis, H.C. & Hansen, P.D. (1999). Cholinesterases of marine teleost fish: enzymological characterization and potential use in the monitoring of neurotoxic contamination. *Mar. Environ. Res.*, **47**, 389–398.

[30] Rodríguez-Fuentes, G. & Gold-Bouchot, G. (2004). Characterization of cholinesterase activity from different tisúes of Nile tilapia (*Oreochromis niloticus*). *Mar. Environ. Res.*, **58**, 505–509.

[31] Varó, I., Navarro, J.C., Amat, F & Guilhermino, L. (2003). Effect of dichlorvos on cholinesterase activity of the European sea bass (*Dicentrarchus labrax*). *Pest. Biochem. Physiol.*, **75**, 61–72.

[32] Bocquené, G., Roig, A. & Fournier, D. (1997). Cholinesterases from the common oyster (*Crassostrea gigas*). Evidence for the presence of a soluble acetylcholinesterase insensitive to organophosphate and carbamate inhibitors. *Fed. Eur. Biochem. Soc. Letters.*, **407**, 261–266.

[33] Chuiko, G.M. (2000). Comparative study of acetylcholinesterase and butyrylcholinesterase in brain and serum of several freshwater fish: specific activities and in vitro inhibition by DDVP, an organophosphorus pesticide. *Comp. Biochem. Physiol., C*, 127, 233–242.

[34] Wogram, J., Sturm, A., Segner, H. & Liess, M. (2001). Effects of parathion on acetylcholinesterase, butyrylcholinesterase, and carboxylesterase in three-spined stickleback (*Gasterosteus aculeatus*) following short-term exposure. *Environ. Toxicol. Chem.*, **20**, 1528–1531.

[35] Whitehead, A., Anderson, S.L., Ramirez, A. & Wilson B.W. (2005). Cholinesterases in aquatic biomonitoring: Assay optimization and species-specific characterization for a California native fish. *Ecotoxicology*, **14**, 597–606.

[36] Hyne, R.V. & Maher, W.A. (2003). Invertebrate biomarkers: links to toxicosis that predict population decline. *Ecotoxicol. Environ. Saf.*, **54**, 366–374.

[37] Barron, M.G., Charron, K.A., Stott, W.T. & Duvall, S.E. (1999). Tissue carboxylesterase activity of rainbow trout. Environ. Toxicol. Chem., 18, 2506–2511.

[38] Barata, C., Solayan, A. & Porte, C. (2004). Role of B-esterases in assessing toxicity of organophosphorus (chlorpyrifos, malathion) and carbamate (carbofuran) pesticides to *Daphnia magna*. *Aquat. Toxicol.*, **66**, 125–139.

[39] Escartín, E. & Porte, C. (1997). The use of cholinesterase and carboxylesterase activities from *Mytilus galloprovinciales* in pollution monitoring. *Environ. Toxicol. Chem.*, **16**, 2090–2095.

[40] Galloway, T.S., Millward, N., Browne, M.A. & Depledge, M.H. (2002). Rapid assessment of organophosphorus/carbamate exposure in the bivalve mollusc *Mytilus edulis* using combined esterase activities as biomarkers. *Aquat. Toxicol.*, **61**, 169–180.

[41] Fourcy, D., Jumel, A., Heydorff, M. & Lagadic, L. (2002). Esterases as biomarkers in *Nereis* (Hediste) *diversicolor* exposed to temephos and *Bacillus thuringiensis* var. *israelensis* used for mosquito control in coastal wetlands of Morbihan (Brittany, France). *Mar. Environ. Res.*, **54**, 755–759.

[42] Küster, E. (2005). Choli- and carboxylesterase activities in developing zebrafish embryos (*Danio rerio*) and their potential use for insecticide hazard assessment. *Aquat. Toxicol.*, **75**, 76–85.

[43] Sheffield, S.R., Sawicka-Kapusta, K., Cohen, J.B. & Rattner, B.A. (2001). Rodentia and Lagomorpha. In: Shore, R.F. & Rattner, B.A. (Eds.), *Ecotoxicology of Wild Mammals*. John Wiley and Sons, West Sussex, U.K., pp. 215-314.

[44] Peakall, D. (1992). *Animal biomarkers as pollution indicators*. Chapman & Hall, London, U.K.

[45] Hart, A.D.M. (1993). Relationships between behavior and the inhibition of acetylcholinesterase in birds exposed to organophosphorus pesticides. *Environ. Toxicol. Chem.*, **12**, 321–336.

[46] Walker, C.H. (2003). Neurotoxic pesticides and behavioural effects upon birds. *Ecotoxicology*, **12**, 307–316.

[47] Fleming, W.J. (1981). Recovery of brain and plasma cholinesterase activities in ducklings exposed to organophosphorus pesticides. *Arch. Environm. Contam. Toxicol.* **10**, 215–229.

[48] Holmes, S.B. & Boag, P.T. (1990). Inhibition of brain and plasma cholinesterase activity in zebra finches orally dosed with fenitrothion. *Environ. Toxicol. Chem.* **9**, 323–334.

[49] McInnes, P.F., Andersen, D.E., Hoff, D.J., Hooper, M.J. & Kinkel, L.L. (1996). Monitoring exposure of nestling songbirds to agricultural application of an organophosphorus insecticide using cholinesterase activity. *Environ. Toxicol. Chem.* **15**, 544–552.

[50] Parsons, K.C., Matz, A.C., Hooper, M.J. & Pokras, M.A. (2000). Monitoring wading bird exposure to agricultural chemicals using serum cholinesterase activity. *Environ. Toxicol. Chem.*, **19**, 1317–1323.

[51] Iko, W.M., Archuleta, A.S. & Knopf, F.L. (2003). Plasma cholinesterase levels of mountain plovers (*Charadrius montanus*) wintering in Central California, USA. *Environ. Toxicol. Chem.*, **22**, 119–125.

[52] Maul, J.D. & Farris, J.L. (2005). Monitoring exposure of northern cardinals, *Cardinalis cardinalis*, to cholinesterase-inhibiting pesticides: enzyme activity, reactivations, and indicators of environmental stress. *Environ. Toxicol. Chem.*, **24**, 1721–1730.

[53] Hall, R.J. & Henry F.P. (1992). Assessing effects of pesticides on amphibians and reptiles: status and needs. *Herpetol. J.* **2**, 65–71.

[54] Lambert, M.R.K. (1997). Effects of pesticides on amphibians and reptiles in Sub-Saharan Africa. *Rev. Environ. Contam. Toxicol.* **150**, 31–73.

[55] Campbell, K.R. & Campbell, T.S. (2000). Lizard contaminant data for ecological risk assessment. *Rev. Environ. Contam. Toxicol.* **165**, 39–116.

[56] Pauli, B.D. & Money, S. (2000). Ecotoxicology of pesticides in reptiles. In: Sparling, D.W., Linder, G. & Bishop, C.A. (eds.), Ecotoxicology of Amphibians and Reptiles. SETAC Press, Pensacola (FL), U.S.A., pp. 269-324.

[57] Hopkins, W.A. Winne, C.T. & DuRant, S.E. (2005). Differential swimming performance of two natricine snakes exposed to a cholinesterase-inhibiting pesticide. *Environ. Pollut.*, **133**, 531–540.

[58] Boone, M.D. & Bridges, C.M. (2003). Effects of carbaryl on green frog (*Rana clamitans*) tadpoles: timing of exposure versus multiple exposures. *Environ. Toxicol. Chem.*, **22**, 2695–2702.

[59] Boone, M.D. & James, S.M. (2003). Interactions of an insecticide, herbicide, and natural stressors in amphibian community mesocosms. *Ecol. Appl.*, **13**, 829–841.

[60] Clark, D.R., Bickham, J.W., Baker, D.L. & Cowman, D.F. (2000). Environmental contaminants in Texa, USA, wetland reptiles: evaluation using blood samples. *Environ. Toxicol. Chem.*, **19**, 2259–2265.

[61] Bain, D., Buttermer, W.A., Astheimer, L., Fildes, K. & Hooper, M.J. (2004). Effects of sublethal fenitrothion ingestion on cholinesterase inhibition, standard metabolism,

thermal preference, and prey-capture ability in the Australian central bearded dragon (*Pogona vitticeps*, Agamidae). *Environ. Toxicol. Chem.*, **23**, 109–116.

[62] Sanchez-Hernandez, J.C., Carbonell, R., Henríquez-Pérez, A., Montealegre, M. & Gómez, L. (2004). Inhibition of plasma butyrylcholinesterase activity in the lizard *Gallotia galloti palmae* by pesticides: a field study. *Environ. Pollut.*, **132**, 479–488.

[63] Van Straalen, N.M. & Van Gestel, C.A.M. (1998). Soil invertebrates and micro-organisms. In: Calow, P. (ed), Handbook of Ecotoxicology, Blackwell Science Ltd, Oxford, UK, pp 251–277.

[64] Cortet, J., Vauflery, A.G., Poinsot-Balaguer, N., Gomot, L., Texier, C. & Cluzeau, D. (1999). The use of invertebrate soil fauna in monitoring pollutant effects. *Eur. J. Soil Biol.*, **35**, 115–134.

[65] Kammenga, J.E., Dallinger, R., Donker, M.H., Köhler, H.R., Simonsen, V., Triebskorn, R., Weeks, J.M. (2000). Biomarkers in terrestrial invertebrates for ecotoxicological soil risk assessment. *Rev. Environ. Contam. Toxicol.* **164**:93–147.

[66] Scott-Fordsmand, J.J. & Weeks, J.M. (2000). Biomarkers in earthworms. *Rev. Environ. Contam. Toxicol.* **165**:117–159.

[67] Sanchez-Hernandez, J.C. (2006). Earthworm biomarkers in ecological risk assessment. *Rev. Environ. Contamin. Toxicol.*, **188**, 85–126.

[68] Pree, D.J., Whitty, K.J., Bittner, L.A. & Pogoda, M.K. (2002). Mechanisms of resistance to organophosphorus insecticides in populations of the obliquebanded leafroller *Choristoneura rosaceana* (Harris) (Lepidoptera: Tortricidae) from southern Ontario. *Pest. Manag. Sci.*, **59**, 79–84.

[69] Stumpf, N., Zebits, C.P.W., Kraus, W., Moores, G.D. & Nauen, R. (2001). Resistance to organophosphates and biochemical genotyping of acetylcholinesterases in *Tetranychus urticae* (Acari: Tetranychidae). *Pest. Biochem. Physiol.*, **69**, 131–142.

[70] Hemingway, J., Hawkes, N.J., McCarroll, L. & Ranson, H. (2004). The molecular basis of insecticide resistance in mosquitoes. *Insect Biochem. Mol. Biol.*, **34**, 653–665.

[71] Coeurdassier, M., Saint-Denis, M., Vaufleury, A.G., Ribera, D. & Badot, P.M. (2001). The garden snail (*Helix aspersa*) as a bioindicator of organophosphorus exposure: effects of dimethoate on survival, growth, and acetylcholinesterase activity. *Environ. Toxicol. Chem.*, **20**, 1951–1957.

[72] Paoletti, M.G. (1999). The role of earthworms for assessment of sustainability and as bioindicators. *Agric. Ecosyst. Environ.* **74**, 137–155.

[73] Stürzenbaum, S.R., Kille, P. & Morgan, A.J. (1998). The identification, cloning and characterization of earthworm metallothioneins. *FEBS Letters* **431**, 437–442.

[74] Svendsen, C., Spurgeon, D.J., Hankard, P.K. & Weeks, J.M. (2004). A review of lysosomal membrane stability measured by neutral red retention: is it a workable earthworm biomarker? *Ecotox. Environ. Saf.*, **57**, 20–29.

[75] Ribera, D., Narbonne, J.F., Arnaud, C., Saint-Denis, M. (2001). Biochemical responses of the earthworm *Eisenia fetida andrei* exposed to contaminated artificial soil, effects of carbaryl. *Soil Biol. Biochem.*, **33**, 1123–1130.

[76] Booth, L.H., Hodge, S. & O'Halloran, K. (2000). Use of cholinesterase in *Aporrectodea caliginosa* (Oligochaeta: Lumbricidae) to detect organophosphate contamination: comparison of laboratory tests, mesocosms, and field studies. *Environ. Toxicol. Chem.*, **19**, 417–422.

[77] Rao, J.V., Pavan, Y.S. & Madhavendra, S.S. (2003). Toxic effects of chlorpyrifos on morphology and acetylcholinesterase activity in the earthworm, *Eisenia foetida. Ecotoxicol. Environ. Saf.*, **54**, 296–301.

[78] Sorour, J. & Larink, O. (2001). Toxic effects of benomyl on the ultrastructure during spermatogenesis of the earthworm *Eisenia fetida. Ecotoxicol. Environ. Saf.*, **50**, 180–188.

[79] Espinosa-Navarro, O. & Bustos-Obregón, E. (2005). Effect of malathion on the male reproductive organs of earthworms, *Eisenia foetida. Asian J. Androl*., **7**, 97–101.

[80] Booth, L.H. & O'Halloran, K. (2001). A comparison of biomarker responses in the earthworm *Aporrectodea caliginosa* to the organophosphorus insecticides diazinon and chlorpyrifos. *Environ. Toxicol. Chem*., **20**, 2494–2502.

[81] Racke, K.D. (1992). Degradation of organophosphorus insecticides in environmental matrices. In: Chambers, J.E.& Levi, E.P. (eds.), Organophosphates: *Chemistry, Fate, and Effects*. Academic, New York, pp. 47–77.

[82] Panda, S. & Sahu, S.K. (2004). Recovery of acetylcholine esterase activity of *Drawida willsi* (Oligochaeta) following application of three pesticides to soil. *Chemosphere* **55**, 283–290.

[83] Dell'Omo, G., Turk, A. & Shore, R.F. (1999). Secondary poisoning in the common shrew (*Sorex araneus*) fed earthworms exposed to an organophosphate pesticide. *Environ. Toxicol. Chem*., **18**, 237–240.

[84] Sanchez, J.C., Fossi, M.C. & Focardi, S. (1997). Serum "B" esterases as a nondestructive biomarker for monitoring the exposure of reptiles to organophosphorus insecticides. *Ecotoxicol. Environ. Saf.*, **38**, 45-52.

[85] Roy, C., Grolleau, G., Chamoulaud, S. & Rivière, J.L. (2005). Plasma B-esterase activities in European raptors. *J. Wild. Dis*., **41**, 184–208.

[86] Thompson, H.M., Walker, C.H. & Hardy, A.R. (1991). Inhibition of avian esterases by organophosphorus insecticides: problems of reactivation and storage. *Arch. Environ. Contam. Toxicol.,* **17**, 379–386.

[87] Bartkowiak, D.J. & Wilson, B.W. (1995). Avian plasma carboxylesterase activity as a potential biomarker of organophosphate pesticides exposure. *Environ. Toxicol. Chem.* **14**, 2149–2153.

[88] Walker, C.H. (1995). Biochemical biomarkers in ecotoxicology – some recent developments. *Sci. Total Environ.,* **171**, 189–195.

[89] Relyea, R.A. & Mills, N. (2001). Predator-induced stress makes the pesticide carbaryl more deadly to gray treefrog tadpoles (*Hyla versicolor*). *Proc. Natl. Acad. Sci*., **98**, 2491–2496.

[90] Rattner, B.A., Becker, J.M. & Nakatsugawa, T. (1987). Enhancement of parathion toxicity to quail by heat and cold exposure. *Pest. Biochem. Physiol*., **27**, 330–339.

[91] Van Straalen, N.M. (2003). Ecotoxicology becomes stress ecology. *Environ. Sci. Technol.* **37**:324A–330A.

[92] Burton, G.A., Batley, G.E., Chapman P.M., Forbes, V.E., Smith, E.P., Reynoldson, T., Schlekat, C.E., den Besten, P.J., Bailer, A.J., Green A.S. & Dwyer, R.L. (2002). A weight-of-evidence framework for assessing sediment (or other) contamination: improving certainly in the decision-making process. Human Ecol. *Risk Assess*., **8**: 1675–1696.

[93] Neuparth, T., Correia, A.D., Costa, F.O., Lima, G. & Costa, M.H. (2005). Multi-level assessment of chronic toxicity of estuarine sediments with the amphipod *Gammarus locusta*: I. Biochemical endpoints. *Mar. Environ. Res.* **60**, 69–91.

[94] Maxwell, D.M., Lieske, C.N. & Brecht, K.M. (1994). Oxime-induced reactivation of carboxylesterase inhibited by organophosphorus compounds. *Chem. Res. Toxicol.*, **7**, 428–433.

[95] Monserrat, J.M. & Bianchini, A. (2000). Methodological and biological aspects to be considered in acetylcholinesterase reactivation assays using 2-PAM. *Environ. Toxicol. Pharmacol.*, **9**, 39–47.

[96] Barahona, M.V. & Sánchez-Fortún, S. (1999). Toxicity of carbamates to the brine shrimp Artemia salina and the effect of atropine, BW284c51, iso-OMPA and 2-PAM on carbaryl toxicity. *Environ. Pollut.*, **104**, 469–476.

[97] Lieske, C., Clarck, J.H., Maxwell, D.M., Zoeffel, L.D. & Sultan, W.E. (1992). Studies of the amplification of carbaryl toxicity by various oximes. *Toxicol. Lett.*, **62**, 127–137.

[98] Yerushalmi, N. & Cohen, E. (2002). Acetylcholinesterase of the California red scale *Aonidiella aurantii* Mask: catalysis, inhibition, and reactivation. *Pest. Biochem. Physiol.*, **72**, 133–141.

[99] Sanchez-Hernandez, J.C. & Moreno Sanchez, B. (2002). Lizard cholinesterases as biomarkers of pesticide exposure: enzymological characterization. *Environ. Toxicol. Chem.*, **21**, 2319–2325.

[100] Binelli, A., Ricciardi, F., Riva, C. & Provini, A. (2006). New evidences for old biomarkers: effects of several xenobiotics on EROD and AChE activities in Zebra mussel (*Dreissena polymorpha*). *Chemosphere*, **62**, 510–519.

[101] Maul, J.D. & Farris, J.L. (2004). Monitoring exposure of passerines to acephate, dicrotophos, and malathion using cholinesterase reactivation. *Bull. Environ. Contam. Toxicol.*, **73**, 682–689.

[102] Sanchez-Hernandez, J.C. (2003). Evaluating reptile exposure to cholinesterase-inhibiting agrochemicals by serum butyrylcholinesterase activity. *Environ. Toxicol. Chem.*, **22**, 296–301.

[103] Escartín, E. & Porte, C. (1996). Acetylcholinesterase inhibition in the crayfish *Procambarus clarkii* exposed to fenitrothion. *Ecotoxicol. Environ. Saf.*, **34**, 160–164.

[104] Moon, A.P. & Smith, T.J. (2005). Organophosphate inhibition of an arylesterase from *Lumbricus terrestris* and reversal by pralidoxime. *Soil Biol. Biochem.*, **37**, 1211–1213.

[105] Raveh, L., Grauer, E., Grunwald, J., Cohen, E. & Ashani, Y. (1997). The stoichiometry of protection against soman and Vx toxicity in monkeys pretreated with human butyrylcholinesterase. *Toxicol. Appl. Pharmacol.*, **145**, 43–53.

[106] Doctor, B.P. & Saxena, A. (2005). Bioscavengers for the protection of humans against organophosphate toxicity. *Chemico-Biological Interactions*, **157**-158, 167–171.

[107] Casida, J.E. & Quistad, G.B. (1998). Golden age of insecticide research: past, present, or future?. *Annu. Rev. Entomol.*, **43**, 1–16.

[108] Amichot, M., Bergé, J.B., Cuany, A., Pasteur, N., Pauron, D. & Raymond, M. (2000). Molecular, genetic and population bases of insect resistance to insecticides. In: Lagadic L., Caquet, T., Amiard, J.C. & Ramade, F. (eds.), Use of Biomarkers for Environmental Quality Assessment, A.A. Balkema, Rotterdam, The Netherlands, pp. 247–268.

[109] Hemingway, J. & Ranson, H. (2000). Insecticide resistance in insect vectors of human disease. *Annu. Rev. Entomol.*, **45**, 371–391.

[110] Mouchès, C., Magnin, M., Bergé, J.B., Silvestri, M., Beyssat, V., Pasteur, N. & Georghiou, G.P. (1987). Overproduction of detoxifing esterases in organophosphate-resistant Culex mosquitoes and their presence in other insects. *Proc. Natl. Acad. Sci.*, **84**, 2113–2116.

[111] Lari, L., Massi, A., Fossi, M.C., Casini, S., Leonzio C. & Focardi S. (1994). Evaluation of toxic effects of the organophosphorus insecticide azinphos-methyl in experimentally and naturally exposed birds. *Arch. Environ. Contam. Toxicol.* **26**, 234–239.

[112] Boutet, A., Tanguy, A. & Morana, D. (2004). Characterisation and expression of four mRNA sequences encoding glutathione S-transferases pi, mu, omega and sigma classes in the Pacific oyster *Crassostrea gigas* exposed to hydrocarbons and pesticides. *Mar. Biol.*, **146**, 53–64.

[113] Peeples, E.S., Schopfer, L.M., Duysen, E.G., Spaulding, R., Voelker, T., Thompson, C.M. & Lockridge, O. (2005). Albumin, a new biomarker of organophosphorus toxicant exposure, identified by mass spectrometry. *Toxicol. Sci.*, **83**, 303–312.

[114] Sogorb, M.A., Carrera, V., Benabent, M. & Vilanova, E. (2002). Rabbit serum albumin hydrolyzes the carbamate carbaryl. *Chem. Res. Toxicol.*, **15**, 520–526.

[115] Sogorb, M.A., Carrera, V. & Vilanova, E. (2004). Hydrolysis of carbaryl by human serum albumin. *Arch. Toxicol.*, **78**, 629–634.

[116] Lagadic, L., Caquet, T., Amiard, J.C. & Ramade, F. (2000). *Use of biomarkers for environmental quality assessment.* A.A. Balkema, Rotterdam, Netherlands.

[117] Walker, C.H., Hopkin, S.P., Sibly, R.M. & Peakall, D.B. (2001). *Principles of ecotoxicology*, 2nd edition, Taylor & Francis, London, UK

[118] Newman, M.C. & Unger, M.A. (2003). *Fundamentals of ecotoxicology*. 2nd edition, Lewis Publishers, Boca Raton, Florida, USA.

[119] Moulton, C.A., Fleming, W.J. & Purnell, C.E. (1996). Effects of two cholinesterase-inhibiting pesticides on freshwater mussels. *Environ. Toxicol. Chem.*, **15**, 131–137.

[120] McHenery, J.G., Linley-Adams, G.E., Moore, D.C., Rodger, G.K. & Davies, I.M. (1997). Experimental and field studies of effects of dichlorvos exposure on acetylcholinesterase activity in the gills of the mussel, *Mytilus edulis* L. Aquat. *Toxicol.* **38**, 125–143.

[121] Abgrall, P., Rangeley, R.W., Burridge, L.E. & Lawton, P. (2000). Sublethal effects of azamethiphos on shelter use by juvenile lobsters (*Homarus americanus*). *Aquaculture* **181**, 1–10.

[122] Sorensen, F.F., Bayley, M. & Baatrup, E. (1995). The effects of sublethal dimethoate exposure on the locomotor behavior of the collembolan *Folsomia candida* (Isotomidae). *Environ. Toxicol. Chem.*, **14**, 1587–1590.

[123] Capowiez, Y., Rault, M., Mazzia, C. & Belzunces, L. (2003). Earthworm behaviour as a biomarker – a case study using imidacloprid. *Pedobiologia*, **47**, 542–547.

[124] Scott, G.R. & Sloman, K.A. (2004). The effects of environmental pollutants on complex fish behaviour: integrating behavioural and physiological indicators of toxicity. *Aquat. Toxicol.*, **68**, 369–392.

[125] Beauvais, S.L., Jones, S.B., Parris, J.T., Brewer, S.K. & Little, E. (2001). Cholinergic and behavioural neurotoxicity of carbaryl and cadmium to larval rainbow trout (*Oncorhynchus mykiss*). Ecotoxicol. *Environ. Saf.*, **49**, 84–90.

[126] Beauvais, S.L., Jones, S.B., Brewer, S.K. & Little, E. (2000). Physiological measures of neurotoxicity of diazinon and malathion to larval rainbow trout (*Oncorhynchus mykiss*) and their correlation with behavioural measures. *Environ. Toxicol. Chem.*, **19**, 1875–1880.

[127] Sparling, D.W., Linder, G. & Bishop, C.A. (2000). *Ecotoxicology of amphibians and reptiles*. SETAC Press, Pensacola, Florida. 877 pp.

[128] Broomhall, S.D. (2005). Measuring chemical impacts on amphibians: ecotoxicity and behavioural data in governmental regulation. *Appl. Herpetol.*, **2**, 259–285.

[129] Caquet, Th. & Lagadic, L. (2000). Consequences of individual–level alterations on population dynamics and community structure and function. In: Lagadic, L., Caquet, Th., Amiard, J.C. & Ramade, F. (eds.), Use of Biomarkers for Environmental Quality Assessment, A.A. Balkema, Rotterdam, Netherlands, pp. 267–302.

[130] Spieser, O.H., Schwaiger, J., Ferling, H & Negele, R.D. (2001). An introduction to behavioral monitoring – effects of nonylphenol and ethinylestradiol on swimming behavior of juvenile carp. In: Butterworth, F.M., Gunatilaka, A. & Gonsebatt, M.E. (eds.), Biomonitors and Biomarkers as Indicators of Environmental Changes. Kluwer Academic/Plenum Publishers, New York, pp. 93–112.

[131] Capowiez, Y., Bastardie, F. & Costagliola, G. (2005). Sublethal effects of imidacloprid on the burrowing behaviour of two earthworm species: modifications of the 3D burrow systems in artificial cores and consequences on gas diffusion in soil. *Soil Biol. Biochem.*, in press.

[132] ISO. (2004). Draft: Soil Quality – avoidance test for evaluating the quality of soils and the toxicity of chemicals. Test with earthworms (*Eisenia fetida/andrei*). Geneve, Switzerland.

[133] Jongmans, A.G., Pulleman, M.M., Balabane, M., van Oort, F. & Marinissen, J.C.Y. (2003). Soil structure and characteristics of organic matter in two orchards differing in earthworm activity. *Appl. Soil Ecol.* **24**, 219–232.

[134] Schaefer, M. (2005). Assessing 2,4,6-trinitrotoluene (TNT)-contaminated soil using three different earthworm test methods. *Ecotox. Environ. Saf.*, **57**, 74–80.

[135] Loureiro, S., Soares, A.M.V.M. & Nogueira, A.J.A. (2005). Terrestrial avoidance behaviour tests as screening tool to assess soil contamination. *Environ. Pollut.*, **138**, 121–131.

[136] Lukkari, T., Aatsinki, M., Väisänen, A. & Haimi, J. (2005). Toxicity of cooper and zinc assessed with three different earthworm tests. *Appl. Soil Ecol.*, **30**, 133–146.

[137] Lukkari, T., Haimi, J. (2005). Avoidance of Cu- and Zn-contaminated soil by three ecologically different earthworm species. *Ecotoxicol. Environ. Saf.*, **62**, 35–41.

[138] Chapman, P.M. & Loehr, R.C. (2003). Relevant environmental science. *Environ. Toxicol. Chem.*, **22**, 2217–2218.

[139] Andreescu, S. & Marty, J.L. (2006). Twenty years research in cholinesterase biosensors: from basic research to practical applications. *Biomol. Engineer.*, **23**, 1–15.

[140] Walker, C.H. (1998). Birds. In: Calow, P. (ed.), *Handbook of Ecotoxicology*, Blackwell Science Ltd, Oxford, United Kingdom, pp. 326–338.

[141] Carr, R.L. & Chambers, J.E. (1991). Acute effects of the organophosphate paraoxon on schedule-controlled behaviour and esterase activity in rats: dose-response relationships. *Pharmacol. Biochem. Behav.* **40**:929–936.

[142] Sanchez-Hernandez, J.C. & Walker, C.H. (2000). *In vitro* and *in vivo* cholinesterase inhibition in lacertides by phosphonate- and phosphorothioate-type organophosphates. *Pest. Biochem. Physiol.*, **67**, 1–12.

[143] Valbonesi, P., Sartor, G. & Fabbri, E. (2003). Characterization of cholinesterase activity in three bivalves inhabiting the North Adriatic sea and their possible use as sentinel organisms for biosurveillance programmes. *Sci. Total Environ.*, **312**, 79–88.

[144] Monserrat, J.M., Bianchini, A. & Bainy, A.C.D. (2006). Kinetic and toxicological characteristics of acetylcholinesterase from the gills of oysters (*Crassostrea rhyzophorae*) and other aquatic species. *Mar. Environ.* Res., *in press*.

[145] Brown, M., Davies, I.M., Moffat, C.F., Redshaw, J & Craft, J.A. (2004). Characterisation of choline esterases and their tissue and subcellular distribution in mussel (*Mytilus edulis*). *Mar. Environ. Res.*, **57**, 155–169.

[146] Mora, P., Michel, X. & Narbonne, J.F. (1999). Cholinesterase activity as potential biomarker in two bivalves. *Environ. Toxicol. Pharmacol.*, **7**, 253–260.

[147] Frasco, M.F., Fournicr, D., Carvalho, F. & Guilhermino, L. (2006). Cholinesterase from the common prawn (*Palaemon serratus*) eyes: catalytic properties and sensitivity to organophosphorus and carbamate compounds. Aquat. Toxicol., *in press*.

[148] Forget, J & Bocquené, G. (1999). Partial purification and enzymatic characterization of acetylcholinesterase from the intertidal marine copepod *Tigriopus brevicornis*. *Comp. Biochem. Physiol. B*, **123**, 345–350.

[149] Forget, J., Livet, S. & Leboulenger, F. (2002). Partial purification and characterization of acetylcholinesterase (AChE) from the estuarine copepod *Eurytemora affinis* (Poppe). Comp. *Biochem. Physiol. C*, **132**, 85–92.

[150] Key, P.B. & Fulton, M.H. (2002). Characterization of cholinesterase activity in tissues of the grass shrimp (*Palaemonetes pugio*). *Pest. Biochem. Physiol.*, **72**, 186–192.

[151] Varó, I., Navarro, J.C., Amat, F & Guilhermino, L. (2002). Characterisation of cholinesterases and evaluation of the inhibitory potential of chlorpyrifos and dichlorvos to *Artemia salina* and *Artemina parthenogenetica*. *Chemosphere*, **48**, 563–569.

[152] Cunha, I., García, L.M. & Guilhermino, L. (2005). Sea-urchin (*Paracentrotus lividus*) glutathione S-transferase and cholinesterase activities as biomarkers of environmental contamination. *J. Environ. Monit.*, **7**, 288–294.

[153] Sturm, A., Wogram, J., Segner, H. & Liess, M. (2000). Different sensitivity to organophosphates of acetylcholinesterase and butyrylcholinesterase from three-spined stickleback (*Gasterosteus aculeatus*): Application in biomonitoring. *Environ. Toxicol. Chem*, **19**, 1607–1615.

[154] Sturm, A., Wogram, J., Hansen, P.D. & Liess, M. (1999). Potential use of cholinesterase in monitoring low levels of organophosphates in small streams: natural variability in three-spined stickleback (*Gasterosteus aculeatus*) and relation to pollution. *Environ. Toxicol. Chem.*, **18**, 194–200.

[155] Osten, J.R., Ortíz-Arana, A., Guilhermino, L. & Soares, A.M.V.M. (2005). In vivo evaluation of three biomarkers in the mosquitofish (*Gambusia yucatana*) exposed to pesticides. *Chemosphere*, **58**, 627–636.

[156] Ma, E., He, Y. & Zhu, K.Y. (2004). Comparative studies of acetylcholinesterases purified from two field populations of the oriental migratory locust (*Locusta migratoria manilensis*): implications of insecticide resistance. *Pest. Biochem. Physiol.*, **78**, 67–77.

In: Environmental Pollution: New Research
Editor: Rachel H. Plattenberg, pp. 47-98
ISBN: 1-60021-285-9
© 2007 Nova Science Publishers, Inc.

Chapter 2

EARTHWORMS FOR MONITORING METAL CONTAMINATION: FROM CELLS TO MOLECULES

***Barbara Plytycz*[1]*, Joanna Homa*[1]*, Azwady nor Aziz*[2]*, László Molnár*[2,3]*, Peter Kille*[2]*, A. John Morgan*[2]**

[1]Department of Evolutionary Immunobiology, Inetitute of Zoology, Jagiellonian University, R. Ingardena 6, 30-060 Krakow, Poland*

[2]Cardiff School of Biosciences, Main Building, P.O. Box 915, Cardiff University, Cardiff CF10 3TL, Wales, UK**

[3]Department of General Zoology and Neurobiology, University of Pécs, H-7624 Pécs, Hungary.

Abstract

Earthworms are widely used to assess the bioavailability and deleterious effects of metals in contaminated soils. This chapter discusses some of the direct and indirect methods that have recently been used to determine the ligand-binding speciation and the extent of the toxicologically-important bio-reactive metal and metalloid fractions in the cells of earthworms as receptor organisms. It proceeds to describe suites of cell-based biomarkers (e.g. morphometrics; neutral red retention time and immuno-competence assays), and their variants, that are more-or-less widely used for reporting chemically-evoked stress in earthworms under laboratory, semi-field, and field exposure conditions. NMR-based metabonomic profiling of the tissues of stressed earthworms is reviewed, and the potential of infra-red microspectroscopy for determining the biochemical profiles of specific cells in different functional states is highlighted. The way that the rapidly expanding genomic database is beginning to inform and propel research designed to further understanding of the fundamental mechanisms underpinning metal trafficking and toxicosis at the level of cells, partly through the provision of immuno-histochemical and in situ hybridization 'tools', is discussed. The chapter concludes with the prediction that the future will see the adoption of high-throughput cellular, as well as molecular-genetic, biomarker techniques in earthworm ecotoxicology, possibly with parallel metal fractionation measurements done as a desirable component of experimental design.

* E-mail address: plyt@zuk.iz.uj.edu.pl; Tel. = 48-12 663 24 28; Fax = 48 12 634 37 16
** E-mail address: morganaj1@cardiff.ac.uk; Tel. = +029 20875872; Fax = +029 20874305

1. Introduction

Earthworms play prominent roles in terrestrial ecotoxicology (Spurgeon et al., 2003), partly because of their acknowledged ecosystem engineering activities (Lavelle and Spain, 2001), and partly because their anatomical arrangements and life-styles create two interfaces, dermal and alimentary, between the organism and the surrounding abiotic and biotic constituents of soil. This functional intimacy between earthworm and soil is reflected in the ways that earthworms accumulate and respond to inorganic and organic pollutants. The present chapter will, however, be restricted to a consideration of the cellular fates and cytotoxic impacts of metals in earthworms.

Earthworms have a prodigious capacity to colonize and, apparently, establish sustainable populations on soils with metal and metalloid burdens exceeding international definitions of 'heavily contaminated' and accepted trigger or soil guidance levels (Sample et al., 1999). Indeed, it is not unusual that polluted field soils inhabited by earthworms contain metal concentrations exceeding lethal concentrations (LC50 values) derived from laboratory acute toxicity tests. In certain cases (e.g. Cd, Pb) severe metal toxicosis is avoided not by exclusion from earthworm tissues, but by efficient 'accumulative sequestration'; in other cases (e.g. As, Cu, Mn, Ni, Zn) toxicosis is minimised through ill-defined exclusion or regulatory processes. Much is beginning to be known about the edaphic factors that modulate the soil solid-to-aqueous phase metal equilibria, and thus the availabilities of metals for uptake into earthworm tissues (Corp and Morgan, 1991; Neuhauser et al., 1995; Posthuma et al., 1998; Saxe et al, 2001; Peijnenburg, 2002; Lofts et al., 2004; Bradham et al., 2006; Spurgeon et al., 2006), but a good deal less is known about the bio-reactivity of the metals accumulated within target cells and tissues (Oste et al., 2001; Vivjer et al., 2004, 2006; Cotter-Howells et al., 2005).The biochemical efficiency of metal trafficking is graphically illustrated by site-specific accumulation of several dry mass percent Pb concentrations in earthworm chloragocytes (Morgan and Morgan, 1989), cells that are considered to synthesise haemoglobin by a highly conserved biochemical pathway whose several enzymes, many of which are Zn-dependant, are inordinately sensitive to Pb inhibition. So vulnerable is the haem pathway that anaemia provides an early diagnostic indicator of Pb toxicity in vertebrates, and inhibition of δ-aminolevulinic acid dehydratase (ALAD) by Pb (Warren et al., 1998) is noted as a highly specific biomarker of Pb exposure. This example highlights a number of issues. First, total accumulated burden of a metal in earthworm tissues is of less ecotoxicological consequence than are estimates of that proportion of the burden which is bio-reactive or 'free' to indiscriminately interfere with biochemical targets. The implication of this point is clear in a statement from the Framework for Metals Risk Assessment (2004): "Linking residues in tissue to adverse effects can be problematic with metals because of complications associated with their internal speciation and accumulation kinetics" (http://www.rsc.org/pdf/scaf/scaf0030104V2.pdf). Second, even if the accumulated metal could be trafficked to, and stored within, the intracellular sequestration compartments with absolute efficiency, the fitness of the organism could still be indirectly reduced through the metabolic cost of ligand synthesis. In reality, of course, the efficiency is unlikely to be absolute because non-essential metals masquerade to some extent as mimics of essential counterparts, and cause toxic damage that induces costly repair and compensatory mechanisms. Third, how individual cells deviate from their phenotypic reaction norms, and how they qualitatively or quantitatively respond to other members of the cellular community of which they are members, is indicative of their exposure to deleterious conditions. When cellular responses, whether genetic, morphological, physiological or behavioural, are objectively measured and compared with appropriate reference states, they comprise what may be referred to as biomarkers of exposure to sub-lethal stress (Huggett et al., 1992; Morgan et al., 1999a; Hyne and Maher, 2003). The use of biomarkers for ecological risk assessments (Walker et al., 2006) is mildly contentious, with one body of opinion such as the 3rd International Workshop on Earthworm Ecotoxicology advocating more research effort in this area (Van Gestel and Weeks, 2004), and others critical of the approach on conceptual and practical grounds (Power and McCarty, 1997; Forbes et al. 2006). It is not our intention to enter into this debate but to selectively review some of the recent work on earthworm chloragocytes and coelomocytes, in the spirit of "in terms of advancing understanding of the modes of stressor action, experimentation has been invaluable", whilst leaving "the interpretation of relevance (requiring) insights into the functioning of ecological systems as a whole" to others (both quotations from Power and McCarty, 1997).

Table 1. Examples of earthworm cells, other than coelomocytes* and chloragocytes (see this chapter, Sections 4, 6-8), that have been 'exploited' as biomarkers of effects (to metals and certain organic residues).

Cell Type	Observations	Reference
oocytes	Cd, laboratory exposure, *Dendrobaena veneta*; [light and electron microscopy]	Siekierska , Urbańska-Jasik (2002)
spermatocytes	Cd and chlordane, laboratory exposure, *Lumbricus terrestris*; [sperm counts]	Cikutovic et al. (1993)
spermatocytes	dieldrin, laboratory exposure, *Eudrilus eugeniae*; [ultrastructure]	Reinecke et al. (1995)
spermatocytes	Pesticide (parathion, malathion, diazinon), laboratory exposures; *Eisenia fetida;* [counts, ultrastructure, apoptosis (TUNEL-test), genotoxicity (comet assay)]	Bustos-Obregón and Goicochea (2002); Bustos-Obregón et al. (2005); Espinoza-Navarro and Bustos-Obregón (2005)
nephridial epithelium	Cd distribution, laboratory exposure, *Eisenia fetida*; [micro- proton induced X-ray emission (PIXE)]	Prinsloo et al. (1999)**
intestinal epithelium	metalliferous volcanic soils, field exposure, *Lumbricus terrestris*; [radial thickness; autometallography; apoptosis by TUNEL-test]	Amaral and Rodriguez, 2005; Amaral et al., 2005
calciferous gland	Cd+Pb+Zn-contaminated, mine-associated soil, field exposure, *Lumbricus rubellus*; [proton probe X-ray microanalysis.]	Morgan et al. (2004)**

*It is noteworthy that Reinecke and Reinecke (2004) used the comet (single cell gel electrophoresis) assay to determine the genotoxic effect of Ni on the coelomocytes of *Eisenia fetida*.

**Strictly, these authors did not study the effect of the detected metal; however, their technique is sufficiently unusual to warrant mention in the context of cellular biomarkers.

This chapter will describe the possibilities of new physical analytical techniques, and draw upon recently acquired genomic, transcriptomic and cytological data on earthworms to address two themes of general ecotoxicological significance:

(i) How do cells, like earthworm chloragocytes and coelomocytes, segregate the toxic metal trafficking and sequestration pathways from potentially vulnerable biochemical pathways driving the core functions of the particular cellular targets?

(ii) How the structure and functions of cells are altered either directly or indirectly by metal exposures? These changes manifest at molecular genetic, structural and whole-cell function levels of organisation can be used as environmental diagnostic biomarkers to relate the health or performance of individual earthworms to the quality of the soil in which the earthworm lived.

2. Cellular Biomarkers: General

The impact of stress on organisms occurs at several levels of biological organisation, from genes to ecologically-relevant traits and processes. Cell-level biomarkers are deemed to offer a number of useful advantages (Hinton et al., 1992; Weeks et al., 1992; Cajaraville et al. 1995; Triebskorn et al. 1997), including the integration of biochemical and physiological perturbations, the detection of toxicosis in a functionally important cohort of cells surrounded by more abundant 'normal' cells, an earlier indication of toxicosis than whole organism-level changes, and the potential for measuring chronic effects in field populations. For the ecotoxicologist engaged in evaluating the effects of contaminant mixtures on non-model systems, in comparison with the toxicologist whose objective frequently is to understand the effects of single compounds on models, the choices of cell type(s) and cell parameter(s) to measure are fraught with potential problems. In most cases cell choice is justified on the basis of involvement in a key metabolic function, with an implied notion that significant impairment may compromise individual fitness. Parameter choices are seemingly more catholic, perhaps faithfully reflecting a laboratory's expertise more than the pursuit of fundamental rationale. Choice may also be driven by valid pragmatic criteria; for example, technical simplicity and low cost, high throughput and reproducibility, low invasiveness allowing repeat measurements and/or the conservation of natural populations. Overlying the spectrum of practical considerations are issues related to the relevance and interpretation of the gathered data, hindrances compounded in many invertebrate cells and tissues by relatively poor basic background knowledge of ontology, homology, and holistic function.

The quantitative measurement of cellular structure (morphometry) has been effectively used in fish ecotoxicology (e.g. Braunbeck, 1998; van der Ven et al., 2003) and aquatic invertebrate ecotoxicology (e.g. Lowe and Pipe, 1986; Soto et al., 2002). Morphometry has been under-exploited in terrestrial invertebrate ecotoxicology. Nevertheless, studies on an isopod (Morgan et al., 1990) and earthworms (Morgan et al., 2002; Morgan and Turner, 2005) revealed significant inter-population differences in the sub-cellular structure of hepatopancreatic epithelial cells and chloragocytes, respectively, in residents of reference and metal-polluted field soils. Structure and function are generally, and self-evidently, linked. The body of morphometric observations in diverse vertebrate and invertebrate systems encourage the view that more subtle stress-mediated changes in the fidelity and performance indicators of target cells are both measurable and potentially informative in ecotoxicology.

A number of different cell types in earthworms have been the subjects of ecotoxicological investigations (**Table 1**). However, most work has been done on chloragocytes and one or more of the recognised immuno-competent cells collectively referred to as coelomocytes. Chloragocytes and coelomocytes will, therefore, be the main cellular subjects of this chapter.

3. Earthworm Chloragocytes: Metal Relationships

The capacity of earthworms to bioaccumulate and resist the effects of soil-borne metal contaminants is exceptional (Morgan et al., 1993). Data for the species *Dendrodrilus rubidus* inhabiting a circumneutral metalliferous soil associated with the spoil heaps at an abandoned Pb/Zn mine are illuminating (**Table 2**).

Table 2. Mean Cd, Pb and Zn concentrations in *Dendrodrilus rubidus* (Oligochaeta; Lumbricidae) and in its native soil in the vicinity of Roman Mine, South Wales, UK.

	Pb (μg/g dry mass)	Zn (μg/g dry mass)	Cd (μg/g dry mass)
Soil*	16300	48800	510
Whole earthworm*	2430	2200	1250
14-day EC_{50} cocoon production (i.e. soil metal conc.)**	1629	357	295

*Metal concentrations (conc. HNO_3 digests) measured by ICP-MS; **the reproductive output data were obtained in standard laboratory assays for the species *Eisenia fetida* by Spurgeon and Hopkin (1995).

Two facts and two inferences of general ecotoxicological significance derive from these site-specific whole-worm data. Fact 1: whilst the tissue burdens of the three metals are appreciable, it is evident that the bioaccumulation efficiencies differ between metals in ways that probably reflect the ligand-binding affinities of the metals rather than whether they are essential or non-essential. Fact 2: the field worms maintain a sustainable population in a soil that is contaminated with a mixture of three metals whose individual concentrations are well in excess of the concentrations that severely compromise reproductive output in spiked test soil. Inference 1: the potential toxicity of a field soil cannot be predicted merely from its measured metal content by extrapolation from laboratory toxicity tests. Inference 2: only a small fraction of the equilibrated metal body-burden of earthworm field populations with multi-generation exposure histories can be bioreactive. But not evident in the tabulated data is the fact that each of the three metals is not homogeneously distributed amongst the major organs of the earthworm (Morgan and Morgan, 1990; Morgan et al., 1993; Morgan et al., 1999b). Indeed, integrating data from tissue analyses, sub-cellular microprobe analyses, and morphometrics leads to the conclusion that the major proportion of the body burdens are located in the chloragogenous tissue and posterior alimentary epithelia (Morgan et al., 2002; Table 2). It follows that determining the cellular fates and effects of metals within earthworm cells with a predilection to accumulate metals is important. [It should be noted that there is no intention here to give the impression that cells with a high capacity to sequester metals are, as a consequence, especially vulnerable to the toxic effects of the metals. It is plausible that the converse is true. However, transiting and capturing excessive loads of non-essential and even essential metals must surely tax the resources of the cell indirectly if not directly.]

The critical body residue (CBR) has been defined by Vivjer et al. (2004) as "the threshold concentration of a substance in an organism that marks the transition between no effect and adverse effect". The relationship between the CBR for a metal and its total accumulated body burden is difficult to predict (Van Straalen, 1996). But so important is the relationship that there is a burgeoning interest, stimulated by elegant studies on the transference of metals from adapted and non-adapted aquatic oligochaete worms to their shrimp predators (Wallace et al., 1998; Wallace and Luoma, 2003), in estimating a non-sequestered metal concentration (- it may be acceptable to term it 'potentially bio-reactive'; but it cannot yet be explicitly related to CBR because this would require parallel measurement of a specified effect -) in fractionated whole earthworm tissues (Andre et al., 2005; Vivjer et al., 2006). Morgan et al. (1999b) had already fractionated whole earthworm tissues to show that significant proportions of the Pb (49%) and Zn (71%) body burdens are located in the pellet fraction but, in contrast, a high proportion of the Cd burden (87%) is associated with the supernatant. Thus, tissue fractionation is a technically feasible and relatively undemanding alternative or adjunct to the conventional bio-monitoring practice of analysing metals in whole earthworms (or other reasonable sized invertebrates), and should yield information reflecting toxic consequences. Advantageous though tissue fractionation may be in principle, it has yet to be confirmed that its estimates of the non-sequestered and sequestered metal pools correlate with specific functional parameters in identified cell type(s). It is also prudent to consider the possibility that the homogenization and centrifugal fractionation of tissues could displace metals from their *in vivo* locations. Nevertheless, it anticipated that this approach will increasingly find favour in ecotoxicological studies.

There is a wealth of studies, some qualitative others quantitative, many involving the technique of electron probe X-ray microanalysis (EPXMA), that have shown that earthworm chloragocytes are major sequestration sites for certain metals, notably Cd, Pb and Zn (e.g. Ireland and Richards, 1997; Morgan and Morris, 1982; Prentø, 1987; Winters and Morgan, 1988; Morgan et al., 1993, 1994, 1995; Morgan and Morgan, 1989a, 1989b; Cotter-Howells et al., 2005). EPXMA provides *in situ* non-destructive multi-elemental analysis, with a spatial resolution capability in thin sections making it possible to both detect and quantify elements within morphologically defined organelles (Morgan, 1985). The main findings of these studies can be selectively summarized:

- The characteristic electron-dense, Ca+PO_4-rich, organelles of the chloragocytes (known as chloragosomes) are the predominant sites of Pb and Zn sequestration.
- Quantitative analyses of the organelles are consistent with simple extrapolations from measured whole-worm and tissue metal concentrations that individual chloragosomes can accumulate Pb and Zn dry mass concentrations exceeding 30,000 μg/g.
- Cd is sequestered in a compositionally distinct, S-rich, organelle referred to as a cadmosome. Sub-cellular metal-specific partitioning based on ligand-affinity differences is a characterisitic phenomenon in chloragocytes.

- The cadmosome has been shown by immuno-histochemistry in light and electron microscopy to contain Cd-metallothionein (Stürzenbaum et al., 2004). The Cd:S ratio in the cadmosome matrix deviates from the known Cd:S ratio in earthworm metallothionein (Stürzenbaum et al., 2004), perhaps indicating that protein is hydrolysed and some of its thiols re-cycled to the cytosol whilst Cd is retained. The lysosomal nature of the cadmosome, demonstrated by combined Neutral Red loading and EPXMA (Morgan et al., 2004), is consistent with this hypothesis.

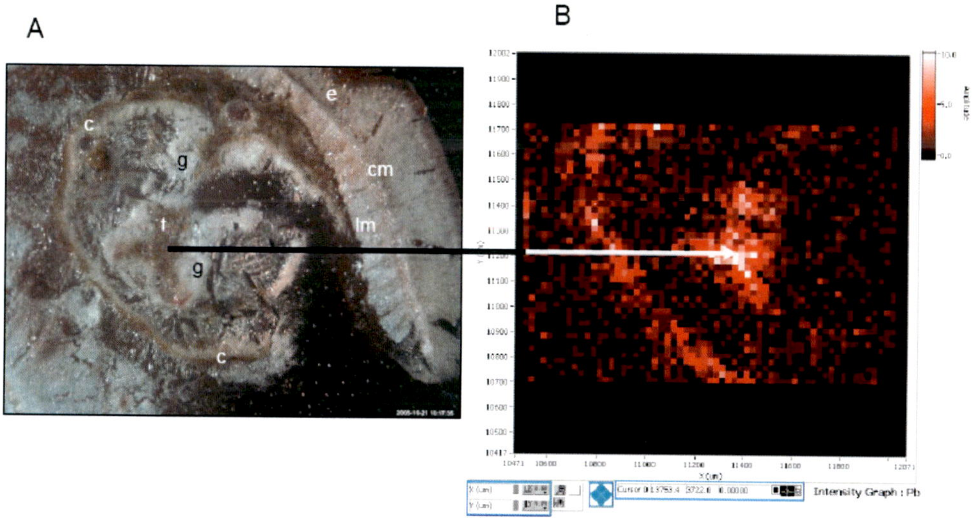

Figure 1. (A) A light micrograph of a transverse, air-dried, cryostat section (nominally 50 μm thick) of an earthworm (*Lumbricus rubellus*) from the Cwmystwyth Stream mine site in South Wales, UK (see Table 3 of this chapter for the soil metal content and pH value). Note the body wall, consisting of the epidermis (e), circular muscle (cm), and longitudinal muscle (lm), and the gut epithelium (g), the brown chloragogenous tissue (c), and the typhlosole fold (t). (B) A Pb distribution map across the area depicted in the corresponding micrograph. Note that Pb is mainly located in the chloragocytes, including the cells 'entrapped' in the typhlosolar fold. Pb L3-edge XANES (data not shown) was performed on the Pb-rich chloragogenous locus indicated by the arrow (see text, Section 3). [Morgan, Bennett, O'Reilly, Charnock – unpublished observations.]

EPXMA detects and localises atoms. A fundamental constraint of the physical technique is its inability to distinguish between the chemical states (of binding or ionisation, for example) of given metals. The nature of the ligands binding detected metals within cells and their constituent organelles can only be deduced from their co-localisation with anionic elements. [It was this constraint that disallowed concluding that cadmosomes contained Cd-metallothionein until an anti-metallothionein antiserum became available for immuno-detecting the protein.] Two studies have recently been published using synchrotron-based X-ray absorption spectroscopy (XAS) for determining the ligand-binding speciation (including coordination number, bond distances, and type of near-neighbours surrounding specific element centres) of As (Langdon et al., 2005) and Cd, Pb, and Zn (Cotter-Howells et al., 2005) in frozen-hydrated, powdered, whole earthworms of the species *Lumbricus rubellus* and *Dendrodrilus rubidus*, respectively. Encouragingly, the XAS findings were consistent with, and complimentary to, EPXMA and immunohistochemical obsevations on chloragocytes and, in the case of the As study, the intestinal epithelium. Advances in synchrotron technology, especially in the targetting of intense fine X-ray beams onto sectioned tissues, is beginning to makes it possible to perform XAS on structurally intact biological specimens (Mikhaylova et al., 2005; Punshon et al., 2005). Although the new technology, referred to as mini-EXAFS (Extended X-ray Absorption Fine Structure) or micro-EXAFS depending on the spatial resolution capability of the system, has evolved into a what might justifiably be described as a histochemical tool with unique metal speciation capabilities, it should be noted that the physical limitations of X-ray compared with electron optics mean that its spatial imaging and analytical resolutions are highly unlikely to match these parameters in EPXMA. However, **Figure 1** shows a Pb distribution map collected by micro-EXAFS from a thick, unfixed, transverse cryo-section of an earthworm sampled from a Pb contaminated soil associated with an abandoned Pb/Zn mine. In principle, each Pb-containing pixel in the map (c. 50x50 μm, which is rather larger than the dimensions of individual chloragocytes) can then be subjected to further analysis of Pb speciation (by Pb L3-edge XANES, X-ray Absorption Near Edge Structure). However, in our limited experience to date with this particular prototype synchrotron facility, even the most concentrated Pb locations yielded noisy signals casting uncertainty on the findings. We can tentatively conclude that the speciation of Pb in the gut contents and in the chloragocytes is different, with Pb probably surrounded by a single shell of S scatterers in the former and a best fit with shells of O and S in the cells. Clearly, more work is required to optimise the performance of this potentially powerful technique which, though likely to provide novel insights into the availabilities of metals in cells (Punshon et al., 2005) whose morphological and compositional fidelities have been maintained through the analytical process, cannot ever be a standard ecotoxicology tool because of its intrinsically specialised nature.

4. Earthworm Chloragocytes: Morphological Plasticity

Oligochaete annelids possess chloragogenous tissue, a tissue unique to the taxon. Chloragocytes have certain hepatocyte-like properties, including prominent roles in carbohydrate storage and, as we have seen, in the accumulation and detoxification of inorganic (and organic) residues (Morgan, 1982). Thus, there are numerous structural and functional reasons for promulgating the notion that chloragocytes and their constituent organelles are potentially good subjects for developing biomarkers. Environmental stressors, including xenibiotics, evoke substantial alterations in earthworm chloragogenous tissue, encompassing the formation and release of the chloragosome granules that are defining features of chloragocytes (Fischer, 1976; Fischer and Molnar, 1992). Transmission electron microscopy readily shows qualitative differences in the morphology of earthworm chloragocytes from metal polluted and reference sites (Ireland and Richards, 1977; Morgan et al., 2002; Morgan and Turner, 2005). Metal exposures tend to disrupt the characteristic concentric internal structure of chloragosome granules, increase the frequency and apparent size of structures variously referred to as debris vesicles or tertiary lysosomes, and reduce the content of glycogen deposits (**Figure 2**). Morphometric analyses (Morgan et al., 2002; Morgan and Turner, 2005; **Table 3**) reinforce the descriptive observations of metal-induced, site-specific, intra- and inter-specific differences in chloragocyte ultrastructure. In principle such plastic alterations in the morphology of these metal-sequestering multi-functional cells could be used as biomarkers of sub-lethal stress in field populations. In practice, transmission electron microscopy combined with manual image analysis of the micrographs (necessitated by the extremely heterogeneous morphologies of the metal-sequestering organelles) is not a technique amenable for most routine ecotoxicological applications because of the associated low sample throughput, high expense, and specialised technical expertise requirements. Light microscopy techniques can on the other hand form the basis of very effective biomarker protocols (Cajaraville et al., 1995; Amaral and Rodriguez, 2005; Amaral et al., 2005). Also, automating the sample preparation, imaging, and image analysis phases of the protocols is readily achieved in light microscopy. In the following section of this chapter a brief review will be given of a widely used method for determining the effects of environmental toxicants on the integrity of lysosomal membranes in earthworm coelomocytes (and, of course, the fluid-suspended cells of other invertebrate taxa). Since this technique known as the Neutral Red Retention Time assay (NRRT) is relatively familiar to terrestrial ecotoxicologists, a derivative technique (the Neutral Red Positive 'Granule' Index, NRPGI) using the same molecular 'probe' (Neutral Red, NR), but applicable to earthworm chloragocytes, will be described. This technique has not previously been described or thoroughly validated, but it is noteworthy that, like the NRRT, histological sectioning is not a prerequisite even though it monitors attached cells, and it also differs from the NRRT in (putatively) measuring the relative volume of the lysosomal pool rather than the structural fidelity of the lysosomal membrane. There is limited evidence indicating that chemical stressors do induce the proliferation of NR-positive vesicles, named "prechloragosomes", in earthworm chloragogenous tissue (Fischer, 1976; Cancio et al., 1995a). Supportive data will be provided here to allow the reader to make a judgment as to the reliability and potential of the new NRPGI biomarker.

Table 3. Inter-population differences in the morphometrics of chloragocyte cells in the epigeic earthworm species, *Dendrodrilus rubidus* and *Lumbricus rubellus*, from one reference site and two metalliferous sites: % Volume Fractions (V_v) of the chloragosome granules and debris vesicles.

Sites*	*Dendrodrilus rubidus*		*Lumbricus rubellus*	
	Chloragososmes	Debris vesicles	Chloragosomes	Debris vesicles
Dinas Powys (reference)	24.4 ± 8.8 (5)[a]	4.3 ± 2.3 (5)[a]	32.0 ± 6.8 (5)[a]	1.5 ± 1.1 (5)[a]
Cwmystwyth (low contamination)	24.6 ±2.1 (5)[a]	9.8 ± 3.5 (5)[a]	30.0 ± 6.5 (5)[a]	7.9 ± 1.7 (5)[a]
Draethen (high contamination)	15.1 ± 5.4 (6)[b]	27.8 ± 12.3 (5)[b]	33.0 ± 9.4 (5)[a]	13.4 ± 4.1 (5)[b]

[See Fig. 2 for illustrations of the two organelle 'types'. Data derived and tabulated from: Morgan et al., 2002; Morgan and Turner, 2005.]

Data expressed as mean ± S.E.; numbers in parentheses = number of true replicates (i.e. number of worms). Statistical analyses by Kruskal-Wallace (a non-parametric equivalent of ANOVA) followed by Mann-Whitney pair-wise comparisons: values with a different lower-case letter within a column are statistically different at $P<0.05$.

*Mean soil metal concentrations (μg/g dry weight): Dinas Powys (Cd, 0.9; Pb, 90; Zn, 420); Cwmystwyth (Cd, 0.1; Pb, 670; Zn, 460); Draethen (Cd, 350; Pb, 5300; Zn, 16400). It is noteworthy that Cwmystwyth soil is acidic(pH = 3.5) compared with Dinas Powys (5.8) and Draethen (6.5).

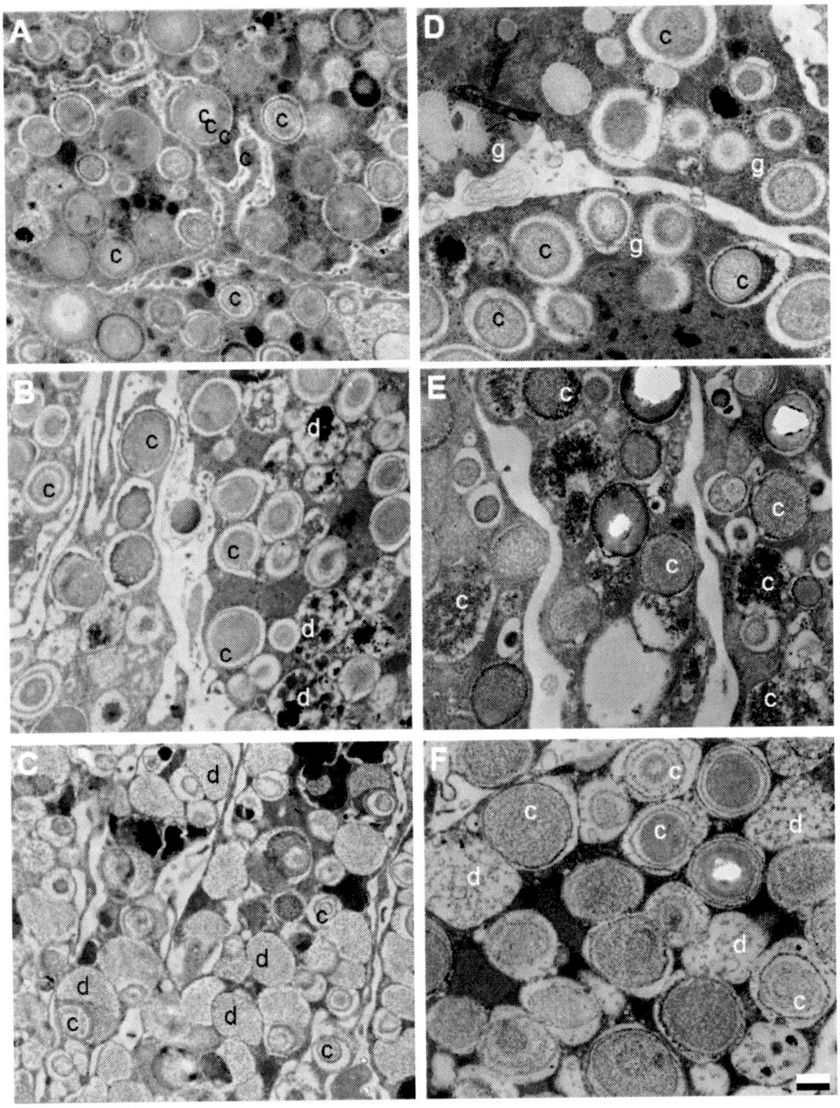

Figure 2. Transmission electron micrographs of representative chloragocytes in the epigeic earthworms, *Dendrodrilus rubidus* (A, B, C) and *Lumbricus rubellus* (D, E, F), collected from their native soils: (A & D) at an uncontaminated reference site at Dinas Powys; (B & E) at a moderately Pb+Zn-contaminated, acidic, mine-associated soil at Cwystwyth Stream; (C & F) at a heavily Pb+Zn+Cd-contaminated, calcareous, soil at Draethen.
Labelling code: c, chloragosome granules; d, debris vesicles; g, glycogen rosettes. All micrographs were taken at the same magnification (- the scale marker represents 1μm).
See Table 3 for a summary of the inter-population morphometric data. [For more details, see: Morgan et al. (2002) and Morgan and Turner (2005).]

Lysosomes are multi-functional organelles (Blott and Griffiths, 2002), and have been shown in a number of different animal taxa to be susceptible to the deleterious effects of a broad spectrum of environmental contaminants (Lowe et al., 1981; Marigomez et al., 1989; Svendsen et al., 1996; Köhler et al., 2002), as well as to non-chemical stressors such as hypoxia, hyposalinity, and starvation (Moore, 1985). They are acidic organelles containing cocktails of hydrolases whose normal functions include the digestion of endocytosed molecules and the autophagic disposal of cytoplasmic components. Thus, the unregulated leakage of lysosomal enzymes through the limiting membrane causes severe, occasionally irreversible, cytosolic damage. Some researchers (Varute and More, 1972, 1973; Cancio et al., 1995b) consider earthworm chloragosomes to be (perhaps specialised) lysosomes, although only a minor cohort is cytochemically positive for lysosome markers. Thus, the homology of earthworm chloragosomes is uncertain, which is not unexpected given that lysosomes are dynamic organelles with complex life-histories, as well as plastic morphologies reflecting their functional status. In light of these factors, it may be prudent not to assign NR-positive granules in chloragocytes to any conventional nomenclature. For this reason, we consider them to be an unidentified subset of lysosomes whose relative numbers appear to be responsive to the degree of stress experienced by an individual earthworm.

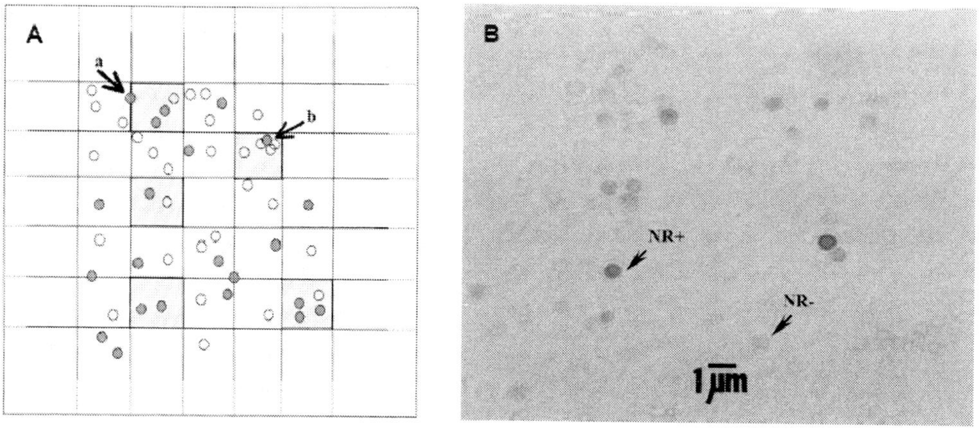

Figure 3. (A) Schematic diagram showing an example of NRPGI scoring. Counts were made in 20 'standard' grid-squares (only 5 such squares are illustrated here, in a representation of part of the eyepiece grid with a total of 100 identical squares), and the percentage of non-overlapping NR-positive granules, the NRPGI value, calculated. In this example the NRPGI value is (8÷12) × 100 = 75%. Note that the granule on the grid boundary (a) and the overlapping NR-positive and NR-negative granules (b) were not counted. (B) Photomicrograph of a typical air-dried chloragogenous smear preparation from *L. rubellus* to illustrate the sizes of the granules. Note that even in this black-and-white example the discrete and dense NR-positive granules (NR+; actually an intense red colour) are easily distinguished from the NR-negative granules (NR-).

Two generic methods of evaluating stress-induced deterioration in the integrity of the lysosomal membrane have been relatively widely used in ecotoxicology. First, enzyme histochemistry has been used to measure increased permeability; damaged lysosomes are maximally stained for a constituent enzyme after shorter incubation periods than intact lysosomes because substrate and capturing molecules can traverse the membrane more readily (e.g. Moore, 1985). Second, Lowe et al. (1992) introduced a Neutral Red retention assay (NRAT) as a means of measuring the effects of sub-lethal stress in marine organisms. The principle of the technique is simple: NR being a cationic, acid-seeking, molecule accumulates in lysosomes, but its retention period is shorter in lysosomes whose membranes have become leaky due to environmental perturbations. Weeks and Svendsen (1996) described a modification of the NRRT, which was subsequently exploited on lysosomes in earthworm coelomocytes for monitoring single and multifactorial toxicant exposures under laboratory (Svendsen and Weeks, 1997a; Scott-Fordsmand et al., 1998; Spurgeon et al., 2000; Booth and O'Halloran, 2001; Booth et al., 2002, 2005; Reinecke et al., 2002; Reinecke and Reinecke, 2003; Robidoux et al., 2002; Hønsi et al., 2003), semi-field (Svendsen and Weeks, 1997b; Booth et al., 2001; Hankard et al., 2004; Maboeta et al., 2004; Robidoux et al., 2004a, 2004b; Reinecke and Reinecke, 2005) and field conditions (Svendsen et al., 1996; Maboeta et al. 2002, 2003). Despite its undoubted utility in experienced hands, the NRRT has probably not been even more widely used as a biomarker tool because scoring is notoriously difficult for untrained observers. This practical constraint prompted Reinecke et al. (1999) to describe an automated image analysis scoring system, but the specialised equipment itself represents a constraint.

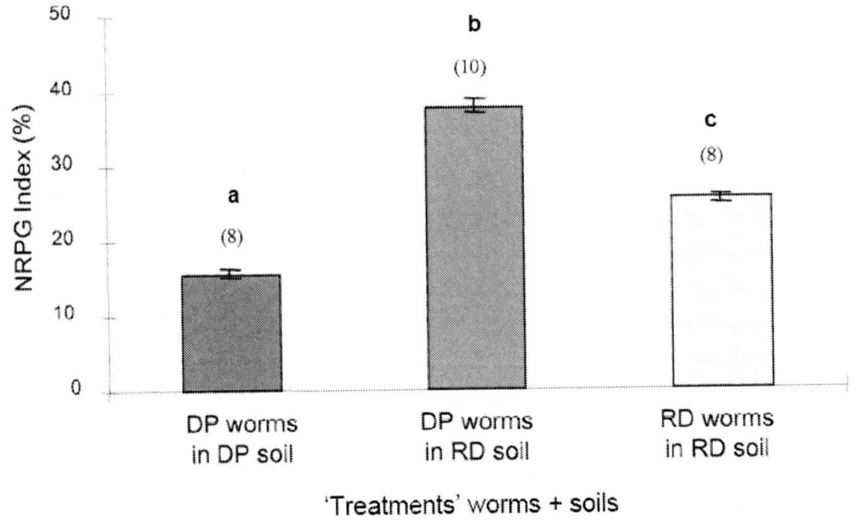

Figure 4. Mean ± S.E. (denoted by verticals) NRPGI values in three experimental groups of earthworms after 3 weeks exposure to two soils under laboratory conditions. Group 1 = DP 'clean' worms on 'clean' DP (reference) soil; Group 2 = DP worms on polluted RD soil; 'polluted' RD worms on RD soil (see: Tables 4 and 5 for soil and earthworm metal concentrations). Different lower case letters above the histogram columns indicate statistically significant differences in measured NRPGI values (at $P<0.05$, ANOVA and Tukey pair-wise comparisons). Numbers in parentheses are the numbers of replicates = number of earthworms.

The NRPGI assay:

Neutral Red (3-amino-7-dimethylamino-2-methylphenzine hydrochloride; Sigma) was injected (0.5% aqueous solution; 30-50µl) using 26G needles (Fischer Scientific Ltd.) into the post-clitellar coelom of depurated worms (*Lumbricus rubellus*). Injected worms were left for 5 min. in a Petri dish with a few drops of deionized water. After the brief dye-loading period, a dorsal longitudinal incision was made in the body wall in the vicinity of the NR injection site. Excess coelomic fluid was removed by careful mopping with filter paper; a triangular piece of clean filter paper was then gently drawn over the exposed, NR-stained, chloragogenous tissue. Cytoplasmic organelles drawn into the filter paper from the mechanically ruptured chloragocytes were then smeared gently onto the central area of glass microscope slides. The smeared preparations were air-dried, mounted, and coverslipped.

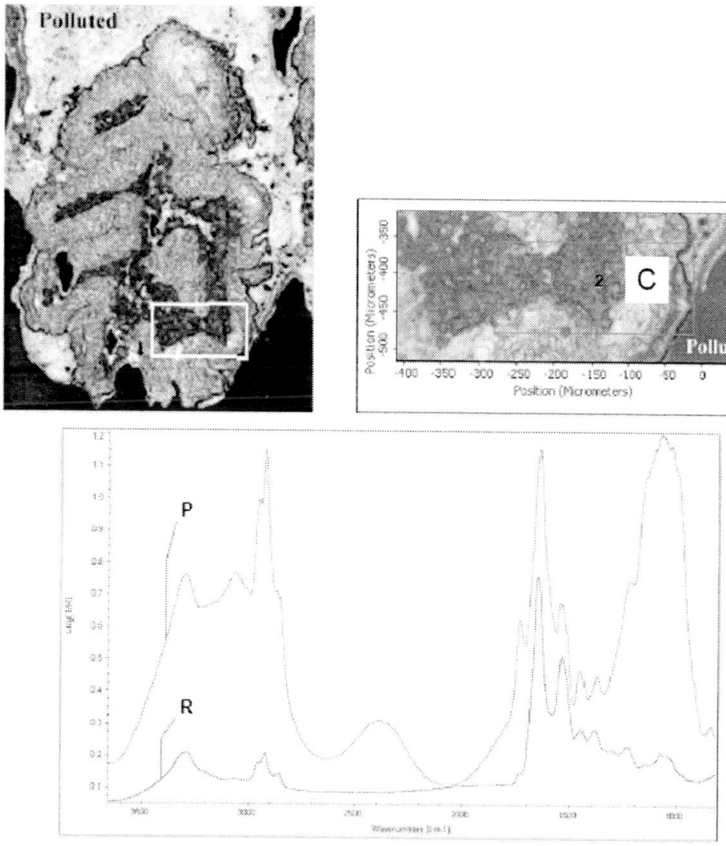

Figure 5. Light micrograph of a transverse cryostat section through the intestinal region of the earthworm *Dendrodrilus rubidus* sampled from its native metal-polluted, acidic, soil at the abandoned Cwmystwyth Pb/Zn mine site; the selected area contains gut contents (dark deposits) and chloragogenous tissue (c). The spectra are synchrotron-FTIR spectra derived from single points within the chloragogenous tissue of a worm from the polluted mine soil (P) and a corresponding region in a worm from an unpolluted reference site called Dinas Powys (R). Note the major qualitative differences in the two spectra in the wavenumber ranges c. 1800 to 1500, and 1250 to 800 cm^{-1}. See Table 3 for soil metal concentration values at the two sites. [Bahrami and Morgan – unpublished data.]

A standardized scoring system was adopted based on a grid (1cm×1cm, divided into 100 equal sized, 0.01cm^2, grid squares) placed in the eyepiece of a light microscope. Each smear was observed under an oil-immersion lens (×1000 magnification) without changing the focal plane after initial focusing. Between 5 and 7 separate microscope fields were scored on each chloragogenous smear from individual worms (sub-sample/nested data). In each microscope field, counts of 'free' NR-positive and 'free' NR-negative granules or vesicles were made in 20 (out of the total of 100) pre-determined grid squares (**Figure 3**). Thus, NRPGI = [number of NR-positive granules per field ÷ (sum of NR-positive and NR-negative granules per field)]

× 100%. The nested data were used to statistically compare NRPGI values between three exposure groups: Group1 - worms sampled from an unpolluted reference site, and maintained on their 'own' soil in the laboratory for 3 weeks; (ii) Group 2 – reference site worms maintained for 3 weeks on a metal polluted (Pb+Zn) mine-site soil; (iii) Group 3 – worms native to the mine site and maintained for 3 weeks on their 'own' metalliferous soil. At least 18 earthworms, kept in pairs, were used in each exposure group. One member of a pair of worms was digested for metal analysis (see **Tables 4 and 5** for soil and earthworm metal concentrations); the other was used in the NRPGI test (**Figure 4**).

Table 4. pH, % loss-on-ignition (LOI), and conc. HNO_3-extractable metal concentrations (μg/g dry weight) in the two field soils used for the NRPGI experiment. Data expressed as mean ± S.E. [See: Tables 5 and 6.]

Site	pH	% LOI	Cd	Cu	Pb	Zn
Reference	6.08±0.03	12.1±0.2	1.2±0.1	22±1	86±4	167±12
(DP)	(3)	(3)	(5)	(5)	(5)	(5)
Mine	6.64±0.05	31.6±2.2	103±3	106±8	11850±406	6155±12
(RD)	(3)	(3)	(5)	(5)	(5)	(5)
RD:DP	-	2.6	86	4.8	138	40

Numbers in parentheses = number of individual samples analysed

Group 1 worms, as expected, had much lower Cd, Pb, and Zn concentrations (but, curiously, higher Cu) than worms native to the polluted soil (Group 3). Group 2 worms (i.e. clean worms transferred to metal-polluted soil) had significantly higher tissue Cd, Pb, and Zn concentrations than their counterparts maintained on their 'own' clean soil (Group 1), but appreciably lower concentrations than Group 3 worms. Significant differences were found in the NRPGI values for the three exposure groups (**Table 5**): Group 2 worms had a mean value 100% higher than 'controls' (Group 1 worms), and about 50% higher than Group 3 worms in their 'own' polluted soil. No significant differences were recorded ($F_{23,142} = 0.86$, $P>0.05$) in the NRPGI values between individuals within any given experimental group.

Table 5. Metal concentrations (μg/g dry weight) in whole worms after 3 week exposures to field soils under laboratory conditions. [Data expressed as mean ± S.E.]

*Exposure Group**	Cd	Cu	Pb	Zn
Group 1	2.4±0.6^{a}	5.7±0.7^{a}	N.D.	386±15^{a}
	(9)	(9)	(9)	(9)
Group 2	94±9^{b}	2.5±0.6^{b}	930±32^{a}	1347±59^{b}
	(8)	(8)	(8)	(8)
Group 3	237±78^{c}	0.5±0.3^{c}	1033±73^{b}	2791±61^{c}
	(8)	(8)	(8)	(8)

*Group 1 = DP worms on DP (i.e. reference) soil; Group 2 = DP worms on RD (i.e. polluted) soil; Group 3 = RD worms on RD soil [see Table 2 for soil values].

Means in a given vertical column followed by different lower-case superscript letters differ significantly (P<0.001).

N.D. = non-detectable with flame atomic absorption spectrophotometry.

The low observed variability in the measured NRPGI values within treatment groups (**Figure 4**) indicated that scoring with this protocol is relatively reliable and objective, especially when the prepared tissue smears were sufficiently thin. The NRPGI assay indicated that the three groups of earthworms examined in the present study were subjected to, or experiencing, different degrees of cellular stress. In the cases of the worms exposed to their 'own' native or polluted soils, respectively, the observed difference could be either an indirect or a direct effect of the inter-population differences in accumulated metal burdens. The elevated NRPGI value in the worms transferred from their 'own' relatively clean soil to the polluted soil could similarly be explained by the upsurge in tissue metal burdens during the three-week exposure period. More problematic to explain is the significantly higher NRPGI in those worms transferred to the polluted soil for a limited period compared with the values recorded in those worms (Group 3) sampled from a population with a multi-generational and protracted individual history of exposure to the metalliferous soil. It is reasonable to hypothesise that the NRPGI is reporting a shock reaction during the early acclimatory phase in the introduced worms. Alternatively, the cellular biomarker may be indicating that the introduced earthworms are hypersensitive to metal stressors compared with an adapted resident population. Reinecke and Reinecke (2003) found, using the NRRT assay, that the lysosomal membrane of coelomocytes from the *Eisenia fetida* with a multi-generational Pb-exposure history in laboratory culture was more stable under Pb challenge than the lysosomal membranes of earthworms with no prior exposure history. Analogous observations were made by Eckwert and Köhler (1997) on the terrestrial isopod, *Oniscus asellus*. They measured higher stress protein 70 (hsp70) levels in isopods transferred from a clean to a polluted mine site compared with hsp70 levels in mine site residents. Notwithstanding the fact that hsp70 is intimately involved in important lysosomal functions, such as protein translocation (Agarraeres and Dice, 2001) and the inhibition of membrane permeabilization (Nylandsted et al., 2004), there is apparent correspondence between the present NRPGI findings and those of Eckwert and Köhler (1997) and Reinecke and Reinecke (2003). However, we must remain circumspect: the pronounced NRPGI difference between introduced (DP, Group 2) and mine-site resident *L. rubellus* (Group 3) could possibly reflect genetic differentiation in the latter population, but to be more certain it would be necessary to establish whether or not the NRPGI value in the DP worms introduced to polluted RD soil will eventually equilibrate at a similar level to that in RD worms on their 'own' soil. In conclusion, the NRPGI possesses a number of practical attributes that meet the general criteria for a Tier 1 biomarker (Mayer et al., 1992): it is quick and easy; it facilitates the rapid throughput of individual samples; it is inexpensive and statistically robust; and it is transferable without making heavy demands on local technical expertise or facilities. However, the NRPGI assay needs to be thoroughly assessed and validated by comparison against, for example, earthworm demographic parameters and independent biomarkers, following the precedents established by the proponents of the NRRT assay (e.g. Spurgeon et al., 2000; Reinecke et al., 2002; Spurgeon et al., 2004; Svendsen et al., 2004).

5. Chloragocytes: Metabonomics

Metabonomics (or metabolomics as it is sometimes called) is a family of techniques based on a number of physical platforms, usually nuclear magnetic resonance spectroscopy (NMR) or mass spectrometry (MS), for determining the metabolic profiles of biological samples (Robertson, 2005). It is an approach that has been intensively used by a few research groups in various biomedical fields and, of particular relevance for this chapter, in a limited number of environmental applications (e.g. Viant et al., 2003). The technique has been used to very good effect to characterise the metabolic responses of earthworms to low temperature (Bundy et al., 2003) and toxic stress (Gibb et al., 1997; Bundy et al., 2001, 2002a, 2004; Lenz et al., 2005), as well as for phenotypically differentiating eco-physiologically similar species belonging to the genus *Eisenia* (Bundy et al., 2002b). Since this chapter is restricted to cell-level responses, it is proposed to give a short account of another physical technique, high-brightness synchrotron Fourier Transform Infrared microspectroscopy (S-FTIR-M), which also has the capacity to measure 'global' biochemical profiles, but can do so in intact cells and tissues with spatial resolutions as fine as 3 to 10 μm (Holman et al., 2003; McNaughton 2005). S-FTIR-M has been fairly widely exploited in a number of biomedical applications, including live cell and time-resolved studies (Dumas and Miller, 2003; Holman et al., 2003; Hamlin and Price, 2004). To our knowledge the technique has not previously been used for tracking the compositional changes in specified cells and tissues in invertebrates. **Figure 5** includes two S-FTIR-M spectra from the chloragogenous tissue of two earthworms (*L. rubellus*), one inhabiting a clean soil the other a heavily metal-contaminated soil. The impact of metal exposure under field conditions on the metabolic fingerprints of chloragocytes is immediately apparent, especially in the wavenumber range 1800-800 cm^{-1}. The full potential of this technique for measuring changes in the cell-specific biochemistry of organisms as they progress through different life stages, or experience stress caused by habitat fluxes, toxicant exposures, pathogen invasions etc., will be realised when the acquired spectra have been subjected to the quantification routines, and multivariate statistical analyses, that have already been exploited in conjunction with bench-top FTIR analyses of bulk, earthworm-related, environmental samples (Scullion et al., 2003a, 2003b; Gwynn-Jones et al., 2003).

6. Coelomocytes: Morphology and Composition

Coelomocytes, earthworm immunocompetent cells, circulate in coelomic fluid constituting the hydrostatic skeleton of oligochaete annelids. Earthworm coelomocytes consist of two apparent cell populations of different morphology and origin, namely amoebocytes (hyaline and granular) and eleocytes (Cooper and Stein, 1981). There are indications that amoebocytes derive from the mesenchymal lining of the coelom (Hamed et al., 2002), whilst eleocytes (chloragocytes) differentiate from the chloragogen cells that cover the coelomic surfaces of the alimentary tract and major blood vessels (Affar et al., 1998). As such, eleocytes seem to be fully differentiated chloragocytes that have sloughed off into the coelomic fluid.

The number and composition of earthworm coelomocytes is species specific (Kurek et al., 2002; Kurek and Plytycz, 2003; Cholewa et al., 2006) (**Table 6**). They fluctuate in an annual cycle (Kurek et al., 2002; Kurek and Plytycz, 2003), and are affected by several environmental stressor, including metal contaminants (Homa et al., 2003, 2005; Olchawa et al., 2006). Their number in the coelomic cavity may be also be modulated by non-chemical stressors; for example, coelomic fluid with suspended coelomocytes may be expelled through dorsal pores and metanephridial openings during the convulsive body movements of worms irritated either naturally (e.g. by predators) or experimentally (e.g. by mild electric current –

Roch, 1979), 5% ethanol (Cooper et al., 1995), or ultrasound stimuli (Hendawi et al., 2004). Under natural conditions, fluid extrusion is used by some earthworm species as an anti-predator reaction (Sims and Gerard, 1985).

Table 6. Number of 'total' coelomocytes and autofluorescent eleocytes extruded by electric shock (1 min, 4.5V) from adult (clitellate) individuals from 12 lumbricid earthworm species.

Earthworms		Coelomocytes	
Species	**b.w. [g]**	**C [x10^6]**	**AF [%]**
Allolobophora chlorotica	**0.22** ±0.01 [e] (7)	**1.3** ±0.1 [bd]	**30** ±4 [a]
Aporrectodea caliginosa	**0.91**±0.12 [bce] (5)	**0.4** ±0.1 [d]	**1.0** ±0.3 [b]
Aporrectodea longa	**1.90** ±0.16 [d] (5)	**1.6** ±0.5 [bd]	**0.2** ±0.1 [b]
Dendrodrilus rubidus	**0.17** ±0.02 [e] (7)	**1.1** ±0.1 [bd]	**21.0** ±6.5 [a]
Eisenia fetida	**0.59** ±0.04 [ce] (5)	**3.0** ±0.3 [ac]	**22** ±1 [a]
Lumbricus castaneus	**0.19**±0.01 [e] (5)	**0.3** ±0.1 [d]	**0.5** ±0.2 [b]
Lumbricus festivus	**0.97**±0.06 [bc] (5)	**2.0** ±0.3 [ab]	**0.05** ±0.03 [b]
Lumbricus rubellus	**0.59** ±0.04 [ce] (5)	**0.8** ±0.1 [bd]	**0.5** ±0.3 [b]
Lumbricus terrestris	**3.96** ±0.27 [a] (5)	**1.9** ±0.3 [ab]	**0.5** ±0.3 [b]
Octolasion cyaneum	**1.54**±0.03 [bd] (3)	**1.5** ± 0.1 [abcd]	**22** ±8 [a]
Octolasion tyrtaeum lacteum	**0.40** ±0.02 [e] (5)	**1.3** ±0.3 [bd]	**11** ±3 [a]
Octolasion tyrtaeum tyrtaeum	**0.70** ±0.07 [bce] (5)	**3.4** ±0.5 [ac]	**35** ±8 [a]

b.w. = body weight; numbers in parentheses = number of individuals; C = coelomocytes; AF = autofluorescent eleocytes; means ± SE.

Statistical analysis by ANOVA followed by *post hoc* Tukey's test; Means in a given vertical column followed by different lower-case superscript letters differ significantly ($P<0.05$).

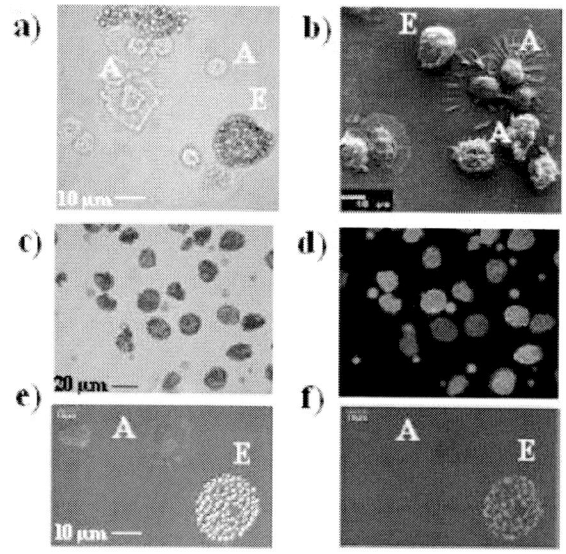

Figure 6. Continued on next page.

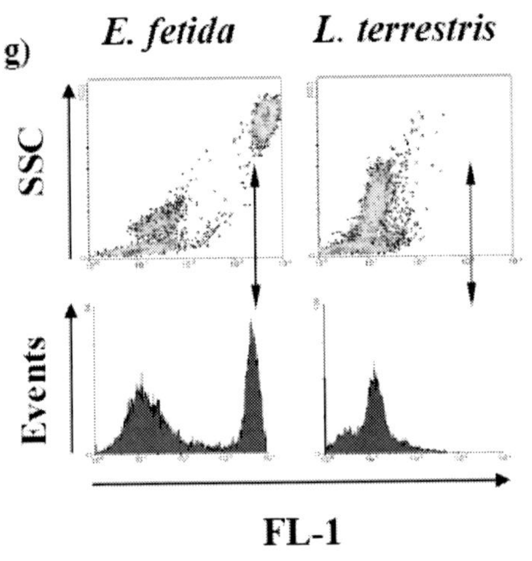

Figure 6. Morphology and characteristics of earthworm coelomocytes : amoebocytes (A) and eleocytes (E). (a) Morphology of fresh coleomocyte suspension in a haemocytometer and (b) in a scanning electron microscope; (c,d) pinocytosis of neutral red (bright field, c, and fluorescence, d, microscopy); (e,f) autofluorescence of granules in eleocytes under scanning confocal microscope; (g) flow cytometric analysis of autofluorescent cells: density plots (upper row) and histograms (lower row); SSC = cell complexity/granularity; note that autofluorescent cells are present in *E. fetida* (arrows) but absent in *L. terrestris* preparations.

Morphological differences between amoebocytes and eleocytes **(Figure 6)** are visible even in a haemocytometer; eleocytes are filled with prominent ball-like granules, whilst amoebocytes, especially hyaline amoebocytes, quickly spread and adhere firmly to the substratum **(Fig. 6a)**. Similar differences may be discerned in scanning electron micrographs **(Fig. 6b)**. More cytological details and cell type-specific features are visible after May-Grunwald-Giemsa stained smears or cytospin preparations, and in cells imaged by transmission electron microscopy. Intracellular granules are apparently absent in hyaline amoebocytes, but prominent granules with electron-dense material are common in granular amoebocytes. In marked contrast, the cytoplasm of eleocytes is filled with electron-lucent granules reminiscent of chloragosomes (Adamowicz, 2005; Kureket al., submitted).

Both amoebocytes and eleocytes avidly pinocytose neutral red; thus, they have been widely used in the NRRT assay (see: Section 4 in this chapter for references). Neutral red uptake is visible under bright field light microscopy, but fluorescence microscopy enhances and facilitates observations of neutral red containing vesicles (lysosomes and/or chloragosomes) **(Figs. 6c, 6d)**. Studies on the application of flow cytometry for NRRT assay, recommended by Svendsen et al. (2004), are currently in progress in our laboratory at the Jagiellonian University.

The coelomocytes of several lumbricid earthworm species contain eleocytes exhibiting autofluorescence detectable by fluorescent microscope and flow cytometry (Cholewa et al., 2006). Confocal scanning microscopy revealed that the autofluorescence is confined to the chloragosome-like granules of eleocytes **(Fig. 6e,f)**. Coelomocytic autofluorescence of makes these cells ideal candidates for analysis by flow cytometry. This is exemplified in the density plots and histograms presented in **Figure 6g**, where a prominent population of cells exhibiting strong autofluorescence (region toward the right on the FL-1 axis) corresponds with a prominent populations of cells with high complexity/granularity (plotted on the SSC axis). The inter-species difference should be noted; strongly autofluorescent cells (i.e. eleocytes) are a major cohort in *Eisenia fetida* but are very scarce in *Lumbricus terrestris* (Cholewa et al., 2006). Parallel analyses by phase contrast/fluorescence microscopy and flow cytometry have demonstrated a high percentage (>10%) of autofluorescent eleocytes in several earthworm species, namely in *Allolobophora chlorotica*, *Dendrodrilus rubidus*, *E. fetida*, *Octolasion spp.* (Cholewa et al., 2006) and *Dendrobaena veneta* (Olchawa et al., 2006), while <1% of autofluorescent coelomocytes was recorded in representatives of *Aporrectodea spp.* and *Lumbricus spp.* (Cholewa et al., 2006) **(Table 6).** We hypothesise that chloragocytes are detached from chloragogenous tissue to form a population of freely-floating eleocytes in the former group of species, while in the latter group free chloragosomes are released from chloragocytes into the coelomic fluid. Detailed comparative studies are necessary to explore the cytological events and functional significance of these species differences in cellular behaviour.

Flow cytometry, spectrofluorometry, and high performance liquid chromatography (HPLC) show that riboflavin (vitamin B2) is the predominant molecular contributor to the autofluorescence displayed by earthworm eleocytes (Koziol et al., 2006). Recently, the riboflavin content of eleocytes has been found to be species-related, in the order: *E. fetida* > *Octolasion spp.* > *A. chlorotica* > *D. rubidus* (Plytycz et al., submitted).

Earthworms experimentally depleted of coelomocytes are fully vital and capable of restoring the total pool of these cells. For instance, electric shock (1 min, 4.5V) induced the expulsion of more then 95% of coelomocytes suspended in the coelomic fluid of *Dendrobaena veneta*, but almost half of the initial cell density was restored within a few hours, presumably by recruitment from the existing resident pool of coelomocytes. Subsequently, the 'normal' coelomocyte density level was gradually attained presumably by the division of proliferating cells. This relatively late, slow, recruitment phase was temperature-dependent. Complete restoration took 3-6 weeks in earthworms kept at ambient room temperature (c.20°C). In contrast, coelomocyte density levels in earthworms kept in the cold (c.10°C) were significantly below 'normal' even 2 months after expulsion **(Fig. 7a)** (Olchawa et al., submitted). Flow cytometric analysis of DNA profiles and immunohistochemical detection of bromodeoxyuridine (BrdU) revealed that proliferating cells were present both among free-floating coelomocytes and among cells located in the typhlosole and metanephridial regions of the posterior body (Olchawa et al., submitted). We postulate that under field conditions the rate of coelomocyte restoration might be affected by various factors (season, temperature, humidity, food availability, etc.), including soil contamination.

Since coelomocytes are directly implicated in immune functions (see Section 7 of this chapter), it would be reasonable expect that earthworms experiencing coelomocyte depletion would consequentially experience higher massive morbidity and mortality. Surprisingly, we have observed that *D. veneta* and *E. fetida* subjected daily for 3 weeks (i.e. 21 times!) to the electrically-induced coelomocyte extrusion were fully vital even with the minimal number of immunocompetent cells. Furthermore, they retained their ability to restore their coelomocyte populations (in preparation).

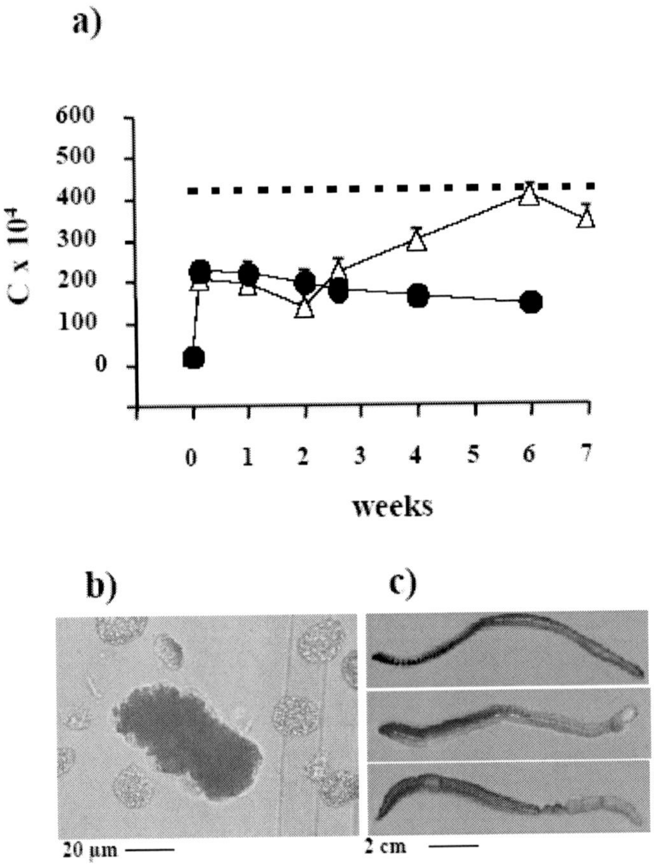

Figure 7. Salient features of *D. veneta* coelomocytes. (a) Temporal restoration of the coelomocyte population after experimental extrusion at t_0 by electrical stimulation (1 min, 4.5 V). At 22°C the number of coelomocytes reached the control level (dashed line) within several weeks (empty symbols), while coelomocyte density remains low in animals kept 10°C (solid symbols). (b) Haemocytometer image showing coelomocytes and a multicellular brown body encapsulating foreign matter probably including invading pathogens. (c) Photographs of *D. veneta*: control (top) and worms exposed to metal-soaked filter papers with body malformations connected with autotomy [see Section 7 of this chapter].

7. Coelomocytes: Immunocompetence and Soil Quality

Both categories of coelomocytes, amoebocytes and eleocytes, are involved in immune functions such as phagocytosis (Adamowicz and Wojtaszek, 2001; Engelmann et al., 2004), cytotoxicity (Suzuki and Cooper, 1995; Cossarizza et al., 1996; Kauschke et al., 2001), and encapsulation (Valembois et al., 1992). They also participate in humoral immunity through their secretions; for example: lysozyme (Çotuk and Dales, 1984), hemolysin (Eue et al., 1998; Koenig et al., 2003), Eiseniapore (Lange et al., 1999; Kauschke et al., 2001), coelomic cytolytic factor (Bilej et al., 1995), fetidins (Milochau et al., 1997). Eleocytes functionally resemble vertebrate hepatocytes in that they participate in the metabolism and storage of glycogen and lipids, in excretion, and they transport nutrients systemically via the coelomic fluid to diverse cells and tissues (Affar et al., 1998). Our recent results (Koziol et al., 2006;

Plytycz et al., submitted) indicate that eleocytes participate in the storage and distribution of vitamins (e.g. vitamin B2 - riboflavin). It is noteworthy that in mammals riboflavin potentiates immune responses. For example, riboflavin increases neutrophil migration (Verdrengh and Tarkowski, 2005), and neutrophil counts and activation in mice and cattle (Osame et al., 1995). Riboflavin also displays a potent ability to incapacitate the human malaria parasite *Plasmodium falciparum* (Akompong et al., 2000).

Under 'normal' conditions the coelomic cavity of healthy earthworms is inhabited by various soil-derived parasites, such as bacteria, gregarines, and fungi (Dales and Kalaç, 1992; Tučková and Bilej, 1996; Reinhart and Dollahon, 2003; Wieczorek-Olchawa et al., 2003). That the populations of extraneous invading organisms are kept in check by the combined activities of coelomocytes and humoral factors is generally accepted. However, the balance between parasites and the host may be tilted in favour of the invading parasites either by the elevation of the pathogenicity of a given source soil or by the impairment of host immunity. In response to large foreign bodies, including massive bacterial invasion, earthworm coelomocytes form so-called "brown bodies". These distinctive coelomic entities are formed when invaders are entrapped within a multicellular capsule, whose encapsulating cells darken with the accumulation brown pigments like lipofuscins and melanins (Cotuc and Dales, 1984; Valembois et al., 1994) **(Fig. 7b).** Lipofuscin (also referred to as "ageing pigment") is an accumulation of insoluble, highly cross-linked, lipoproteins formed as a consequence of the powerful activity of reactive oxygen species on polyunsaturated lipids, an important mechanism accompanying the immune response (Valembois and Lassegues, 1995). Valembois et al. (1994) demonstrated that in *E. fetida andrei*, the amount of lipofuscin is negligible in free coelomocytes but increases in parallel with the formation and maturation of the brown bodies. Fischer (1977), in comparison, found that the synthesis of lipofuscin occurs earlier in the life-cycle of *L. terrestris* coelomocytes. In any case, brown body formation accompanied by lipofuscin accumulation could be viewed as a biomarker of the status of the earthworm immune system. Indeed, brown body formation was significantly higher, and free coelomocyte counts were lower, in *D. veneta* exposed to a heavy metal contaminated soil (Wieczorek-Olchawa et al., 2003). Brown bodies, once formed, are moved towards the caudal segments; these terminal segments may then become necrotic and are self-amputated. This strategy, beginning with encapsulation and ending with autotomy (Sims and Gerard, 1985), is an efficient mechanism in an organism with high regenerative potential for periodically eliminating foreign invaders **(Fig. 7c).**

Earthworm coelomocytes are, therefore, fluid-suspended cells that are readily sampled without severe intrusion either by direct withdrawal or by different extrusion techniques. The cells themselves are responsive to a wide spectrum of chemical and physical stress-inducing factors. Changes in the population structure of coelomocytes, individual cell morphologies, and in individual cell composition and functional parameters may all be used as more-or-less effective biomarkers of sub-lethal response to environmental conditions. Because these coelomocytes are so central to cell-based earthworm ecotoxicology, and because the range and depth of knowledge of their responses to metal induced toxicosis is rapidly expanding, the following sections focuses on the new information emerging largely from the Jagiellonian University laboratory.

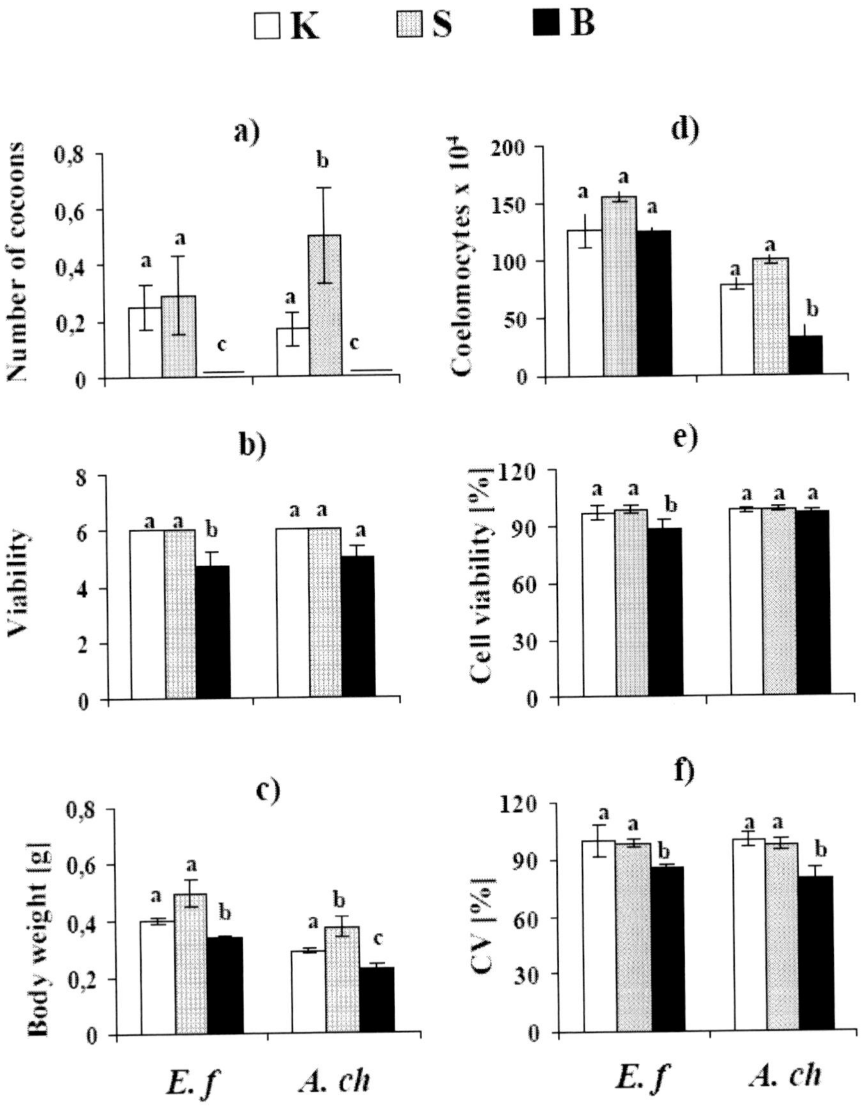

Figure 8. The responses of earthworms and their coelomocytes after 2-week exposure to soils samples from Krakow ('K', relatively clean), Sierbowice ('S', metal-free) and Bukowno ('B', heavily contaminated). [see Table 7 for soil metal concentrations]

Whole-worm exposures to metal-contaminated soil followed by cell extrusion

In a series of experiments, adult *A. chlorotica*, *E. fetida*, and *D. veneta* were maintained in either relatively unpolluted ('K' samples), or metal-free ('S' samples), or a soil heavily polluted with Zn, Pb, Cd ('B' samples) due to local industrial activities. Soils 'K' and 'S' possess a rich bacterial flora, while the 'B' soil is almost bacteria-free **(Table 7)** (Homa et al. 2003; Wieczorek-Olchawa et al. 2003).

Two earthworm species, *A. chlorotica* and *E. fetida*, were transferred from 'K' soil ('reference') to either 'S' ('unpolluted') or 'B' ('Polluted') soils, respectively **(Fig. 8).** After exposure to 'B' soil the reproduction of both species (i.e., cocoon production) was completely inhibited **(Fig. 8a),** whilst mortality was recorded (Fig. 8b), the body weight was significantly reduced **(Fig. 8c)**, and the number, viability and plastic-adherence of their coelomocytes were reduced **(Fig. 8d-f)**. In contrast, worms maintained in the clean 'S' soil displayed enhanced reproductive output **(Fig. 8a)**, they gained body weight **(Fig. 8c)**, whilst the numbers and properties of their coelomocytes were unaltered by transference. Some *A. chlorotica* individuals were kept on the metalliferous 'B' soil for up to 8 weeks. It was observed that the survivors possessed even higher number of coelomocytes than their counterparts kept on their native 'K' soil (Homa et al. 2003), indicating an ability to cope successfully with the coelomic bacterial load even in heavily polluted environment. This phenomenon of pathogen resistance under seemingly immuno-compromised circumstances clearly needs further investigation.

Table 7. Physico-chemical and microbiological characteristics of soil samples from Krakow ('K'), Sierbowice ('S') and Bukowno ('B'). N = 3-4 per sample.

Characteristics	Soil samples		
	K	B	S
pH H_2O	7.5	7.6	5.9
pH KCl	7.2	7.4	5.7
O.C (%)	2.9	4.4	1.4
O.M (%)	5.0	7.7	2.5
Metals (μg/g)			
Zn	331.2±2.2	9820.5±345.8	172.6±14.5
Pb	94.1±2.7	1174.8±126.8	3.2±2.9
Cd	1.4±0.1	72.2±1.7	3.4±0.2
Cu	40.44±0.34	18.61±1.30	7.61±1.19
Bacteria;			
Number of colonies	420±12.83	7±0.82	Nt

O.C. = organic carbon; O.M. = organic matter; means ± SE; Nt - not tested.

Coelomocytes retrieved from *A. chlorotica* maintained for 4 weeks in soils ('K', original uncontaminated; 'S', very clean; or 'B', heavily polluted) were tested *ex vivo* by various methods, including detection of heat shock protein 70 (HSP70) on cytospins/smears, or dot-blot analysis of cell lysates. Both assays gave complementary results: they revealed that HSP70 expression was moderate in the 'K' treatment, but was down-regulated in the 'S' treatment, and upregulated appreciably in 'B' soil treatment **(Fig. 9)** (Homa et al., submitted). Therefore, exposure to the metal-contaminated 'B' soil up-regulates stress protein expression in the free-floating coelomocytes and reduces their number (compare **Figs 8** and **9**). Perhaps coelomocytes die and/or participate in formation of parasite-encapsulating multicellular brown bodies which are too large to be expelled through pores in body wall of animals subjected to the electric shock (unpublished observations).

Metal contaminats can potentially exert deleterious effects on earthworms and bacteria (Wieczorek-Olchawa 2003; Olchawa et al. 2006). This notion is illustrated by the observed cascade of effects experienced by *D. veneta* when experimentally transferred for 4 weeks from uncontaminated soil from the commercial supplier to the 'B' soil at either high (22°C) or low (10°C) ambient temperatures. At the higher temperature, Zn, Cd, and Pb were each accumulated efficiently, leading to high mortality rates. Increased morbidity and mortality in the worms maintained at a relatively high temperature correlated with a decreased free coelomocyte count, increased numbers of brown bodies, and a significant increase in the coelomic bacterial load despite the low bacterial content of polluted 'B' soil (Wieczorek-Olchawa et al., 2003). This response pattern was not observed in transferred worms maintained at 10°C. These observations are consistent with the hypothesis that soil-derived or coelom-inhabiting bacteria proliferate in the favourable nutrient-rich coelomic environment of the host as a consequence of the earthworm's immune system being impaired by metal intoxication.

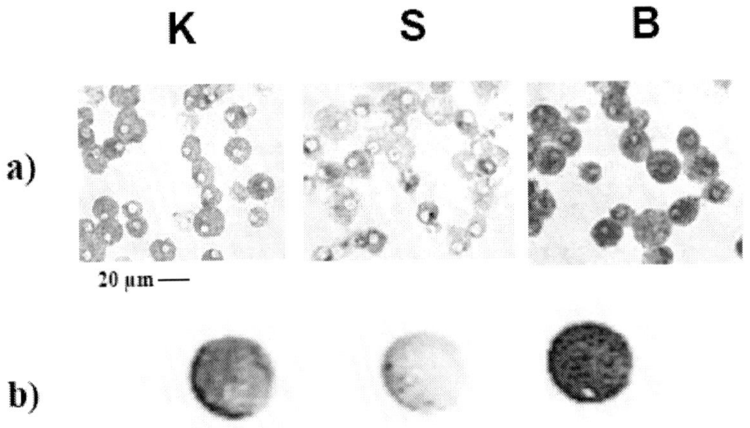

Figure 9. Expression of heat shock protein, HSP70, in coelomocytes extruded from the *A. chlorotica* after 4-week laboratory exposure to soils from Krakow ('K', relatively clean), Sierbowice ('S', metal-free) and Bukowno ('B', heavily contaminated). (a) immunohistochemistry; (b) dot-blot analysis of cell lysates. [see Table 7 for soil metal concentrations]

It seems clear that elevated metal exposures can tilt the balance between invasive microbes and the earthworm's immune system toward the pathogen. However, it is unclear which particular factor(s) was/were critical for compromising immunocompetence. There are a number of possibilities: (i) extensive proliferation of bacteria derived from the polluted soil in the favourable internal environment of the coelomic cavity of newly introduced earthworms; (ii) suboptimal edaphic factors, such as pH and organic matter content/quality lowering the general condition of the introduced worms; (iii) hitherto unappreciated factors such as the synergistic interactions between the various components in the mixed-pollution industrial soil. Even though the interpretation of findings may be difficult, it is evident that the number, composition, and activity of coelomocytes extruded from earthworms transferred to and maintained on complex contaminated soils can serve as sound biomarkers of stressful conditions. Data from single-chemical laboratory exposures may provide the background information necessary for ascribing particular cellular effects to at least a category of chemical stressors in field soils.

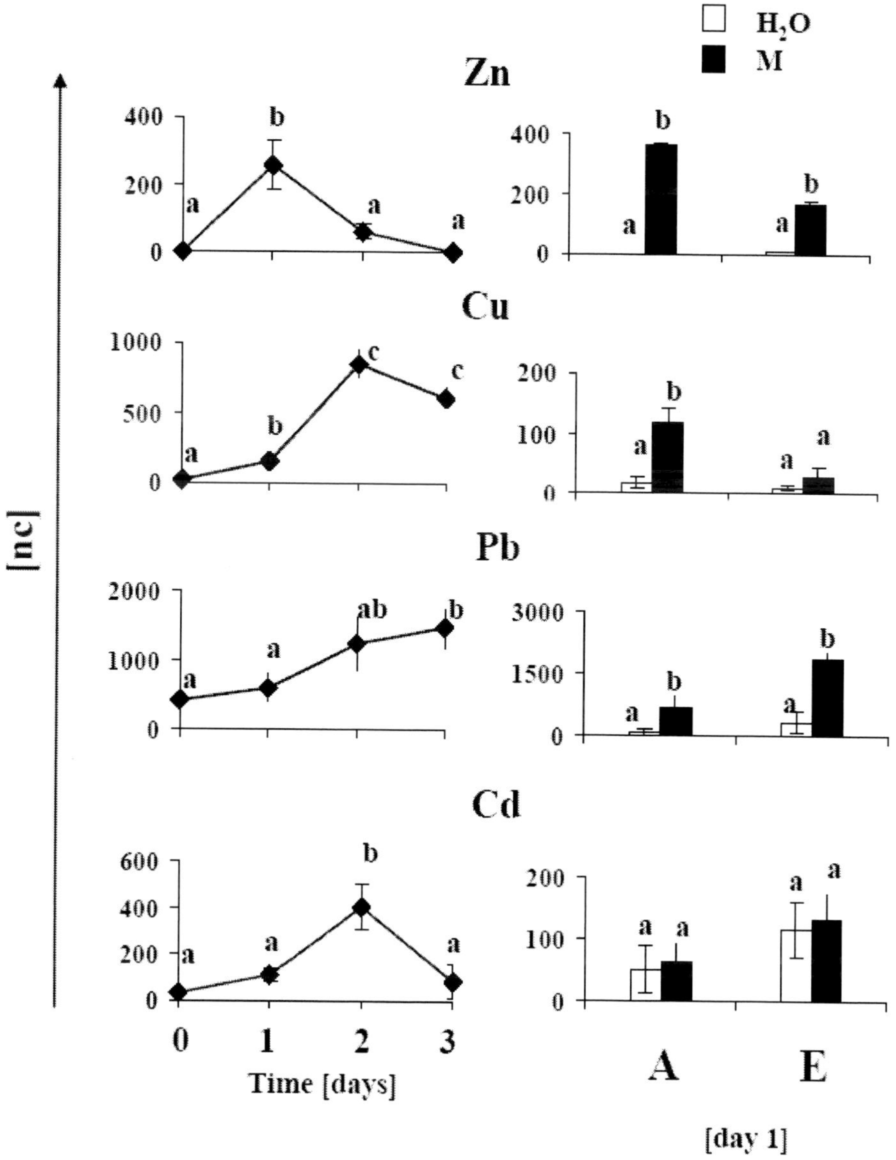

Figure 10. X-ray microanalysis in a scanning electron microscope of metal accumulation in individual coelomocytes: amoebocytes (A) and eleocytes (E), retrieved from *E. fetida* exposed dermally to metal chlorides (M) or water (H_2O) for 1, 2, or 3 days. Nc = Net counts. Methanol-fixed (100%) freeze-dried (Edwards ETD4, Edwards High Vacuum International, UK) and carbon coated (JEE4B, JEOL, Tokyo, Japan) cytospin preparations were analysed in scanning electron microscope (JSM-5410, JEOL, Tokyo, Japan) with a NORAN 679A-3SES energy dispersed spectrometer system (Noran Instruments, Inc., Middletown, WI, USA) [Homa et al., submitted].

Whole-worm contact filter paper (i.e., dermal) exposure

In order to check the effects of a known dose (44 mg/L, corresponding to 1.32 ug/cm^2) of certain metals (Zn, Cu, Cd, or Pb chlorides) on the immune system, earthworms (*E. fetida*, *D. veneta*, and *A. chlorotica*) were dermally exposed in contact filter paper tests as recommended by the OECD (OECD, 1984; Rao et al., 2003) except that the exposure period was extended to 3 days. Under these exposure conditions, Cu was lethal to *A. chlorotica* after 3 days, tissue Zn concentrations were efficiently regulated by all three species, whilst Cu, Cd, and Pb accumulated in species-related patterns. Specifically, the morphologically and ecologically similar species, *E. fetida* and *D. veneta,* accumulated metals with the following order of efficiency: Cd>>Pb>Cu; *A. chlorotica* displayed an entirely different pattern: Cu>Cd (with no discernible Pb accumulation) **(Table 8).**

Since it is evident that metals accumulate in and disrupt the elemental chemistry of chloragocytes (Morgan and Morgan, 1998; Fischer and Molnar, 1992; see: Section 3, this chapter), cells that are related to if not precursors of coelomocytes, it was deemed appropriate to use EPXMA (Banach et al., 2004) to determine whether metals were detectable in the immune cells after 3 day contact filter paper exposures. The following patterns were found in *E. fetida*: Cu accumulated gradually over the exposure period, but mainly in amoebocytes, not eleocytes; Pb was detectable in both amoebocytes and eleocytes; Zn appeared transiently in both eleocytes and amoebocytes on day 1, and its concentration subsequently declined to 'background level; Cd was only detectable on day 2 **(Fig. 10).** Each metal, to different extents and in different ways, affected the Ca/P and S/P ratios of coelomocytes **(Fig. 11).**

Table 8. Accumulation factors (AF) of heavy metals in whole earthworm bodies after 3-day exposure to filter paper soaked with metal chlorides (M) at 1.32 μg/cm^2 or 0.66 μg/cm^2 or to water only (controls, C).

μg/cm^2	*E. fetida*	*D. veneta*	*A. chlorotica*
	M / C	M / C	M / C
$Zn_{1.32}$	1.1	0.9	1.1
$Cu_{1.32}$	3.4	4.34	lethal
$Cu_{0.66}$	nt	nt	10.6
$Pb_{1.32}$	10.4	10.9	0.6
$Cd_{1.32}$	58.3	35.74	4.70

AF = mean metal concentration in whole worms exposed to metal (M) ÷ mean metal concentration in whole worms exposed to water (C).

The disrupted cation homeostasis in *E. fetida* coelomocytes leads to cell deletions, as evidenced by decreased cell density and increased Trypan blue uptake indicative of decreased viability **(Table 9),** either necrosis or the late apoptotic stage of cell death (Kirk et al., 1975). The apoptotic pathway, as reported by caspase-3 upregulation (Pang and Chau, 1999; Jankowski-Hennig et al. 2000; Raes et al., 2000), was activated by Cd and Pb exposures, and especially strongly by Cu exposure, but neither by Zn or water exposures (Homa et al. 2005; submitted). HSP-72 and MT-2 levels were upregulated by each test metal, with HSP72 appearing to be more sensitive to Pb than to either Cd or Cu exposures. The constitutive expression of HSP70 in 'control' cells make this chaperone protein less potentially useful as a biomarker of metal contamination than HSP72, caspase-3, and MT-2 (e.g. Dallinger 1996; Kohler et al. 1996; 1999a; Roesijadi 1996; Spurgeon et al., 2004) **(Table 9, Fig. 12).**

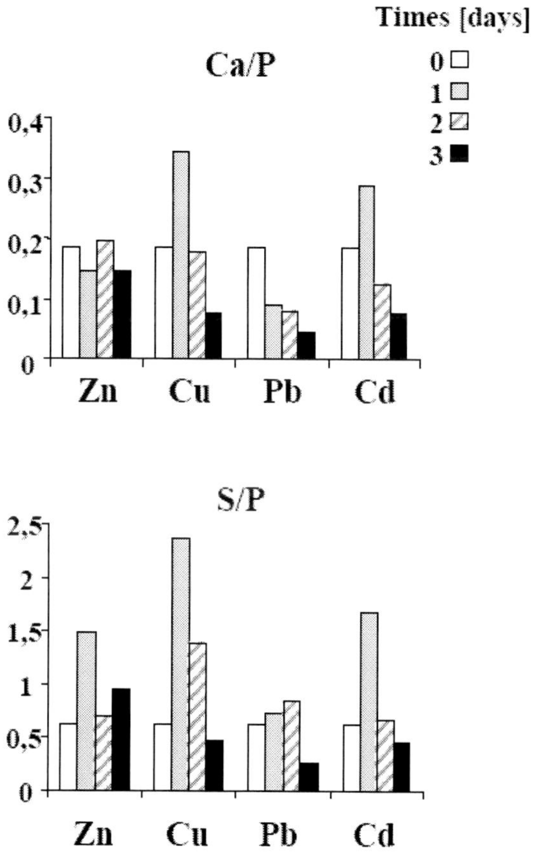

Figure 11. Relative elemental contents of individual coelomocytes (expressed semi-quantitatively as Ca/P and S/P ratios) of *E. fetida* exposed dermally to metal chlorides (M) or water (H_2O) for 1, 2, or 3 days., measured by electron probe X-ray microanalysis in a scanning electron microscopy.

Similar experiments were performed on *A. chlorotica* **(Table 10).** In general, it may be concluded that *A. chlorotica* is more Cu-sensitive then *E. fetida*, and that *A. chlorotica* coelomocytes provide sensitive biomarkers of Cu, Cd, and Pb dermal exposures, at least under laboratory conditions (compare **Table 9** and **Table 10).**

Dermal exposure of *D. veneta* to metals at 22°C or 10°C add more information on the disrupted balance between earthworm immunocompetence and pathogen invasion (Olchawa at al., 2006**) (Fig. 13, Table 11).** The main findings can be summarized as follows:

- Tissue Zn concentration was regulated; Pb and Cd concentrations accumulated in a temperature-dependent manner (7-times more efficient at 22°C compared with at 10°C); a 4-fold increase in Cu concentration at both treatment temperatures (**Table 11, Fig. 13a**).
- Coelomocytes retrieved from the coelomic cavity were affected by all four metals at both temperatures, as evidenced by up-regulated w-MT2 (**Table 11**).
- The number of free coelomocytes was drastically reduced after Cu treatment at both temperatures, but only at 22°C under Cd exposure **(Fig. 13b).**

- Contrary to expectations, the number of bacteria in the coelom was decreased in Cd-treated animals, despite the lowered coelomocyte number **(Fig. 13c).** This paradox was resolved by the results of an experiment on isolated coelom-derived bacteria which showed that Cd completely inhibited bacterial proliferation **(Fig. 13d).**
- We may conclude that Cd is more toxic toward bacteria than it is to coelomocytes, but Cu is apparently much more damaging the immune cells than to bacteria, thus resulting in impaired immunity leading to the morbidity and mortality characteristic of exposure to overdose of this essential transition metal **(Fig. 13).**

Table 9. Number, viability (trypan blue exclusion), and protein content (caspase 3, HSP70, HSP72, and w-MT2) in coelomocytes extruded by electric shock (1 min, 4.5 V) from adult (clitellate) *E. fetida* maintained in unpolluted K soil samples (see Table 7), or exposed for 1, 2 or 3 days to water or metal-soaked filter papers (1.32 μg/cm^2).

Exposure	Days	Number C (x 10^4)	Viability % V	Protein expression [#]			
				Cas -3	HSP70	HSP72	w-MT2
Soil		274.6±19.67	97.14±0.52	+	++	±	+
H2O	1d	323.5±14.78	97.3±0.38	+	++	±	+
	2d	313.2±30.78	96.5±0.50	+	++	±	Nt.
	3d	243.1±37.01	96.8±0.76	+	++	±	+
Zn	1d	263.6±15.83	95.5±1.45	+	+++	+	++
	2d	285.3±13.56	92.0±1.71	+	+++	++	Nt.
	3d	249.9±8.93	94.5±1.42	+	++++	++	+++
Cu	1d	269.3±9.30 *	79.1±1.79 *	++	+++	++	+++
	2d	246.8±24.68 *	61.9±3.38 *	+++	+++	+++	Nt.
	3d	80.0±8.54 *	77.5±5.98 *	+++	+++	+++	+++
Pb	1d	215.8±8.05	80.9±2.68 *	++	+++	+++	++
	2d	181.5±18.47 *	75.7±4.34 *	++	+++	++++	Nt.
	3d	127.2±15.60 *	79.4±4.94 *	++	+++	++++	+
Cd	1d	280.4±20.06	84.3±4.74 *	++	+++	++	+++
	2d	293.2±20.76	63.7±1.66 *	++	+++	+++	Nt.
	3d	165.3±17.95 *	81.1±3.78 *	++	+++	+++	+++

Means ± SE. *mean values significantly different from those in the water-treated group (Mann-Whitney test, $p<0.05$); protein expression absent or negligible (±), weak (+), moderate (++), strong (+++), or very strong (++++); nt - not tested. [#] protein expression levels were estimated by qualitative examination of immuno-peroxidase stained cytospin preparations (Homa et al. 2005)

Flow cytometry has been exploited as a high-throughput technique for monitoring the responses of differenent coelomocyte cohorts to dermal metal exposures. The relatively large size and intense autofluorescent properties of *D. veneta* coelomocytes facilitated this analytical approach. Cd and (especially) Cu exposures drastically reduced the proportion of the coelomocyte population represented by eleocytes (Olchawa et al., 2006). Moreover, such changes were reversible after the transfer of animals from metal-contaminated to clean substrates (Olchawa et al., 2006). In conclusion, flow cytometry is a valuable tool for rapidly assessing the numbers and properties of cells in fluid suspension, especially if some cell types have distinct features such as autofluorescent contents and 'granularity'.

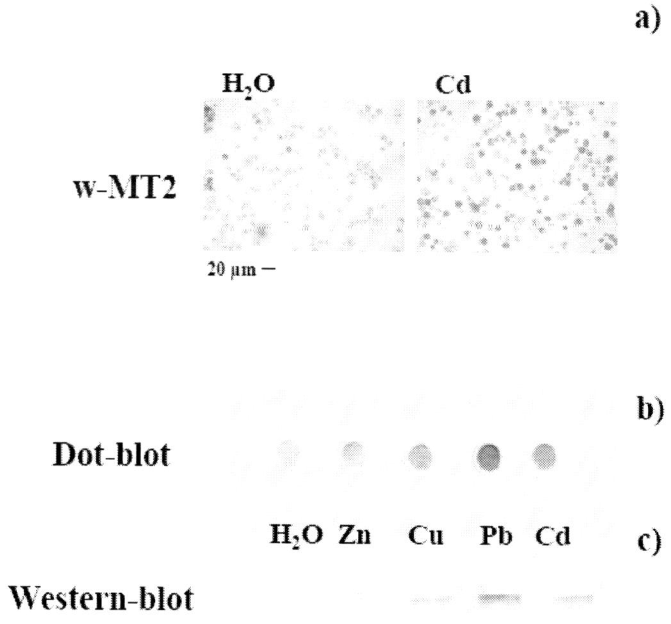

Figure 12. Stress proteins (HSP72 and w-MT2) expression in coelomocytes extruded by electric shock (1 min, 4.5 V) from *E. fetida* exposed for 3 days to filter-paper soaked with H_2O (controls) or metal chloride aqueous solutions. (a) immuno-peroxidase histochemical detection by polyclonal antibodies raised against worm metallothionein isoform 2 (w-MT2) of *L. rubellus* [Stürzenbaum et al. 2001, 2004]; (b,c) dot-blot and Western blot detection of HSP72 in coelomocyte lysates; Stress protein expression levels were negligible in coelomocytes from the control animals while dot-blot and Western blot analysis show the up-regulation of HSP72 in coelomocyte lysates in the induction series, Pb > Cd,Cu > Zn [Homa et al. 2005 and unpublished data].

In vitro post-extrusion exposure of coelomocytes to metals

The sensitivity of isolated coelomocytes to single metal-exposures may be tested under optimised *in vitro* conditions (Kurek at al., 2002; Stankiewicz and Plytycz, 1998). This has been done on extruded coelomocytes from *A. chlorotica*, *D. veneta*, *E. fetida*, and *L. terrestris* by incubating them for 3 hrs in growth medium supplemented with a range of Zn, Cu, Pb, or Cd concentrations **(Fig. 14).** It was found that Zn did not affect cell viability (measured by Trypan blue exclusion - Kirk et al., 1975), MTT reduction to formazan (Plytycz and Jozkowicz, 1994; Stankiewicz and Plytycz, 1998), and coelomocyte functions, including plastic adherence (measured by staining with crystal violet - Plytycz et al., 1992) and pinocytic activity (measured by neutral red (NR) uptake - Plytycz et al., 1992). In contrast, all the assays revealed severe negative effects of Cu on coelomocyte viability and functioning, whilst Pb and Cd toxicities were less severe. Thus it turned out, that the effects of an *in vitro* metal exposure corresponded to those revealed in coelomocytes extruded from animals dermally exposed to particular metals (compare **Fig. 14**, **Fig. 13**, **Tables 9** and **10**).

In conclusion, several complementary methods revealed that coelomocytes homeostatis is significantly disrupted, at least transiently, in the earthworms exposed to metal stressors under laboratory conditions and, thus, these easily accessible cells may provide the basis for developing sensitive and relatively non-intrusive biomarkers of soil quality in an ecologically-relevant organism.

Comparative studies on coelomocytes from various earthworm species should be performed, especially those, which differ in biology, environmental demands and coelomocyte composition. Perhaps some species will be especially suitable for the subsequent studies due to peculiar sensitivity to some toxic factors and/or as most convenient for experimental manipulations (like easy handling in the laboratory, high reproduction rate, or high number of coelomocytes). We expect that studies on earthworm coelomocytes will give the results valuable from both theoretical and practical point of view as they may find a practical application in biomonitoring.

8. Earthworm Genomics

The burgeoning Expressed Sequence Tag database for earthworms, LumbriBASE (www.earthworms.org; Stürzenbaum et al., 2003), currently populated with 17,225 ESTs and 7629 gene objects for the species *Lumbricus rubellus* and 1108 ESTs for *Eisenia andrei*, provides a major resource for diverse earthworm biology studies. It has already spawned a number of field and semi-field ectoxicology studies, each exploiting the technique of quantitative real time PCR (Q-RT-PCR) (Stürzenbaum and Kille, 2001; Galay-Burgos et al., 2003; Van Straalen and Roelofs, 2006) to measure toxicant-induced perturbations of the transcription levels of a small number of earthworm target genes (Galay-Burgos et al., 2003; Ricketts et al., 2004; Stürzenbaum et al., 2004; Galay-Burgos et al., 2005; Spurgeon et al., 2005, 2006). At this relatively early stage in the evolution of ecotoxicogenomics (Snape et al., 2004; Lettieri 2006) it is not easy to comprehend the value of single-gene transcription measurements as faithful reporters of the environmental conditions impinging upon the test organism. The difficulty is illustrated by Spurgeon et al.'s (2006) conclusion that the expression of the metal-binding protein metallothionein-2 in earthworms (*L. rubellus*), exposed under laboratory conditions to metalliferous field soils, was weakly correlated with soil metal concentrations. Transient response kinetics, as reported by Köhler et al., 1999b) for hsp70 in isopods dosed with particular organic toxicants, may be important albeit confounding factors. Other, largely overlooked, confounders are the 'background' fluxes in the expression of a chosen genetic biomarker in response to intrinsic changes (e.g. mating condition) or extraneous variables (e.g. climatic extremes, food availability and quality). A more optimistic or progressive view of the potential of genomics in ecotoxicology is that it will furnish knowledge about mechanistic aspects of toxicosis, a view succinctly presented by Snape et al. (2004):

"Whilst…phenomenological approaches ('endpoints' e.g. median lethal concentrations, no observed effect concentrations, etc.) are useful for identifying chemicals of potential concern they provide little understanding of the mechanism of chemical toxicity. Without this understanding, it will be difficult to address some of the key challenges that currently face ... ecotoxicology, e.g. predicting toxicant responses across a very broad range of...phylogenetic groups present in...ecosystems; estimating how changes at one ecological level or organisation will affect other levels (e.g. predicting population-level effects); predicting the influence of time-varying exposure on toxicant responses."

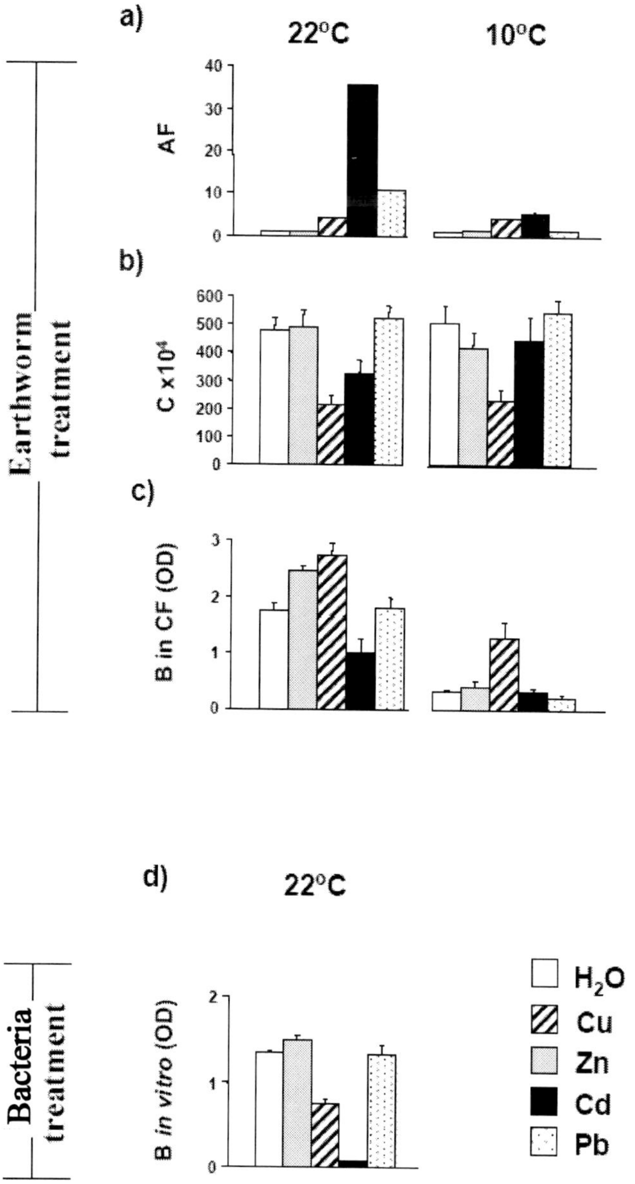

Figure 13. Effects on *D. veneta* of 3-day exposures at 22°C or 10°C to water or metal chlorides: (a) accumulated whole-body tissue metal concentrations; (b) coelomocyte number per gram body weight; (c) bacterial load in coelomic cavity; (d) proliferation of coelom-derived bacteria in a growth medium (bacterial broth) spiked with metals – the bacterial proliferation was measured by the MTT/formazan reduction test [Olchawa et al. 2006].

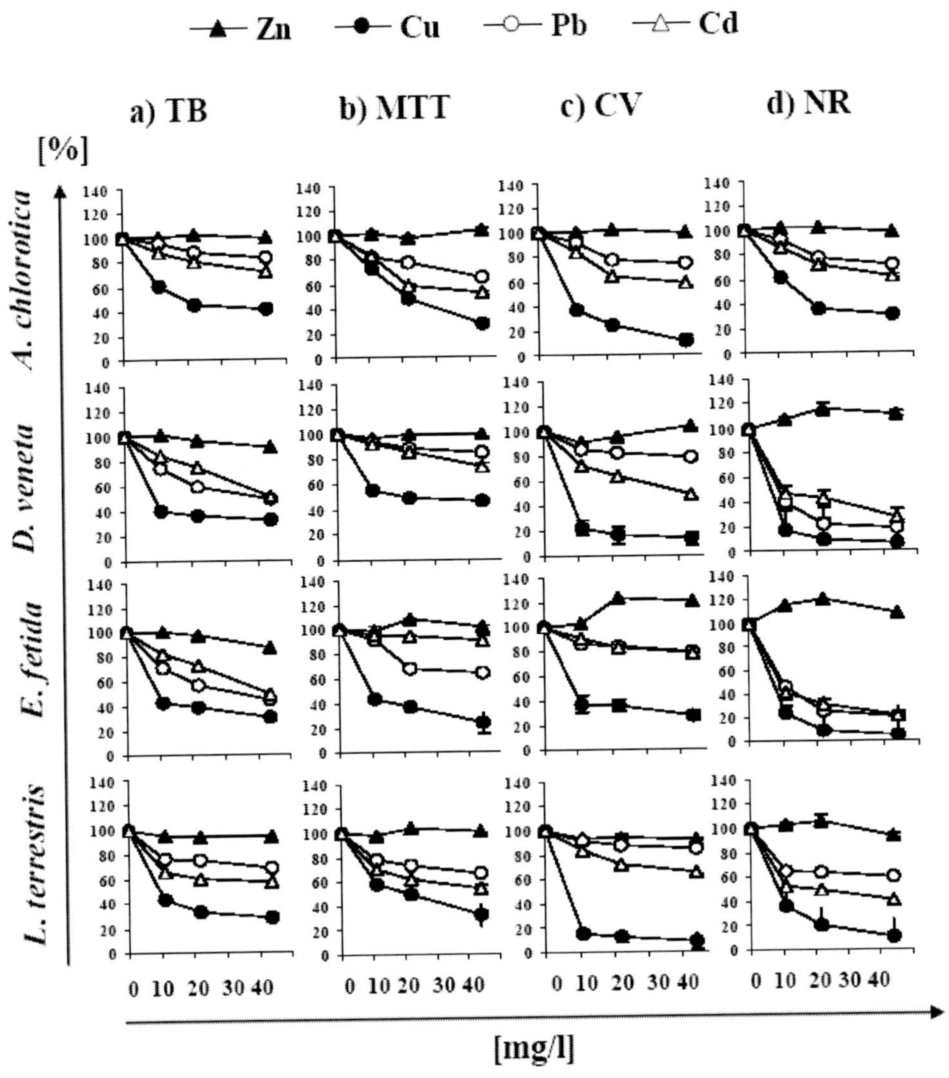

Figure 14. Viability and activity of earthworm coelomocytes after *in vitro* incubation for 3 hours (a,b) or 1 hour (c,d,) in media supplemented with different concentrations of different metal (Cd, Cu, Pb, and Zn) chlorides. Cell viability assessed by Trypan blue exclusion (a) or MTT reduction to formazan (b); plastic adherence assessed by staining with crystal violet (CV, c), and pinocytosis of neutral red (NR, d) by adherent cells [Homa, unpublished].

The reader may be forgiven for feeling that this section of the chapter has lost sight of the cellular perspective. However, not only has the commitment to earthworm genomics lead to Q-RT-PCR-based biomarker assays, but it has been paralleled by studies using microscopy techniques to detect and localize functionally important gene products by immuno-histochemistry, and gene activity by in situ hybridization, in earthworm cells and tissues. Some of the studied proteins are directly or indirectly implicated in anti-microbial responses (Ohata et al., 2000; Wang et al., 2003; Engelmann et al., 2004), whilst heat shock proteins play multifaceted roles in combating physiological stress (Mariño and Morgan, 1998; Mariño et al., 1999; Homa et al., 2005), and metallothioneins have isoform-dependent roles in metal homeostasis, sequestration and trafficking (Stürzenbaum et al., 2001, 2004; Morgan et al., 2004). In many cases genomic observations have led to cellular studies in an effort to understand the function of a given gene product, from secretion, to target sites, and eventual fate. Metallothionein trafficking in earthworm chloragocytes is a relatively complete example of this approach (Stürzenbaum et al., 2004), but it may be instructive to summarize salient facts in the research history of another gene-protein system in earthworms. Annetocin, a member of the oxytocin/vasopressin (OT/VP) peptide super-family, was isolated from *E. fetida* (Ukena et al., 1995), and subsequently found to be a potent inducer of egg-laying behaviour (Oumi et al., 1996). More recently, an annetocin receptor was isolated and sequenced (Kawada et al., 2004). It was found to have a high sequence similarity with

Table 10. Number, viability (trypan blue exclusion), and protein content (caspase 3, HSP70 and HSP72, and w-metallothioneine 2) in coelomocytes extruded by electric shock (1 min, 4.5 V) from adult (clittellate) animals of *A. chlorotica* maintained in unpolluted K soil samples (see Table 7), or exposed for 1, 2 or 3 days to water or metal chloride -soaked filter papers (1.32 μg/cm²).

Exposure	Days	Number C (x 10^4)	Viability % V	Protein expression Cas -3	HSP70	HSP72	w-MT2
Soil		179.4±10.03	98.17±1.82	+	++	±	+
H2O	1d	189.3±13.58	97.70±0.72	+	++	±	+
	2d	212.6±19.23	96.37±1.06	+	++	±	+
	3d	181.4±17.45	93.64±1.32	+	+	±	+
Zn	1d	207.3±13.17	95.89±1.63	+	++	+	++
	2d	219.6±15.72	95.53±3.09	+	++	++	+++
	3d	177.3±11.57	91.95±4.60	+	+	++	+++
Cu	1d	93.33±5.50*	73.94±4.23 *	+++	++	+	++
	2d	112.8±16.27*	56.51±5.06 *	++++	+	+	+/-
	3d	nt.	nt.	nt.	nt.	nt.	nt.
Pb	1d	135.6±22.10 *	81.48±1.19 *	++	+++	++	++
	2d	131.0±11.88 *	73.13±1.29 *	++	++	++	++
	3d	124.2±12.25 *	71.98±1.75 *	++	+	++	++
Cd	1d	108.0±10.60 *	77.48±4.56 *	++	++	+++	+++
	2d	121.8±12.97 *	67.08±4.07 *	++	+++	+++	++++
	3d	119.7±10.57 *	66.97±4.62 *	++	++	+++	+++

Means ± SE. *mean values significantly different from those in respective water group (Mann-Whitney test, $p<0.05$); protein expression absent or negligible (±), weak (+), moderate (++), strong (+++), or very strong (++++); nt - not tested. # protein expression levels were estimated by qualitative examination of immuno-peroxidase stained cytospin preparations (Homa et al. 2005) mammalian OT/VP receptors, and to be expressed specifically in nephridia in the clitellum region of *E. fetida*. These aggregated observations are novel because they demonstrate the hitherto unsuspected regional specialization of nephridia and, from an ecotoxicology viewpoint, because they provide convincing evidence linking characterised gene products (i.e., annetocin and its receptor) to a clearly defined facet of reproductive behaviour. Ricketts et al. (2004) used Q-RT-PCR to show the distribution of annetocin transcripts in the anterio-clitellar segments, and further showed that earthworms exposed to metal-contaminated soils have lower annetocin expression and cocoon (i.e. egg capsule) production. Thus, annetocin is a very promising molecular-genetic biomarker: not only have its physiological mode and cellular sites of action been elucidated, but its function is explicitly and mechanistically linked with a significant demographic parameter. Very recently, valosine-containing proteins (VCP) were sequenced in *E. fetida* (Suzuki et al., 2005). Q-RT-PCR showed that one VCP isoform (evcp-2) was expressed in the anterior segments; in situ hybridization localized evcp-2 expression to the cytophores surrounded by spermatids in the seminal vesicles of sexually mature worms, thus implicating the protein in some capacity in spermatogenesis (Suzuki et al., 2005). It remains to be seen whether VCP expression is altered by environmental toxicants and, if so, whether this impacts upon cocoon production and/or viability.

Table 11. Metal concentrations, metal accumulation factors (see Table 8 for definition of AF), and relative expression of coelomocyte w-MT2 in *D. veneta* exposed for 3 days to filter paper soaked either with water (W) or metal (M) ions (1.32 μg/cm²) at 22°C or 10°C.

Metals	Temp. °C	Metal concentrations in *D. veneta* [μg/g b.w.]		Accumulation Factors [AF]		C w-MT2
		Water exposure [W]	Metal exposure [M]	M/W	22°C/10°C	
H_2O	22					+
	10					+
Zn	22	102.07±7.54	92.35±6.59 NS	0.9	0.7	+++
	10	89.85±5.77	114.98±17.91 NS	1.28		++
Cu	22	6.32±0.58	27.48±0.31 S	4.34	0.98	+++
	10	4.84±0.82	21.40±6.44 S	4.42		++++
Cd	22	1.41±0.14	50.40±11.43 S	35.59	6.56	++++
	10	1.19±0.09	6.49±0.64* S	5.42		++
Pb	22	0.33±0.06	3.62±0.25 S	10.9	7.12	+++
	10	0.31±0.05	0.48±0.11* NS	1.53		++++

Means ± SE. Differences between W and M, and between groups treated with the same metal at 22°C or 10°C, were deemed significant (S, *p<0.05) or not significant (NS) by the Mann-Whitney 'U' test.

Metallothionein (w-MT2) expression was assessed in cytospine preparations after immuno-peroxidase staining (Morgan et al., 2004; Homa et al., 2005; Olchawa et al., 2006) and scored weak qualitatively: (+), moderate (++), strong (+++), very strong immuno-staining (++++).

The reader may now be further forgiven for considering that a cell biologist's involvement in ecotoxicology begins with a sequenced gene and ends with the microscopic localization of the active gene or of its encoded product. Moreover, it might be supposed that whether the impact of toxicants at field-relevant doses on the chosen gene or the protein reduces the fitness of the individual organism is a mere happenstance. Fortuitously and perhaps paradoxically, genomic technologies propel ecotoxicology toward more holistic rather than increasingly reductionist approaches. Genomics can also screen efficiently those genes that are both responsive to particular toxicants at given dose levels, and (through sequence conservation, and knowledge derived from model organisms) are conspicuously involved in fundamental fitness-related processes. A visit to LumbriBASE (www.earthworms.org) or similar database reveals that it is increasingly possible to assess genome-wide stress effects in non-model, non-target, 'sentinel' organisms like *L. rubellus* and *E. fetida*. The technology facilitating this approach is DNA microarrays and accompanying statistical procedures (Schulze and Downward, 2001; Lettieri, 2005; Van Straalen and Roelofs, 2006). Microarray slides have already been fabricated, optimised and deployed for monitoring the transcription profiles of *L. rubellus* challenged by Cd, atrazine, and fluoranthene in relation to the control state (Kille et al., unpublished). Two important ecotoxicological conclusions emerge from these data: (i) a significant proportion of the sampled genome show residue-specific response patterns (gene groupings signifying involvements in metabolic pathways or cascades), such that the microarray output is itself a diagnostic tool; (ii) by correlating the activities (highly significant up- or down-regulation) of individual genes in the microarrays with contemporaneous toxicant-induced changes in fitness indicators of the whole organism, it should be possible to identify a suite of key genes (perhaps including annetocin and evcp-2, but probably a number of novel ones) whose transcriptome levels could be measured in Q-RT-PCR assays. In this way molecular-genetic biomarkers are logically and efficiently screened for biomarker potential by demonstrating a significant linkage between a specific transcriptional change and (putative) ecologically-significant performance indicator. Note that this is not flagged as a mechanistic link: such a linkage may, however, be inferred subsequently by knowing a gene's identity and through comparison with its function in other taxa. Mechanistic knowledge, as shown by the metallothionein and annetocin examples (see above), would be enhanced by determining the cellular sites of a particular gene's activity and the protein's cellular target. In those cases where genes are well annotated, EST sequences can be used to generate primers for in situ hybridization and to design and make, usually through commercial providers, monoclonal antibodies. **Figure 15** provides four ontological diagrams, constructed by informatically interrogating LumbriBASE. It gives a flavour of some of the key processes for which there are EST sequences which can be used to design Q-RT-PCR primers. as well as cellular and sub-cellular molecular-genetic 'probes', for various uses in earthworm ecotoxicology.

Many of the organisms of interest to the ecotoxicologist are, of course, non-model organisms. Indeed, an ecotoxicology laboratory almost by definition may be interested in the impacts of chemicals on several different species, representing different phylogenetic groups and ecological functions. Even within a taxon like the lumbricid earthworms there are a number of recognisable eco-physiological groupings of species, and the differences in distribution within the soil profile and other dimensions of their niches may render them more or less susceptible to chemical stressors under field conditions even if, under laboratory conditions, they have similar responses. It is unlikely that genetic resources as comprehensive as LumbriBASE will become available for many of these 'other' species in the near future (see: Storey and McMullan, 2004). Therefore, it is noteworthy that the technique of 'heterologous screening', where cDNA microarrays fabricated for one species are screened with cDNA from another species, has proved rather effective for characterising the gene expression profile underpinning insect adaptations to stressful environments (Eddy and Storey, 2002; Storey and McMullan, 2004). Whether this pragmatic approach will prove to be widely applicable in ecotoxicogenomics is uncertain because the theoretical objections to it are profound, as Storey and McMullan (2004) readily acknowledged. But, as they also state, the technique probably can yield a number of useful gene sequences in 'new' species despite the problems because of the substantial degrees of genetic conservation across fairly wide phylogenetic distances.

9. Conclusion

This chapter attempted to give the reader an accurate, sometimes idiosyncratic, synthesis of what is the current state-of-the-art in cell-based earthworm ecotoxicology. Idiosyncratic, in that the account was self-restricted to studies where 'the cell' (if the expression may be pardoned) is treated simultaneously as a morphological and a biochemical entity i.e., the cell as a biological system. A notable feature of this approach is that the identities of individual cells must first be imaged or otherwise ascertained prior to subsequent compositional or functional analyses. Implicit to the approach is the notion that the cell complement of metazoans form communities with an integral 'ecology' i.e., the members of the community respond to their surroundings, they have distinctive life-cycle traits, as well as 'professions' or niches. The metaphorical comparison with the more widely accepted use of 'ecology' may have some merits, but should not conceal an important difference. Communities of animals and plants in natural environments are essentially communities of genotypes or of gene pools, whereas individual organisms are comprised of cells all possessing identical genotypes, and yet they differentiate into morphologically (i.e. phenotypically) distinct, often inter-mixed, cohorts. As Kirschner and Gerhart (2005) state, "a cell type is a compartment in 'gene expression space'". Thus, the approaches that we describe in this chapter may be said to examine the effects of chemical stressors impinging upon 'gene expression spaces' in earthworms. For now we shall refer to cells and cell types in the more conventional way, whilst recognising the heuristic insight that the Kirschner and Gerhart descriptive phrase offers.

The preamble in the opening paragraph of this concluding section leads to the question: what is the future of cell-based earthworm ecotoxicology? It may be that attempting to answer such a question is both futile and hazardous but, nevertheless, we shall not be deterred. Some pointers to the future may be found in the history of the field. A broad overview recognises three dominant themes. Firstly, the exploitation of easily-sampled cell systems, containing stress-responsive parameters that can be measured by cost-effective, technically undemanding, methods. The NRRA and various immunocompetence assays on coelomocytes furnish good examples of such biomarkers of sub-lethal toxicosis effects. Secondly, there have been efforts to exploit high-throughput technology when laboratory provision and nature of the bio-sample allow. For example, flow cytometry has been used as a platform for assaying stress indicators in free-floating coelomocytes (Fugere et al., 1996; Brousseau et al.,

1997; Diogene et al., 1997; Massicotte et al., 2003, 2004). Thirdly, genomic and metabolomic technologies are informing fresh initiatives, encompassing immunohistochemical and in situ hybridization protocols, to understand the mechanisms of toxicant metabolism and effects at the level of cells. This brings us to the present, but whither the future? Clinical diagnostics and the drug discovery industry have embraced certain high-throughput technologies that, to the best of our knowledge, have not yet been exploited in ecotoxicology. Perhaps automated fluorescence microscopy for cytometric dose-response profiling of very large cell populations (Perlman et al., 2004) may find favour in some laboratories, but we draw attention to a particular 'family' of high-throughput techniques that the advent of LumbriBASE, for example, makes accessible. Tissue microarrays (TMA) allow for the simultaneous sectioning of hundreds if not thousands of tiny tissue cores (typically about 0.6 mm in diameter). The cored samples can be derived from formalin-fixed or fresh soft-tissue samples, and can be subjected to simultaneous 'staining' by fairly standard immunohistochemistry, fluorescence in situ hybridization (FISH), or RNA in situ hybridization protocols. Thus, a very large experiment can be performed on a single slide, making the approach both cost effective and highly informative when coupled with multivariate statistical tools (Andersen et al., 2001; Liu et al., 2002; Packeisen et al., 2003; Sauter et al., 2003).

Finally, authors have rightly drawn attention to the core value of the 'eco' component of the term 'ecotoxicology' (e.g., Cairns 1988; Chapman 2002). In Chapman's words, "ecological understanding must be integrated into toxicology for a better, more coherent whole". We also note Snape et al.'s (2004) statement, "...the challenge in ecotoxicology for the 21st century is to understand the mechanisms of toxicity to different wildlife species". The fact that this chapter addresses the challenge alluded to in the second quotation does not provide reason to disengage from commitment to the sentiment expressed in the first.

Acknowledgements

This work was supported by the State Committee for Scientific Research, DS/BiNoZ/IZ/773 (awarded to BP and JH), and the British Council (Visiting Scientist Scholarship awarded to LM). It was completed during the tenure a Natural Environment Research Council grant (GST/02/1782) awarded to PK and AJM.

References

[1] Adamowicz A. (2005). Morphology and ultrastructure of the earthworm *Dendrobaena veneta* (Lumbricidae) coelomocytes. *Tissue and Cell* **37**, 125-133.

[2] Adamowicz A.,Wojtaszek J. (2001). Morphology and phagocytotic activity of coelomocytes in *Dendrobaena veneta* (Lumbricidae). *Zoologica Poloniae* **46**, 91-104.

[3] Affar E.B., Dufor M., Poirier G.G., Nadeau D. (1998). Isolation, purification and partial characterization of chloragocytes from the earthworm species *Lumbricus terrestris*. *Mol. Cell. Biochem.* **185**, 123-133.

[4] Agarraberes F.A., Dice J.F. (2001). A molecular chaperone complex at the lysosomal membrane is required for protein translocation. *J. Cell Sci.* **114**, 2491-2499.

[5] Akompong T., Ghori N., Haldar K. (2000). In vitro activity of riboflavin against the human parasite Plasmodium falciparum. *Antimicrob Agents Ch.* **44**, 88-96.

[6] Amaral A.F.S., Rodriguez A.S. (2005). Metal accumulation and apoptosis in the alimentary canal of *Lumbricus terrestris* as a metal biomarker. *BioMetals* **18**, 199-206.

[7] Amaral A.F.S., Soto M., Cunha R., Marigomez I., Rodriguez A. (2005). Bioavailability and cellular effects of metals on *Lumbricus terrestris* inhabiting volcanic soils. *Environ. Pollut.*

[8] Andersen C.L., Hostetter G., Grigoryan A., Sauter G., Kallioniemi A. (2001). Improved procedure for fluorescence in situ hybridization on tissue microarrays. *Cytometry* **45**, 83-86.

[9] Andre J., Hodson M.E. Morgan A.J., Stürzenbaum S.R. (2005). 'Bioavailable' versus 'sequestered' Pb fractions in earthworms: inter-population and exposure-history differences. SETAC Europe (Chronic & Diffuse Pollution), University of Newcastle, October 2005, p. 48.

[10] Ashburner M., Ball C.A., Blake J.A., Botstein D., Butler H., Cherry J.M., Davis A.P., Dolinski K., Dwight S.S., Eppig J.T., Harris M.A., Hill D.P., Issel-Tarver L., Kasarskis A., Lewis S., Matese J.C., Richardson J.E., Ringwald M., Rubin G.M., Sherlock G. (2000). Gene ontology: tool for the unification of biology. The Gene Ontology Consortium. *Nat. Genet.* **25**, 25-29.

[11] Banach Z., Borowska J., Pyza E., Tylko G. (2004). X-Ray microanalysis method of elemental content evaluation in the brain of *Musca domestica* exposed to heavy metals. *Rad. Phys. Chem.,* **71**, 1001-1002.

[12] Bilej M., Brys L., Beschin A., Lucas R., Vercauteren E., Hanušová R., De Baetselier P. (1995). Identification of a cytolytic protein in the coelomic fluid of *Eisenia foetida* earthworms. Immunol Lett. 45, 123-126.

[13] Blott E.J., Griffiths G.M. (2002). Secretory lysosomes. *Nature Rev. Mol. Cell Biol.* **3**, 122-131.

[14] Booth L.H., Heppelthwaite V.J., O'Halloran K. (2005). Effects-based assays in the earthworm *Aporrectodea caliginosa*: their utilisation for evaluation of contaminated sites before and after remediation. *J. Soils and Sediments* **5**, 87-94.

[15] Booth L.H., Heppelthwaite V.J., Webster R., O'Halloran H. (2001). Lysosomal neutral red retention time as a biomarker of organophophate exposure in the earthworm *aporrectodea caliginosa*. Laboratory and semi-field experiments. *Biomarkers* **6,** 77-82.

[16] Booth L.H., O'Halloran K. (2001). A comparison of biomarker responses in the earthworm *Aporrectodea caliginosa* to the organophosphorus insecticides diazinon and chloropyrifos. *Environ. Toxicol. Chem.* **20**, 2494-2502.

[17] Booth L.H., Palasz F., Darling C., Lanno R., Wickstrom M. (2002). The effects of lead-contaminated soil from Canadian prairie skeet ranges on the neutral red retention assay and fecundity in the earthworm Eisenia *fetida. Environ. Toxicol. Chem.* **22**, 2446-2453.

[18] Bradham K.D., Dayton E.A., Basta N.T., Schroder J., Payton M., Lanno R. (2006). Effect of soil properties on lead bioavailability and toxicity to earthworms. *Environ. Toxicol. Chem.* **25**, 769-775.

[19] Braunbeck T. (1998). Cytological alterations in fish hepatocytes following *in vivo* and *in vitro* sublethal exposure to xenobiotics – structural biomarkers of environmental contamination. In: Fish Ecotoxicology. Eds. T. Braunbeck, D.E. Hinton, B. Streit, Birkhäuser Verlag, Basel, pp. 59-140.

[20] Brousseau P., Fugere N., Bernier J., Coderre D., Nadeau D., Poirier G., Fournier M. (1997). Evaluation of earthworm exposure to contaminated soil by cytometric assay of coelomocytes phagocytosis in *Lumbricus terrestris* (Oligochaeta). *Soil Bio. Biochem.* **29**, 681-684.

[21] Bundy J.G., Lenz E.M., Bailey N.J., Gavaghan C.L., Svendsen C., Spurgeon D., Hankard P.K., Osborn D., Weeks J.M., Trauger S.A., Speir P., Sanders I., Lindon J.C., Nicholoson J.K., Tang H. (2002a). Metabonomic assessment of toxicity of 4-fluoroaniline, 3,5-difluoroaniline and 2-fluoro-4-methylaniline to the earthworm *Eisenia veneta* (Rosa): identification of new endogenous biomarkers. *Environ. Toxicol. Chem.* **21**, 1966-1972.

[22] Bundy J.G., Osborn D., Weeks J.M., Lindon J.C., Nicholoson J.K. (2001). An NMR-based metabonomic approach to the investigation of coelomic fluid biochemistry in earthworms under toxic stress. *FEBS Lett.* **500**, 31-35.

[23] Bundy J.G., Ramlov H., Holmstrup M. (2003). Multivariate metabolic profiling using ^{1}H nuclear magnetic resonance spectroscopy of freeze-tolerant and freeze-intolerant earthworms exposed to frost. *Cryo Lett.* **24**, 347-358.

[24] Bundy J.G., Spurgeon D.J., Svendsen C., Hankard P.K., Osborn D., Lindon J.C., Nicholson J.K. (2002b). Earthworm species of the genus *Eisenia* can be phenotypically differentiated by metabolic profiling. *FEBS Lett.* **521**, 115-120.

[25] Bundy J.G., Spurgeon D.J., Svendsen C., Hankard P.K., Weeks J.M., Osborn D., Lindon J.C., Nicholson J.K. (2004). Environmental metabonomics: Applying combination biomarker analysis in earthworms at a metal contaminated site. *Ecotoxicology* **13**, 797-806.

[26] Bustos-Obregón E., Goicochea R.I. (2002). Pesticide soil contamination mainly affects earthworm male reproductive parameters. *Asian J. Androl.* **4**, 195-199.

[27] Bustos-Obregón E., Gonzalez J.R., Espinoza O. (2005). Melatonin as protective agent for the cytotoxic effects of diazinon in the spermatogenesis in the earthworm *Eisenia foetida*. *Ital. J. Anat. Embryol.* **110**, 159-165.

[28] Cairns Jr. J. (1988). Putting the eco in ecotoxicology. *Regul. Toxicol. Pharmacol.* **8**, 226-238.

[29] Cajaraville M.P., Robledo Y., Etxeberria M., Marigómez I. (1995). Cellular biomarkers as useful tools in the biological monitoring of environmental pollution: molluscan digestive lysosomes. In: Cajaraville MP (ed) Cell biology in environmental toxicology. University of the Basque Country Press Service, Bilbao, Spain, pp. 29-55.

[30] Cancio I., Gwynn ap I., Ireland M.P., Cajaraville M.P. (1995b). The effect of sublethal lead exposure on the ultrastructure and on the distribution of acid phosphatase activity in chloragocytes of earthworms (Annelida, Oligochaeta). *Histochem. J.* **27**, 965-973.

[31] Cancio I., Gwynn ap I., Ireland M.P., Cajaraville M.P. (1995a). Lysosomal origin of the chloragosomes in the chloragogenous tissue of the earthworm *Eisenia foetida*: cytochemical demonstration of acid phosphatase acivity. *Histochem. J.* **27**, 591-596.

[32] Chapman P.M. (2002). Integrating toxicology and ecology: putting the "eco" into ecotoxicology. *Mar. Pollut Bull.* **44**, 7-15.

[33] Cholewa J., Feeney G.P., O'Reilly M., Stürzenbaum S.R., Morgan A.J., Plytycz B. (2006). Species specific autofluorescence in earthworm coelomocytes. *Folia Hist. Cytobiol.* **44**, 65-71.

[34] Cikutovic M.A., Fitzparick L.C., Venables B.J., Goven A.J. (1993). Sperm count in earthworms (*Lumbricus terrestris*) as a biomarker for environmental toxicology: effects of cadmium and chlordane. *Environ. Pollut.* **81**, 123-125.

[35] Cooper E.L. (2002). The earthworm: a new model with biomedical applications. In A. Beschin, M. Bilej, E.L. Cooper (Eds.). New Model for Analyzing Antimicrobial Peptides with Biomedical Applications (pp 3-26). IOS Press, Ohmsha.

[36] Cooper E.L., Cossarizza A., Suzuki M.M., Salvioli S., Capri M., Quaglino D., Franceschi C. (1995). Autogeneic but not allogeneic earthworm effector coelomocytes kill the mammalian tumor cell target K562. *Cell Immunol.* **166**, 113-122.

[37] Cooper E.L., Roch P. (2003). Earthworm immunity: a model of immune competence. *Pedobiologia* **47**, 676-688.

[38] Cooper E.L., Stein E.A. (1981). Oligochaetes: In N. A. Ratcliffe & A. F. Rowley (Eds.), *Invertebrate blood cells* (pp.75-140). Academic Press, London, San Francisco.

[39] Corp N., Morgan A.J. (1991). Accumulation of heavy metals from polluted soils by the earthworm *Lumbricus rubellus*: can laboratory exposure of `control' worms reduce biomonitoring problems? *Environ. Pollut.* **74**, 39-52.

[40] Cossarizza A., Cooper E.L., Suzuki M.M., Salvioli S., Capri M., Gri G., Quaglino D., Franceschi C. (1996). Earthworms leukocytes that are not phagocytic and cross-react with several human tumor cell lines. *Exp. Cell Res.* **224**, 174-182.

[41] Cotter-Howells J., Kille P., Charnock J.M., Winters C., Fry J.C., Morgan A.J. (2005). Metal coordination in earthworm cells and tissues: correlative electron probe X-ray mapping and EXAFS analyses. *Environ. Sci. Technol.* **39**, 7731-7740.

[42] Çotuk A., Dales R.P. (1984). Lysozyme activity in the coelomic fluid and coelomocytes of the earthworm *Eisenia foetida* Sav. in relation to bacterial infection. *Comp. Biochem. Physiol.* **78 A**, 469-474.

[43] Dales R.P., Kalaç Y. (1992). Phagocytic defence by the earthworm *Eisenia foetida* against certain pathogenic bacteria. Comp. Biochem. Physiol. 101 A, 487-490.

[44] Dallinger R. (1996). Methallothionein research in terrestrial invertebrates: Synopsis and perspectives. *Comp. Biochem. Physiol.* **113C**, 125-133.

[45] Diogene J., Dufour M., Poirier G.G., Nadeu D. (1997). Extrusion of earhworm coelomocytes: comparison of the cell populations recovered from the species *Lumbricus terrestris*, *Eisenia fetida*, and *Octolasion tyrtaeum*. *Lab. Anim.* **31**, 326-336.

[46] Dumas P., Miller L. (2003). Biological and biomedical applications of synchrotron infrared microspectroscopy. *J. Biol. Phys.* **29**, 201-2003.

[47] Eckwert H., Köhler H.-R. (1997). The indicative value of the hsp70 stress response as a marker of metal effects in *Oniscus asellus* (Isopoda) field popultions: variability between populations from metal-polluted and uncontaminated sites. *Appl. Soil Ecol.* **6**, 275-282.

[48] Eddy S.F., Storey K.F. (2002). Dynamic use of cDNA arrays: heterologous probing for gene discovery and exploration of animal adaptations to stressful environments. In: Cell and Molecular Responses to Stress. Vol. 3. Eds. K.B. Storey, J.M. Storey, Elsevier Press, Amsterdam, pp. 315-325.

[49] Engelmann P., Molnár L., Pálinkás L., Cooper E.L., Németh P. (2004). Earthworm leukocyte populations specifically harbor lysosomal enzymes that may respond to bacterial challenge. *Cell Tissue Research* **316**, 391-401.

[50] Espinoza-Navarro O., Bustos-Obregón E. (2005). Effect of malathion on the male reproductive organs of earthworms, *Eisenia foetida*. *Asian J. Androl.* **7**, 97-101.

[51] Eue I., Kauschke E., Mohrig W., Cooper E.L. (1998). Isolation and characterization of earthworm hemolysins and agglutinins. *Dev. Comp. Immunol.* **22**, 13-25.

[52] Field S.G., Kurtz J., Cooper E.L., Michiels N.K. (2004). Evaluation of an innate immune reaction to parasites in earthworms. J. Invert Pathol. 86, 45-49.

[53] Fischer E. (1976). Chloragocyte-eleocyte transformation induced by benomyl and carbofuran toxication of earthworms (Oligochaeta). *Zool. Anz.* **197**, 225-233.

[54] Fischer E. Horvath I. (1977). In vivo cumulation and discharge of azine, thiazine, and xanthene dyes and their effect on the chloragogen cells of Lumbricidae (Oligochaeta). *Acta Biologica Academiae Scientiarium hungaricae*. **28**, 33-47.

[55] Fischer, E., Molnar, L. (1992). Environmental aspects of the chloragogenous tissue of earthworms. *Soil Biol. Biochem.*, **24**, 1723-1727.

[56] Fitzpatrick L.C., Muratti-Oritz J.F., Venables B.J., Goven A.J. (1996). Comparative toxicity in earthworms *Eisenia fetida* and *Lumbricus terrestris* exposed to cadmium nitrate using artificial soil and filter paper protocols. *Bull Environ. Contam. Toxicol.* **57**, 63-68.

[57] Forbes V.E., Palmqvist A., Bach L. (2006). The use and misuse of biomarkers in ecotoxicology. *Environ. Toxicol. Chem.* **25**, 272-280.

[58] Fugere N., Brousseau P., Krzystyniak K., Coderre D., Fournier M. (1996). Heavy metal-specific inhibition of phagocytosis and different in vitro sensitivity of heterogeneous coelomocytes from *Lumbricus terrestris* (Oligochaeta). Toxicology 109, 157-166.

[59] Galay Burgos M., Winters C., Stürzenbaum S.R., Randerson P.F., Kille P., Morgan A.J. (2005). Cu and Cd effects on the earthworm *Lumbricus rubellus* in the laboratory: multivariate statistical analysis of relationships between exposure, biomarkers, and ecologically-relevant parameters. *Environ. Sci. Technol.* **39**(6), 1757-1763.

[60] Galay-Burgos M., Spurgeon D.J., Weeks J.M., Sturzenbaum S.R., MorganA.J., Kille P. (2003). Developing a new method for soil pollution monitoringusing molecular-genetic biomarkers. *Biomarkers* **8**, 229-239.

[61] Gibb J.O.T., Svendsen C., Weeks J.M., Nicholson J.K. (1997). 1H NMR spectroscopic investigations of tissue metabolite biomarker responses to Cu II exposure in terrestrial invertebrates: identification of free histidine as a novel biomarker of exposure to copper in earthworms. *Biomarkers* **2**, 295-302.

[62] Gwynn-Jones D., Huang W., Easton G., Goodacre R., Scullion J. (2003). UV-B radiation induced changes in litter quality affects earthworm growth and cast characteristics as determined by metabolic fingerprinting. *Pedobiologia* **47**, 784-787.

[63] Hamed S.S., Kauschke E., Cooper E.L. (2002). Cytochemical properties of earthworm coelomocytes enriched by Percoll. In A. Beschin, M. Bilej, E. L. Cooper (Eds.). A New Model for Analyzing Antimicrobal Peptides with Biomedical Applications (pp. 29-37). IOS Press, Ohmsha.

[64] Hamlin N.J., Price P.A. (2004). Mineralization of decalcified bone occurs under cell culture conditions and requires bovine serum but not cells. *Calcif. Tisue Int.* **75**, 231-242.

[65] Hankard P.K., Svendsen C., Wright J., Wienberg C., Fishwick S.K., Spurgeon D.J., Weeks J.M. (2004). Biological assessment of contaminated land using earthworm biomarkers in support of chemical analysis. *Sci. Total Environ.* **330**, 9-20.

[66] Hendawi M., Sauve S., Ashour M., Brousseau P., Fournier M. (2004). A new ultrasound protocol for extrusion of coelomocyte cells from the earthworm *Eisenia fetida*. *Ecotoxicol Environ Saf.* **59**, 17-22.

[67] Hinton D.E., Baumann P.C., Gardner G.R., Hawkins W.E., Hendricks J.D., Murchelano R.A., Okihiro M.S (1992). *Histopathologic biomarkers*. In: Huggett R.J., Kimerle R.A., Mehrle Jr. P.M., Bergman H.L. (eds) Biomarkers. Biochemical, physiological, and histological markers of anthropogenic stress. Lewis Publishers, Boca Raton, pp 155-209.

[68] Holman H.-Y. N., Martin M.C., McKinney W.R. (2003). Tracking chemical changes in a live cell: Biomedical applications of SR-FTIR spectromicroscopy. *Spectroscopy* **17**, 139-159.

[69] Homa J., Morgan A.J., Plytycz B. Disrupted homeostasis in coelomocytes of *Eisenia fetida* exposed dermally to heavy metals. Submitted.

[70] Homa J., Morgan A.J., Plytycz B. Effects of heavy metals on coelomocytes of *Allolobophora chlorotica*. Submitted.

[71] Homa J., Niklinska M., Plytycz, B. (2003). Effects of heavy metals on coelomocytes of the earthworm *Allolobophora chlorotica*. *Pedobiologia* **47**, 640-645.

[72] Homa J., Olchawa E., Stürzenbaum S.R., Morgan A.J., Plytycz B. (2005). Early-phase immunodetection of metallothionein and heat shock proteins in extruded earthworm coelomocytes after dermal exposure to metal ions. *Environ. Pollut* **135**, 275-280.

[73] Hønsi T.G., Stubberud H.E., Andersen S., Stenersen J. (2003). Lysosomal fragility in earthworms (*Eisenia veneta*) exposed to heavy metal contaminated soils from two abandoned pyrite ore mines in Southern Norway. Water, *Air, & Soil Pollution* **142**, 27-37.

[74] Hopkin S.P. (1989). Ecophysiology of metals in terrestrial invertebrates. Elsevier Applied Science Publishers. Ltd., London and New York, p. 365.

[75] Hsu S.M., Soban E. (1982). Colour modification of diaminobenzidine precipitation by metallic ions and its application to double immunocytochemistry. *J. Histochem. Cytochem.* **30**, 1079-1082.

[76] Huggett R.J., Kimerle R.A., Mehrle Jr. P.M., Bergman H.L. (eds) (1992). Biomarkers. Biochemical, Physiological, and Histological Markers of Anthropogenic Stress. Lewis Publishers, Boca Raton, pp. 247.

[77] Hyne R.V., Maher W.A. (2003). Invertebrate biomarkers: links to toxicosis that predict population decline. *Ecotoxicol. Environ. Saf.* **54**, 366-374.

[78] Ireland M.P., Richards K.S. (1977). The occurrence and localization of heavy metals and glycogen in the earthworms *Lumbricus rubellus* and *Dendrobaena rubida* from a heavy metal site. *Histochem.* **51**: 153-166.

[79] Jankowski-Hennig M.A., Clegg M.S., Daston J.M.R., Keen C.L. (2000). Zinc-deficient rat embryos have increased caspase 3-like activity and apoptosis. *Biochem. Physiol. Res. Comm.* **271**, 250-256.

[80] Kauschke E., Komiyama K., Moro I., Eue I., König S., Cooper E.L. (2001). Evidence for perforin-like activity associated with earthworm leukocytes. *Zoology* **104**, 13-24.

[81] Kawada T., Kanda A., Minakata H., Matsushima O., Satake H. (2004). Identification of a novel receptor for an invertebrate oxytocin/vasopressin superfamily peptide: molecular and functional evolution of the oxytocin/vasopressin superfamily. *Biochem. J.* **382**, 231-237.

[82] Kirk C.J., Peel R.N., James K.R., Kershaw Y. (1975). Basic medical laboratory technology. John Wiley and Sons, New York.

[83] Kirschner M.W., Gerhart J.C. (2005). The Plausibility of Life. Resolving Darwin's Dilemma. p. 216. Yale University Press, New Haven and London.

[84] Koenig S., Wagner F., Kauschke E., Peter-Katalinie J., Cooper E.L., Eue I. (2003). Mass spectrometric analyses of CL39, CL41 and H1, H2, H3 confirm identity with fetidin and lysenin produced by earthworm leukocytes. *Dev. Comp. Immunol.* **27**, 513-520.

[85] Köhler H.R., Rahman B., Gräff S., Berkus M., Triebskorn R. (1996). Expression of the stress-70 protein family (hsp70) due to heavy metal contamination in the slug, *Deroceras reticulatum*: an approach to monitor sublethal stress conditions. *Chemosphere* **33**, 1327-1340.

[86] Köhler H.R., Eckwert H., Triebskorn R., Bengtsson G. (1999a). Interaction between tolerance and 70 kDa stress protein (hsp70) induction in collembolan populations exposed to long-term metal pollution. *App. Soil Ecol.* **11**, 43-52

[87] Köhler H.R., Knödler C., Zanger M. (1999b). Divergent kinetics of hsp70 induction in *Oniscus asellus* (Isopoda) in response to four environmentally relevant organic chemicals (B[a]P, PCB52, %-HCH, PCP): suitability and limits of a biomarker. *Arch. Environ. Contam. Toxicol.* **36**, 179-185.

[88] Köhler A., Wahl E., Söffker K. (2002). Functional and morphological changes of lysosomes as prognostic biomarkers of toxic liver injury in a marine flatfish (*Platichthys flesus* (L.). *Environ. Toxicol. Chem.* **21**, 2434-2444.

[89] Koziol B., Markowicz M., Kruk J., Plytycz B. (2006). Riboflavin as a source of autofluorescence in *Eisenia fetida* coelomocytes. Photochem. Photobiol., in press.

[90] Kurek A., Homa J., Kauschke E., Bigaj J., Plytycz B. Comparative studies on morphology, flow cytometric characteristics, and activity of coelomocytes extruded from adult individuals of four earthworm species: *Allolobophora chlorotica*, *Dendrobaena veneta*, *Eisenia fetida*, and *Lumbricus terrestris*. Submitted.

[91] Kurek A., Homa J., Plytycz B. (2002). Earthworm coelomocytes: convenient model for basic and applied sciences. In A. Beschin, M. Bilej, E. L. Cooper (Eds.). A New Model for Analyzing Antimicrobial Peptides with Biomedical Applications (pp. 38-46). IOS Press, Ohmsha,

[92] Kurek A., Plytycz B. (2003). Annual changes in coelomocytes of four earthworm species. *Pedobiologia* **47**, 689-701.

[93] Langdon C.J., Winters C., Stürzenbaum S.R., Morgan A.J., Charnock J.M., Meharg A.A., Piearce T.G., Lee P.H., Semple K.T. (2005). Ligand arsenic complexation and immunoperoxidase detection of metallothionein in the earthworm *Lumbricus rubellus* inhabiting arsenic-rich soil. *Environ. Sci. Technol.* **39**, 2042-2048.

[94] Lange S., Kauschke E., Mohrig W., Cooper E.L. (1999). Biochemical characteristics of Eiseniapore, a pore-forming protein in the coelomic fluid of earthworms. *European Journal of Biochemistry* **262**, 547-556.

[95] Lavelle P., Spain A.V. (2001). Soil Ecology, pp. 288-289. Kluer Academic Publishers, Dordrecht, The Netherlands.

[96] Lenz E.M., Weeks J.M., Lindon J.C., Osborn D., Nicholoson J.K. (2005). Qualitative high field ^{1}H-NMR spectroscopy for the characterization of endogenous metabolites in earthworms with biochemical biomarker potential. *Metabolomics* **1**, 123-136.

[97] Lettieri T. (2005). Recent applications of DNA microarray technology to toxicology and ecotoxicology. *Environ. Health Perspect.* **114**, 4-9.

[98] Liu C.L., Prapong W., Natkunam Y., Alizadeh A., Montgomery K., Gilks C.B., Rijn van de M. (2002). Software tools for high-throughput analysis and archiving of immunohistochemistry staining data obtained with tissue microarrays. *Am. J. Pathol.* **161**, 1557-1565.

[99] Lofts S., Spurgeon D.J., Svendsen C., Tipping E. (2004). Deriving soil critical limits for Cu, Zn, Cd, and Pb: a method based on free ion concentrations. *Environ. Sci. Technol.* **38**, 3623-3631.

[100] Lowe D.M., Moore M.N., Clarke K.R. (1981). Effects of oilon digestive cells in mussels: quantitative alterations in cellular and lysosomal structure. *Aquat. Toxicol.* **1**, 213–226.

[101] Lowe D.M., Moore M.N., Evans B.M. (1992). Contaminant impact on interactions of molecular probes with lysosomes in hepatocytes from dab *Limanda limanda. Mar. Ecol. Prog.* **91**, 135-140.

[102] Lowe D.M., Pipe R.K. (1986). Hydrocarbon exposure in mussels: a quantitative study of the responses in the reproductive and nutrient storage cell systems. *Aquat. Toxicol.* **8**, 265-272.

[103] Maboeta M.S., Reinecke S.A., Reinecke A.J. (2002). The relation between lysosomal biomarker and population responses in a field population of *Microchaetus* sp. (Oligochaeta) exposed to the fungicide copper oxychloride. *Ecotoxicol. Environ. Saf.* **52**, 280-287.

[104] Maboeta M.S., Reinecke S.A., Reinecke A.J. (2003). Linking lysosomal biomarker and population responses in a field population of *Aporrectodea caliginosa* (Oligochaeta) exposed to the fungicide copper oxychloride. *Ecotoxicol. Environ. Saf.* **56**, 411-418.

[105] Maboeta M.S., Reinecke S.A., Reinecke A.J. (2004). The relationship between lysosomal biomarker and organismal responses in acute toxicity test with *Eisenia fetida* (Oligochaeta) exposed to the fungicide copper oxychloride. *Environ. Res.* **96**, 95-101.

[106] Marigómez J.A., Vega M.M., Cajaraville M.P., Angulo E. (1989). Quantittive responses of the lysosomal vacuolar system of winkles to sublethal concentrations of cadmium. *Cell. Mol. Biol.* **35**, 555-562.

[107] Marino F., Morgan A.J. (1998). Immunohistochemical detection of heat shock proteins (hsp60, hsp70, hsp90) expression in stressed earthworms. In: Advances in Earthworm Ecotoxicology. Eds. S. Sheppard, J. Bembridge, M. Holmstrup, L. Posthuma, Society of Environmental Toxicology and Chemistry (SETAC), Pensacola FL., pp.199-214.

[108] Marino F., Winters C., Morgan, A.J. (1999). Heat shock protein (hsp60, hsp70, hsp90) expression in earthworms exposed to metal stressors in the field and laboratory. *Pedobiologia*, **43**, 615-624.

[109] Martin D.M., Berriman M., Barton G.J. (2004). GOtcha: a new method for prediction of protein function assessed by the annotation of seven genomes. *BMC Bioinformatics* **5**, 178.

[110] Massicotte R., Robidoux P.Y., Sauvé S., Flipo D., Fournier M., Trottier B. (2003). Immune response of earthworms (*Lumbricus terrestris, Eisenia Andrei* and *Aporrectodea tuberculata*) following *in situ* soil exposure to atmospheric deposition from a cement factory. *J. Environ. Monit.* **5**, 774-779.

[111] Massicotte R., Robidoux P.Y., Sauvé S., Flipo D., Mathiot A., Fournier M., Trottier B. (2004). Immunotoxicological response of the earthworm *Lumbricus terrestris* following exposure to cement kiln dust. *Ecotoxicol. Environ. Saf.* **59**, 10-16.

[112] Matozzo V., Ballarin L., Pampanin D.M., Marin M.G. (2001). Effects of copper and cadmium exposure on functional responses of hemocytes in the clam, *Tapes philippinarum. Arch. Environ. Contam. Toxicol.* **41**, 163-170.

[113] Mayer F.L., Versteeg D.J., McKee M.J., Folmar L.C.,.Graney R.L, McCume D.C., Rattner B.A. (1992). Physiological and nonspecific biomarkers. In: *Biomarkers: Biochemical, Physiological, and Histological Markers of Anthropogenic Stress.* (Ed. R.J. Huggett), Lewis Publishers, Chelsea, MI. pp. 5-86.

[114] McNaughton D. (2005). Synchrotron infrared spectroscopy in biology and biochemistry. *Australian Biochemist* **36**, 55-58.

[115] Mikhaylova A., Davidson M., Toastmann H., Channell J.E.T., Guyodo Y., Batich C., Dobson J. (2005). Detection, identification and mapping of iron anomalies in brain tissue using X-ray absorption spectroscopy. *J. Royal Soc. Interface* **2**, 33-37.

[116] Milochau A., Lassègues M., Valembios P. (1997). Purification, characterization and activities of two hemolytic and antibacterial proteins from coelomic fluid of the annelid *Eisenia fetida andrei. Acta Bioch. Bioph.* **1337**, 123-132.

[117] Moore M.N. (1985). Cellular responses to pollutants. *Mar. Pollut. Bull.* **16**, 134-139.

[118] Morgan A.J. (1982). The elemental composition of the chloragosomes of nine species of British earthworms in relation to calciferous gland activity. *Comp. Biochem. Physiol.* **73A**, 207-216.

[119] Morgan A.J. (1985). X-ray Microanalysis in Electron Microscopy for Biologists, pp.79. Oxford University Press/Royal Microscopical Society, Oxford.

[120] Morgan A.J., Brock J.M., Winters C., Lewis G.H.J. (1994). Mapping metal distributions in thin cryosections without STEM with the Philips electron beam and image deflection (EBID) unit. *Scanning Electron Microsc.* **8**, 231-243.

[121] Morgan A.J., Gregory Z.D.E., Winters C. (1990). Responses of the hepatopancreatic "B" cells of the isopod, *Oniscus asellus*, to metals accumulated from a contaminated habitat: a morphometric analysis. *Bull. Environ. Contam. Toxicol.* **44**, 363-368.

[122] Morgan A.J., Morgan J.E., Turner M., Winters C., Yarwood A. (1993). Metal relationships in earthworms. In: Ecotoxicology of Metals in Invertebrates. Eds. R. Dallinger, P.S. Rainbow, Lewis, Boca Raton FL., pp. 333-358.

[123] Morgan A.J., Morris B. (1982). The accumulation and intracellular compartmentation of cadmium, lead, zinc and calcium in two earthworm species *(Dendrobaena rubida and Lumbricus rubellus)* living in highly contaminated soil. *Histochem.* **75**, 269-285.

[124] Morgan A.J., Stürzenbaum S.R., Kille P. (1999a). A short review of molecular biomarker strategies with particular regard to recent developments in earthworms. *Pedobiologia* **43**, 574-584.

[125] Morgan A.J., Stürzenbaum S.R., Winters C., Grime G.W., Aziz N.A., Kille P. (2004). Differential metallothionein expression in earthworm (*Lumbricus rubellus*) tissues. *Ecotoxicol. Environ. Saf.* **57**, 11-19.

[126] Morgan A.J., Stürzenbaum S.R., Winters C., Grime G.W., Aziz N.A., Kille P. (2004). Differential metallothionein expression in earthworm (*Lumbricus rubellus*) tissue. Ecotoxicol. *Environ. Saf.* **57**, 11-19.

[127] Morgan A.J., Stürzenbaum S.R., Winters C., Kille P. (1999b). Cellular and molecular aspects of metal sequestration and toxicity in earthworms. *Invert. Reprod. Devel.* **36**, 17-24.

[128] Morgan A.J., Turner M. (2005). Quantitative ultrastructure of metal sequestering cells reflect inter-site and inter-species differences in earthworm metal burdens. *Arch. Environ. Contam. Toxicol.* **49**, 45-52.

[129] Morgan A.J., Turner M.P., Morgan J.E. (2002). Morphological plasticity in metal-sequestering earthworm chloragocytes: morphometric electron microscopy provides a biomarker of exposure in field populations. *Environ. Toxicol. Chem.* **21**, 610-618.

[130] Morgan A.J., Winters C., Yarwood Y., Wilkinson N. (1995). *In vivo* metal substitutions in metal sequestering subcellular compartments: X-ray mapping in cryosections. *Scanning Electron Microsc.* **9**, 1041-1060.

[131] Morgan J.E., Morgan A.J. (1989a). Zinc sequestration by earthworm (Annelida: Oligochaeta) chloragocytes. An *in vivo* investigation using fully quantitative electron probe X-ray micro-analysis. *Histochem.* **90**, 405-411.

[132] Morgan J.E., Morgan A.J. (1989b). The effect of lead incorporation on the elemental composition of earthworm (Annelida: Oligochaeta) chloragosome granules. *Histochem.* **92**, 237-241.

[133] Morgan J.E., Morgan A.J. (1990). The distribution of cadmium, copper, lead, zinc and calcium in the tissues of the earthworm *Lumbricus rubellus* sampled from one uncontaminated and four polluted soils. *Oecologia* **84**, 559-566.

[134] Morgan J.E., Morgan A.J. (1998). The distribution and intracellular compartmentation of heavy metals in the endogeic earthworm *Aporrectodea caliginosa* sampled from an unpolluted and a metal-contaminated site. *Environ. Pollut.* **99**, 167-175.

[135] Neuhauser E.F., Cukic Z.V., Malecki M.R., Loehr R.C., Durkin P.R. (1995). Bioconcentration and biokinetics of heavy metals in the earthworm. *Environ. Pollut.* **89**, 293-301.

[136] Nylandsted J., Gyrd-Hansen M., Danielewicz A., Fehrenbacher N., Lademann U., Høyer-Hansen M., Weber E., Multhoff G., Rohde M., Jäättelä M. (2004). Heat shock protein 70 promotes cell survival by inhibiting lysosomal membrane permeabilization. *J. Exp. Med.* **200**, 425-435.

[137] OECD (1984). Guidelines for the testing of chemicals. No. 207. Earthworm acute toxicity tests. Adopted April 4.

[138] Ohta N., Shioda S., Sekizawa Y., Nakai Y., Kobayashi H. (2000). Sites of expression of mRNA for lysenin, a protein isolated from the coelomic fluid of the earthworm *Eisenia foetida*. *Cell Tiss. Res.* **302**, 263-270.

[139] Olchawa E., Bzowska M., Stürzenbaum S.R., Morgan A.J., Plytycz B. (2006). Dermal exposure affects the coelomocyte-bacteria balance in earthworms: Environmental interactions between abiotic and biotic stressors. *Environ. Pollut.*, in press.

[140] Olchawa E., Swiatkiewicz I., Homa J., Bzowska M., Plytycz B. Restoration of coelomocytes in *Dendrobaena veneta* after experimental extrusion. Submitted.

[141] Osame A.S.S., Kimura M. (1995). Effects of vitamin B2 on neutrophil functions in cattle. *J. Vet. Med. Sci.* 57, 493-495.

[142] Oste L.A., Dolfing J., Ma W-C, Lexmond T.M. (2001). Cadmium uptake by earthworms as related to the availability in the soil and the intestine. *Environ. Toxicol. Chem.* **20**, 1785-1791.

[143] Oumi T., Ukena K., Matsushima O., Ikeda T., Fujita T., Minakata H., Nomoto K. (1996). Annetocin, an annelid oxytocin-related peptide, indices egg-laying behaviour in the earthworm, *Eisenia foetida. J. Exp. Zool.* **276**, 151-156.

[144] Packeisen J., Korsching E., Herbst H., Boecker W., Buerger H. (2003). Demystified... Tissue microarray technology. *J. Clin. Pathol.: Mol. Pathol.* **56**, 198-204.

[145] Pang J.H.S., Chau L.Y. (1999). Cooper-induced apoptosis and immediate early gene expression in macrophages. *Atherosclerosis* **146**, 45-52.

[146] Peijnenburg W.J.G.M. (2002). Bioavailability of metals to soil invertebrates. In: Bioavailability of Metals in Terrestrial Ecosystems: Importance of Partitioning for Bioavailability to Invertebrates, Microbes, and Plants. Ed. H.E. Allen, Society of Environ. Toxicol. Chem. (SETAC), Pensacola FL., pp. 89-112.

[147] Perlman Z.E., Slack M.D., Feng Y., Mitchison T.J., Wu L.F., Altschuler S.J. (2004). Multidimensional drug profiling by automated microscopy. *Science* **306**, 1194-1198.

[148] Plytycz B., Jozkowicz A. (1994). Differential effects of temperature on macrophages of ectothermic vertebrates. *Journal of Leukocytes Biology* **56**, 729-731.

[149] Plytycz B., Rozanowska M., Seljelid R. (1992). Quantification of Neutral Red pinocytosis by small numbers of adherent cells: comparative studies. *Folia biologica (Cracow)* **40**, 3-9.

[150] Plytycz B., Homa J., Koziol B., Kruk J., Rozanowska M., Morgan A. J. Riboflavin-derived autofluorescence of eleocytes from several earthworm species. Submitted.

[151] Posthuma L., Notenboom J., De Groot A.C., Peijnenburg W.J.G.M. (1998). Soil acidity as a major determinant of zinc partitioning and zinc uptake in 2 oligochaete worms exposed to contaminated field soils. In: Sheppard SC, Bembridge JD, Holmstrup M, Posthuma L (eds) Advances in earthworm ecotoxicology, Society of Environ. Toxicol. Chem. (SETAC), Pensacola FL, pp 111-127.

[152] Power M., McCarty L.S. (1997). Fallacies in Ecological Risk Assessment Practices. *Environ. Sci. Technol/News* **31**, 8, 370-375.

[153] Prento P. (1987). Distribution of 20 enzymes in the midgut region of the earthworm, Lumbricus terrestris L., with particular emphasis on the physiological role of the chloragog tissue. *Comp. Biochem. Physiol.* **87A**, 135-142.

[154] Prinsloo M.W., Reinecke S.A., Przybylowicz W.J., Mesjasz-Przybylowicz J., Reinecke A.J. (1999). Micro-PIXE studies of Cd distribution in the nephridia of the earthworm *Eisenia fetida* (Oligochaeta). *Nuc. Instru. Meth. Phys. Res. Section B: Beam Interactions with Materials and* Atoms **158**, 317-322.

[155] Punshon T., Jackson B.P., Lanzirotti A., Hopkins W.A., Bertsch P.M., Burger J. (2005). Application of synchrotron X-ray microbeam spectroscopy of metal distribution and speciation in biological tissues. *Spectroscopy Lett.* **38**, 343-363.

[156] Quaglino D., Cooper E.L., Salvioli S., Capri M., Suzuki M.M., Ronchetti I.P., Franceschi C., Cossarizza A. (1996). Earthworm coelomocytes *in vitro*: cellular features and "granuloma" formation during cytotoxic activity against the mammalian tumor cell target K562. *Eur J Cell Biol.* **70**, 278-8.

[157] Raes H., Braeckman B. P., Criel G. R., Rzeznik U., Vanfletern J. R. (2000). Copper induces apoptosis in Aedes C6/36 cells. *J. Exp. Zool.*, **286**, 1-12.

[158] Rao J.V., Pavan Y.S., Madhavendra S.S. (2003). Toxic effects of chlorpyrifos on morphology and acetylcholinesterase activity in the earthworm, *Eisenia fetida*. *Ecotoxicol. Environ. Saf.* **54**, 296-301.

[159] Reinecke A.J., Reinecke S.A. (2003). The influence of exposure history to lead on the lysosomal response in *Eisenia fetida* (Oligochaeta). *Ecotoxicol. Environ. Saf.* **55**, 30-37.

[160] Reinecke S.A., Helling B., Reinecke A.J. (2002). Lysosomal response of earthworm (*Eisenia fetida*) coelomocytes to the fungicide copper oxychloride and relation to life-cycle parameter. *Environ. Toxicol. Chem.* **21**, 1026-1031.

[161] Reinecke S.A., Reinecke A.J. (1999). Lysosomal response of earthworm coelomocytes induced by long-term environmental exposure to heavy metals. *Pedobiologia* **43**, 585-593.

[162] Reinecke S.A., Reinecke A.J. (2004). The comet assay as biomarker of heavy metal genotoxicity in earthworms. *Arch. Environ. Contam. Toxicol.* **46**, 208-215.

[163] Reinecke S.A., Reinecke A.J. (2005). Biomarker response and biomass change of earthworms exposed to chlorpyrifos in microcosms. *Ecotoxicol. Environ. Saf.* ?????????

[164] Reinecke S.A., Reinecke A.J., Froneman M.L. (1995). The effects of dieldrin on the sperm ultrastructure of the earthworm *Eudrilus eugeniae* (Oligochaeta). *Environ. Toxicol. Chem.* **14**, 961-965.

[165] Reinhart M., Dollahon N. (2003). Responses of coelomocytes from *Lumbricus terrestris* to native and non-native eukaryotic parasites. *Pedobiologia* **47**, 710-716.

[166] Ricketts H.J., Morgan A.J., Spurgeon D.J., Kille P. (2004). Measurement of annetocin gene expression: a new reproductive biomarker in earthworm ecotoxicology. *Ecotox. Environ. Saf.* **57**, 4-10.

[167] Robertson D.G. (2005). Metabonomics in toxicology: a review. *Toxicological Sciences* **85**, 809-822.

[168] Robidoux P.Y., Dubois C., Hawari J., Sunahara G.I. (2004a). Assessment of soil toxicity from an antitank firing range using *Lumbricus terrestris* and *Eisenia andrei* in mesocosms and laboratory studies. *Ecotoxicology* **13**, 603-614.

[169] Robidoux P.Y., Svendsen C., Sarrazin M., Hawari J., Thiboutot S., Ampleman G., Weeks J.M., Sunahara G.I. (2002). Evaluation of tissue and cellular biomarkers to assess 2,4,6-trinitotoluene (TNT) exposure in earthworms: effects-based assessment in laboratory studies using *Eisenia andrei*. *Biomarkers* **7**, 306-321.

[170] Robidoux P.Y., Svendsen C., Sarrazin M., Thiboutot S., Ampleman G., Hawari J., Weeks J.M., Sunahara G.I. (2004b). Assessment of a 2,4,6-trinitrotoluene-contaminated site using *Aporrectodea caliginosa* and *Eisenia andrei* in mesocosms. *Arch. Environ. Contam. Toxicol.* **48**, 56-67.

[171] Roch P. (1979). Protein analysis of earthworm coelomic fluid: I. Polymorphic system of natural hemolysin of *Eisenia fetida andrei*. *Dev. Comp. Immunol.* **3**, 599-608.

[172] Roesijadi G. (1996). Metallothionein and its role in toxic metal regulation. *Comp. Biochem. Physiol.* **113/2**, 117-123.

[173] Sample B.E., Suter G.W., Beauchamp J.J., Efroymson R.A. (1999). Literature-derived bioaccumulation models for earthworms: development and validation. *Environ. Toxicol. Chem.* **18**, 2110-2120.

[174] Sauter G., Simon R., Hillan K. (2003). Tissue microarrays in drug discovery. *Nature Rev. Drug Discovery* **2**, 962-972.

[175] Saxe J.K., Impellitteri C.A., Peijnenburg W., Allen H.E. (2001). Novel model describing trace metal concentrations in the earthworm, *Eisenia andrei. Environ. Science Tech.* **35**, 4522-4529.

[176] Schulze A., Downward J. (2001). Navigating gene expression using microarrays – a technology review. *Nat Cell Biol.* **3**, E190-E195.

[177] Scott-Fordsmand J.J., Weeks J.M., Hopkin S.P. (1998). Toxicity of nickel to the earthworm and the applicability of the neutral red retention assay. *Ecotoxicol.* **7**, 291-295.

[178] Scullion J., Elliott G.N., Huang W.E., Goodacre R., Worgan H., Darby R., Bailey M.J., Gwynn-Jones D., Griffith G.W., Winson M.K., Williams P.A., Clegg C., Draper J. (2003a). Use of earthworm casts to validate FT-IR spectroscopy as a 'sentinel' technology for high-throughput monitoring of global changes in microbial ecology. *Pedobiologia* **47**, 440-446.

[179] Scullion J., Goodacre R., Elliott G., Huang W., Worgan H., Gwynn-Jones D., Griffith G., Darby R., Bailey M., Clegg C., Draper J. (2003b). Food quality and microbial succession in ageing earthworm casts: standard microbial indices and metabolic fingerprinting. *Pedobiologia* **47**, 888-894.

[180] Siekierska E., Urbańska-Jasik D. (2002). Cadmium effect on the ovarian structure in earthworm *Dendrobaena veneta* (Rosa). *Environ. Pollut.* **120**, 289-297.

[181] Sims R.W., Gerard B.M. (1985). Earthworms. Keys and Notes for the Identification and Study of the Species. *Synopses of the British Fauna* No. **31**. The Linnean Society and The Estuarine and Brackish-Water Sciences Association, London, p. 171.

[182] Snape J.R., Maund S.J., Pickford D.B., Hutchinson T.H. (2004). Ecotoxicogenomics: the challenge of integrating genomics into aquatic and terrestrial ecotoxicology. *Aquat. Toxicol.* **67**, 143-154.

[183] Soto M., Zaldibar B., Cancio I., Taylor M.G., Turner M., Morgan A.J., Marigómez I. (2002). Subcellular distribution of cadmium and its cellular ligands in mussel digestive gland cells as revealed by combined autometallography and X-ray microprobe analysis. *Histochem. J.* **34**, 273-280.

[184] Spurgeon D.J., Hopkin S.P. (1995). Extrapolation of the laboratory-based OECD earthworm test to metal-contaminated field sites. *Ecotoxicol.* **4**, 190-205.

[185] Spurgeon D.J., Lofts S., Hankard P.K., Toal M., McLellan D., Fishwick S., Svendsen C. (2006). Effect of pH on metal speciation and resulting metal uptake and toxicity for earthworms. *Environ. Toxicol. Chem.* **25**, 788-796.

[186] Spurgeon D.J., Stürzenbaum S.R., Svendsen C., Hankard P.K., Morgan A.J., Weeks J.M., Kille P. (2004). Toxicological, cellular and gene expression responses in earthworms exposed to copper and cadmium. *Comp. Biochem. Physiol.* **138**, 11-21.

[187] Spurgeon D.J., Svendsen C., Rimmer V.R., Hopkin S.P., Weeks J.M. (2000). Relative sensitivity of life-cycle and biomarker responses in four earthworm species exposed to zinc. *Environ. Toxicol. Chem.* **19**, 1800-1808.

[188] Spurgeon D.J., Weeks J.M., Van Gestel C.A. (2003). A summary of eleven years progress in earthworm ecotoxicology. *Pedobiologia* **47**, 588-606.

[189] Stankiewicz A., Płytycz B., (1998). Effects of *in vitro* conditions and *in vitro* thermal adaptation on viability of the earthworm (*Eisenia fetida*) coelomocytes. *Folia Biologica (Cracow)* **46**, 3-4.

[190] Storey K.B., McMullan D.C. (2004). Insect cold hardiness: new advances using gene screening technology. In: Life in the Cold: Evolution, Mechanisms, Adaptation, and Application. 12th International Hibernation Symposium. Eds. B.M. Barnes, H.V. Carey, Biological Papers of the University of Alaska, #27, Fairbanks. Pp. 275-281.

[191] Stürzenbaum S.R., Georgiev O., Morgan A.J., Kille P. (2004). Cadmium detoxification in earthworms: from genes to cells. *Environmental Science and Toxicology* **38**, 6283-6289.

[192] Stürzenbaum S.R., Kille P. (2001). Control genes in quantitative molecular biological techniques: the variability of invariance. *Comp. Biochem. Physiol.* **130B**, 281-289.

[193] Stürzenbaum S.R., Parkinson J., Blaxter M., Morgan A.J., Kille P., Georgiev O. (2003). The earthworm Expressed Sequence Tag project. *Pedobiologia* **47**, 447-451.

[194] Stürzenbaum S.R., Winters C., Galay M., Morgan A.J., Kille P. (2001). Metal ion trafficking in earthworms: identification of cadmium-specific metallothionein. *J. Biol. Chem.* **276**, 34013-34018.

[195] Suzuki M.M., Cooper E.L. (1995). Killing of intrafamilial leukocytes by earthworm effector cells. *Immunol. Lett.* **44**, 45-49.

[196] Suzuki T., Honda M., Matsumoto S., Stürzenbaum S.R., Gamou S. (2005). Valosine-containing proteins (VCP) in an annelid: identification of a novel spermatogenesis related factor. *Gene* **362**, 11-18.

[197] Svendsen C., Meharg A. A., Freestone P., Weeks J. M. (1996). Use of an earthworm lysosomal biomarker for the ecological assessment of pollution from an industrial plastics fire. *App. Soil Ecol.*, **3**, 99-107.

[198] Svendsen C., Spurgeon D. J., Hankard P. K., Weeks J. M. (2004). A review of lysosomal membrane stability measured by neutral red retention: is it a workable earthworm biomarker? *Ecotoxicol. Environ. Saf.*, **57**, 20-29.

[199] Svendsen C., Weeks J.M. (1997a). Relevance and applicability of a simple earthworm biomarker of copper exposure: I. Links to ecological effects in a laboratory study with *Eisenia Andrei. Ecotoxicol. Environ. Saf.* **36**, 72-79.

[200] Svendsen C., Weeks J.M. (1997b). Relevance and applicability of a simple earthworm biomarker of copper exposure: II. Validation and applicability under field conditions in a mesocosm experiment with *Lumbricus rubellus. Ecotoxicol. Environ. Saf.* **36**, 80-88.

[201] Triebskorn R., Köhler H-R, Honnen W., Schramm M., Adams S.M., Müller E.F. (1997). Induction of heat shock proteins, changes in liver ultrastructure, and alterations of fish behaviour: are these biomarkers related and are they useful to reflect the state of pollution in the field? *J. Aquat. Ecosystem Stress Recov.*, **6**, 57-73.

[202] Tučková L., Bilej M. (1996). Mechanism of antigen processing in invertebrates: are there receptors? In E. L. Cooper (Eds.). Comparative and Environmental Physiology (pp. 41-72). Springer-Verlag, Berlin, Heidelberg, New York.

[203] Ukena K., Oumi T., Matsushima O., Ikeda T., Fujita T., Minakata H., Nomoto K. (1995). Effects of annetocin, an oxytocin-related peptide isolated from the earthworm *Eisenia foetida*, and some putative neurotransmitters on gut motility of the earthworm. *J. Exp. Zool.* **272**, 184-193.

[204] Valembois P., Lasségues M. (1995). *In vitro* generation of reactive oxygen species by free coelomic cells of the annelid *Eisenia fetida andrei*: an analysis by chemiluminescence and nitro blue tetrazolium reduction. *Dev. Comp. Immunol.* **19**, 195-204.

[205] Valembois P., Lassègues M., Roch P. (1992). Formation of brown bodies in the coelomic cavity of the earthworm *Eisenia fetida andrei* and attendant changes in shape and adhesive capacity of constitutive cells. *Dev. Comp. Immunol.* **16**, 95-101.

[206] Valembois P., Seymour J., Lassègues M. (1994). Evidence of lipofuscin and melanin in the brown body of the earthworm *Eisenia fetida andrei. Cell Tissue Res* **227**, 183-188.

[207] Van Gestel C.A.M., Weeks J.M. (2004). Recommendations of the 3rd International Workshop on Earthworm Ecotoxicology, Aarhus, Denmark, August 2001. *Ecotoxicol. Environ. Saf.* **57**, 100-105(6).

[208] Van Straalen N.M. (1996). Critical body concentrations: their use in bioindication. In: *Bioindicator Systems for Soil Pollution*, Ed. By N.M. Van Straalen, D.A. Krivolutsky, pp. 5-16, Kluwer, Dordrecht.

[209] Van Straalen N.M., Roelofs D. (2006). *An Introduction to Ecological Genomics*. Oxford University Press, Oxford.

[210] Varute A.T., More N.K. (1972). Are chloragosomes in earthworm chloragogen cells lysosomes? *Acta Histochem.* **44S**, 144-151.

[211] Varute A.T., More N.K. (1973). Lysosomal acid hydrolases in the chloragogen cells of earthworms. *Comp. Biochem. Physiol.* **45A**, 607-635.

[212] Ven van der LTM, Wester PW, Vos JG (2003) Histopathology as a tool for the evaluation of endocrine disruption in zebrafish (*Danio rerio*). *Environ Toxicol Chem* **22**: 908-913.

[213] Verdrengh M., Tarkowski A. (2005). Riboflavin in innate and acquired immune responses. *Inflamm. Res.***54**, 390-393.

[214] Viant M.R., Rosenblum E.S., Tieerdema R.S. (2003). NMR-based metabolomics: A powerful approach for charaterizing the effects of environmental stressors on organism health. *Environ. Sci. Technol.* **37**, 4982-4989.

[215] Vijver M.G., Van Gestel C.A.M., Lanno R.P., Van Straalen N.M., Peijnenburg W.J.G.M. (2004). Internal metal sequestration and its ecotoxicological relevance: a review. *Environ. Sci. Technol.* **38**, 4705-4712.

[216] Vijver M.G., Van Gestel C.A.M., Van Straalen N.M., Lanno R., Peijnenburg W.J.G.M. (2006). Biological significance of metals partitioned to subcellular fractions within earthworms (*Aporrectodea caliginosa*). *Environ. Toxicol. Chem.* **25**, 807-814.

[217] Walker C.H., Hopkin S.P., Sibly R.M., Peakall D.B. (2006). *Principles of Ecotoxicology*, 3rd edition. Taylor & Francis, Boca Raton, FL. pp. 149-163.

[218] Wallace W.G., Lopez G.R., Levinton J.S. (1998). Cadmium resistance in an oligochaete and its effect on cadmium trophic transfer to an omnivorous shrimp. *Marine Ecology Progress Series* **172**, 225-237.

[219] Wallace W.G., Luoma S.N. (2003). Subcellular compartmentalization of Cd and Zn in two bivalves. II. Significance of trophically available metal (TAM). *Marine Ecology Progress Series* **257**, 125-137.

[220] Wang X., Wang X., Zhang Y., Qu X., Yang S. (2003). An antimicrobial peptide of the earthworm *Pheretima tschiliensis*: cDNA cloning, expression and immunolocalization. *Biotechnol. Lett.* **25**, 1317-1323.

[221] Warren M.J., Cooper J.B., Wood S.P., Shoolingin-Jordan P.M. (1998). Lead poisoning, haem synthesis and δ-aminolaevulinic acid dehydratase. *Trends in Biochem. Sci.* **23**, 217-221.

[222] Weeks B.A., Anderson D.P., DuFour A.P., Fairbrother A., Goven A.J., Lahvis G.P., Peters G. (1992). Immunological biomarkers to assess environmental stress. In: Huggett R.J., Kimerle R.A., Mehrle Jr. P.M., Bergman H.L. (eds) Biomarkers. Biochemical, Physiological, and Histological Markers of Anthropogenic Stress. Lewis Publishers, Boca Raton, pp 211-233.

[223] Weeks J.M., Svendsen C. (1996). Neutral red retention by lysosomes from earthworm (*Lumbricus rubellus*) coelomocytes: a simple biomarker of exposure to soil copper. *Environ. Toxicol. Chem.* **15**, 1801-1805.

[224] Wieczorek-Olchawa E., Niklinska M., Miedzobrodzki J., Plytycz B. (2003). Effects of temperature and soil pollutions on the presence of bacteria, coelomocytes and brown bodies in coelomic fluid of *Dendrobaena veneta*. *Pedobiologia* **47**, 702-709.

[225] Winters C., Morgan A.J. (1988). Quantitative electron probe X-ray microanalysis of lead-sequestering organelles in earthworms: technical appraisal of air-dried smears and freeze-dried cryosections. *Scanning Microsc.* **2**, 947-958.

[226] Yang Y.L., Ye Y.M., Li F.M., Li Y.F., Ma P.Z. (1987). Characteristic autofluorescence for cancer diagnosis and its origin. *Lasers Surg Med*, **7**, 528-532.

In: Environmental Pollution: New Research
Editor: Rachel H. Plattenberg, pp. 99-119
ISBN: 1-60021-285-9
© 2007 Nova Science Publishers, Inc.

Chapter 3

TRANSFORMATION OF LEAD BULLETS AND PHOSPHATE-INDUCED LEAD IMMOBILIZATION IN SHOOTING RANGE SOILS

Xinde Cao [1,2*] ***and Lena Q. Ma*** [1]

[1] Soil and Water Science Department, University of Florida, Gainesville, FL 32611

[2] Center for the Environmental Systems, Stevens Institute of Technology, Hoboken, NJ 07030

Abstract

Lead (Pb) contamination at shooting range sites is under increasingly environmental concern. This study investigated Pb bullet transformation, Pb contamination, and phosphate-induced Pb immobilization in shooting range soils. Soil samples were collected from three shooting ranges, referred to as SR1, SR2, and SR3 ranges. Mineralogical analysis by powder X-ray diffraction (XRD) and scanning electron microscopy-energy dispersive X-ray spectroscopy (SEM-EDS) revealed that spent Pb bullets was transformed into soluble cerussite ($PbCO_3$) or hydrocerussite ($Pb_3(CO_3)_2(OH)_2$) in both SR1 and SR2 soils, while in the SR3 soil Pb was transformed into insoluble hydroxypyromorphite ($Pb_5(PO_4)_3OH$) due to the presence of high level of soil phosphorus (P). The Pb transformation is proposed to follow a four-stage process: Oxidation, carbonation, dissolution, and precipitation. As a result of Pb transformation, the three soils were heavily contaminated with Pb, with Pb concentrations ranging from 12 to 57 g kg^{-1}. In SR1 and SR2 soils, as much as 68 % and 26 % of Pb was extracted by Toxicity Characteristic Leaching Procedure (TCLP), whereas SR3 soil leached only 5.9 % of Pb by TCLP due to the formation of hydroxypyromorphite. Amendment with phosphate rock (PR), phosphoric acid (PA), or monocalcium phosphate (MCP) to SR1 and SR2 soils significantly reduced TCLP-Pb concentrations. The treatments of PR+PA or MCP reduced TCLP-Pb to be below USEPA for non-hazadous regulatory limit of 5 mg L^{-1} at P:Pb molar ratios $\geq$3.0 after 28-d reaction in both two soils. Lead immobilization in the phosphate-treated SR1 and SR2 soils resulted mainly from the formation of insoluble hydroxypyromorphite at the expense of soluble Pb carbonates. Modeling showed that Pb leachability in the SR3 soil was controlled by hydroxypyromorphite, while cerussite

* E-mail address: xindecao@ufl.edu. Corresponding author: X. Cao. Tel: (352) 392-1951. Fax: (352) 392-3902. Soil and Water Science Department, University of Florida, Gainesville, FL 32611

determined Pb leaching in the SR1 and SR2 soils. Upon application of PR+PA or MCP to the SR1 and SR2 soils, Pb activity was controlled by the formed hydroxypyromorphite. Our study indicated that shooting ranges are significantly contaminated with Pb due to the transformation (weathering) of Pb bullets. The end products of the transformation depend on soil chemical conditions and environment conditions. Application of phosphate is an effective alternative for stabilizing Pb to minimize adverse impacts of Pb to the environment. Among the amendments studied, the treatment combining phosphate rock and phosphoric acid is more cost-effective in immobilizing Pb in shooting range soils and poses a lower risk of secondary P leaching.

Keywords: shooting range; lead; contamination; phosphate; transformation; immobilization.

1. Introduction

Soil contamination with lead (Pb) is a widespread problem and can pose serious threats to human and ecosystem health (Adriano 2001). The U.S. Environmental Protection Agency (USEPA) considers Pb as the number one environmental threat to children's health. Even at low levels, it is linked to subtle developmental delays and reduced mental function in children (USEPA 2003).

Lead contamination in soils originates from various sources, including mining and smelting of ores, deposition or recycling of spent batteries, and waste incineration (Davies 1995, Manceau et al. 1996). Recently, shooting activities are being considered as one of the largest contributors of Pb to the environment (USGS 2002). Annual deposition of metallic Pb by shooting varies between 200 to 6,000 tons in the Netherlands, Denmark, Canada, and England (Jorgensen and Willems 1987; VanBon and Boersema 1988, Mellor and McCartney 1994, Scheuhammer and Norris 1996). In the US alone, there are more than 3,000 active ranges for military training and more than 9,000 non-military outdoor ranges for recreation shooting (Houlihan and Wiles 2002, IRTC 2003). Total amount of lead expended as munitions in hunting and recreational shooting has exceeded 3 million metric tons in the 20th century and is presently increasing at a rate of approximately 60,000 metric tons per year (Craig et al. 1999).

When Pb bullets are fired onto a soil berm, intact metallic Pb can be rapidly transformed into reactive Pb species. The most abundant secondary Pb phases include cerussite ($PbCO_3$) and hydrocerussite ($Pb_3(CO_3)_2(OH)_2$) (Jorgensen and Willems 1987, Lin 1996, Cao et al. 2003a,b, Hudison Jr. et al. 2004). In addition, small amounts of litharge (α-PbO), massicot (β-PbO), and anglesite ($PbSO_4$) may also occur, depending on soil and environment conditions (Lin et al. 1995). These Pb compounds may readily find their way into the surrounding environment (e.g. soil and water), posing a risk to the ecological system. Lead concentrations as high as 5% have been documented in shooting range soils, while Pb concentrations of up to 300 $\mu g\ L^{-1}$ Pb have been reported in the retention pond in a Florida shooting range (Cao et al. 2003a). Downward migration of Pb can be significant in shooting ranges, especially with the presence of high organic matters and high alkalinity (Cao et al. 2003a). Heavy Pb-loading from spent Pb bullets also makes it possible for Pb to move down in shooting range soils (Deiss et al. 2004, Richards et al. 2000).

In addition, the bioaccessibility of Pb may increase as a result of the weathering of bullets (Cao et al. 2003a, Migliorini et al. 2004). Lead concentrations as high as 800 mg kg^{-1} Pb were found in the shoots of Bermuda grass collected from a Florida shooting range (Cao et al. 2003a), i.e. considerable amounts of Pb were transported to the aboveground biomass. Migliorini et al. (2004) indicated that a significant portion of metallic Pb from spent pellets accumulated in the predator *Ocypus olens* (Coleoptera). The health of individuals who work and participate in recreational shooting at these ranges may also be at risk due to the secondary Pb minerals that are present in Pb contaminated soils (USEPA 2001). The USEPA reported that Pb in shooting ranges meets the statutory definition of solid waste, and if the Pb is not reclaimed it can be labeled as hazardous waste, which subjects to the Resource Conservation and Recovery Act (USEPA 2001). Therefore, controlling Pb leachability from shooting range soils is an important step to minimize Pb exposure to effectively protect public health and the environment.

Chemical stabilization is a process by which one or more additives are mixed with the soils to render contaminants less susceptible to dissolution and subsequent migration. The effectiveness of this process s has been very well documented (Basta et al. 2001, Brown et al. 2004). Phosphate addition has proven to be an effective means of immobilizing Pb in contaminated soils (Ma et al. 1995). This immobilization is accomplished via the formation of less soluble or insoluble lead phosphate compounds such as pyromorphite-like minerals ($Pb_5(PO_4)_3X$, X=F, Cl, OH), which is induced by phosphate at the expense of more soluble Pb compounds such as cerussite ($PbCO_3$) and litharge (PbO) (Laperche et al. 1997, Zhang et al. 1999, Yang et al. 2001, Cao et al. 2002). Due to the significant reduction of Pb bioavailabilty and the improvement of soil properties (e.g., the permeability or porosity), phosphate-treated soils can be reused. A pilot-scale application of phosphate amendment to a Pb-contaminated site has been demonstrated at an abandoned battery recycling site by Cao et al. (2002) who found that phosphate significantly reduced Pb leachability and phytoavailability and Pb immobilization in the soil showed a long-term stability. Although a significant effort has been made to evaluate the effectiveness of phosphate-based in situ remediation of contaminated soils (Brown et al. 2004, Yang et al. 2001, Cao et al. 2002), little information has been reported for application of phosphate to shooting range soils.

The purpose of the present research is (1) to determine the transformation of spent Pb bullets in shooting range soils; (2) to assess total soil Pb concentrations and Pb leachability; and (3) to investigate application of phosphate for Pb immobilization in the soils.

2. Materials and Methods

2.1. Soil Sampling and Characterization

Soil samples were collected from the surface (top 20cm) of berms (backstops) at three shooting ranges in Florida and New Jersey. They were referred to as SR1, SR2 and SR3 range (SR1 and SR3 in New Jersey, SR2 in Florida). After air-dry, the SR1 and SR3 soils were passed through a 2 mm sieve, and the SR2 soils through a 0.075 mm sieve. Preliminary experiments indicated that these soil fractions have the highest Pb content and need to be treated before disposed to the environment (Data not shown). Therefore, these fractions were the focus of this research. Soil pH was measured at a 1:1 soil:water ratio. Soil organic carbon was determined using the Walkley-Black procedure (Nelson and Sommers 1982). Soil particle separation was done using the ASTM C136-96A (ASTM 2000). The selected properties of the three soils are shown in Table 1.

Table 1. Selected physic-chemical properties of three range soils

Range	Soil pH 1:1	Organic C %	Sand %	Caly+silt %	P[a] %	Al %	Ca %	Fe %	Pb mg kg^{-1}	TCLP-Pb mg L^{-1}
SR1	7.22	0.59	87.8	12.2	0.20	1.22	0.32	1.21	57,000	1,940
SR2	8.47	0.50	0	100	0.21	2.24	0.68	2.54	35,868	466
SR3	5.76	1.27	21.9	78.1	4.24	1.62	2.81	1.69	12,710	37.2

[a] Total concentrations.
[b] pH of TCLP leachate after18-h reaction

2.2. Phosphate Treatments

Several P additives were used as soil amendments. The mineral phosphate rock (PR) was obtained from PCS Phosphate Mining Company (White Springs, Florida). The PR sample was classified as ground concentrated. Its preparation included air-drying, passing through a 0.425-mm sieve, and rinsing with de-ionized water. Fluoroapatite ($Ca_{10}(PO_4)_6F_2$) has been identified as a major component of this PR (Singh et al. 2001). The characteristics of the PR were presented in Singh et al. (Singh et al. 2001). Analytical grade monocalcium phosphate ($Ca(H_2PO_4)_2\,H_2O$) (MCP) and phosphoric acid (H_3PO_4) (PA) were also used as P sources.

Since the SR3 soil was rich in P (containing 4.24% P) (Table 1), phosphate amendments were only applied to the SR1 and SR2 soils. The SR1 soil was treated by addition of powdered PR alone as well as a combination of PR and PA at P:Pb molar ratios of 2:1, 3:1, 4:1, 5:1, and 6:1. For the treatment of PR+PA combination, PR and PA were simultaneously added into the soil at an equal P molar ratio. The SR2 soil was treated with soluble MCP at P:Pb molar ratios of 1.5:1, 3:1, 4.5:1, 6:1, and 7.5:1. Different sources of P were applied to two soils, because of the differences in site hydrology and soil properties. In Florida, groundwater is generally shallow (Cao et al. 2003a), and the SR1 was a sandy soil containing 88% sand (Table 1), which tends to exhibit low P sorption capacity (He et al. 1999). Thus, in order to avoid P leaching into the groundwater and surface runoff, less soluble P source (PR) was added to the SR1 soil, whereas soluble P source (MCP) was applied to the SR2 soil, which contained no sand. For quality control purposes each batch of experiments was run in duplicate.

Specifically, 200 g of the soil were placed in a stainless steel container and homogenized for 30 minutes. Pre-weighed amendments and 40 wt% deionized water were added to the container simultaneously. The materials were mixed well and incubated in sealed bags at room temperature. The treated soil samples were collected at the beginning (1 day) and the end of the treatment (28 days).

2.3. Toxicity Characteristic Leaching Procedure

Lead leachability in the untreated and treated soils was evaluated by using the modified toxicity characteristic leaching procedure (TCLP) (USEPA 1992). Specifically, 25 g soil was mixed with 500 g of the number one leaching fluid (pH=4.93) in 500 ml high-density polyethylene (HDPE) bottles. The mixture was shaken at a speed of 30 rpm for 18 hrs. At the end of shaking, 20ml of the leachate solution was collected for pH measurement and the rest was then filtered through a 0.45 μm membrane filter. The filtrate was acidified with cocncentrated nitric acid to a pH < 2 prior to metal determination by using ICP-OES (Thermo Varian Vista-MPX, Varian, CA).

2.4. Modeling of Pb Leaching

The concentrations of cations and anions in the TCLP leachates were determined. The results of the ion analyses were used as inputs for chemical speciation using Visual MINTEQ (Visual MINTEQ, 2002). Activity ratio diagram for Pb was based on the dissolution equilibrium of Pb minerals including cerussite and lead phosphates. Chemical speciation data obtained from Visual MINTEQ were plotted on the diagram to identify mineral phases potentially controlling Pb solubility in the untreated and P-treated soils.

2.5. Chemical Analysis

The concentrations of solution P was colorimetrically measured with a Shimadzu 160U spectrometer using the molybdate/ascorbic acid method (Olsen and Sommers 1982). Concentrations of soluble anions (e.g., Cl, Br, F, SO_4, and PO_4) were determined by using ion chromatography (Waters 2690 Separations Module, Waters Corporation, USA). Concentrations of solution metals (e.g., Pb, Ca, Fe, Al, and Mn) were determined by ICP-OES. Concentrations of total soil metals were determined by first digesting soil solids according to the USEPA Method 3050B (USEPA 1995) and the leachates were then analyzed by ICP-OES. The quality assurance/quality control (QA/QC) program included a blank, a replicate, and a spike every 20 samples (USEPA 1995). The QC soil sample included 2710 Montana soil Standard Reference Materials (US NIST, Gaithersburg, MD)

2.6. Mineralogical Analyses

The minerals in the selected samples of interest were determined by powder X-ray Diffraction (XRD) using a Rigaku DXR-300 computer-automated diffractometer. Step scanned data were collected using Bragg-Brentano geometry. The X-ray source was from a Cu anode operating at 40 kV and 30 mA using CuKα radiation with a diffracted beam graphite-monochromator. Data were collected between 5° and 65° in 2θ with a step size of 0.05° and count time of 5 seconds per step. Soil samples for XRD analysis were prepared by sieving the selected soils through a 75 μm sieve and the sieved particles were mounted into a quartz sample holder. To analyze the corrosion products in the weathering crust of bullets, the bullets collected from the >2mm soil particles were placed in a beaker, submerged in deionized water, and ultrasonicated until no further crust material was removed. The resulting suspensions with the crust materials were then centrifuged and the sediments were air-dried and ground with an agate mortar for XRD analysis.

The selected soil and crust samples were further subjected to scanning electron microscopy-energy dispersive x-ray spectroscopy (SEM-EDS) analysis using a LEO-810 Zeiss microscope equipped with an ISIS-LINK system. The samples were prepared using double-sided carbon tape.

3. Results and Discussion

3.1. Soil Characteristics and Lead Contamination

Three soils show a wide range of physical and chemical characteristics (Table 1). Soil pH ranged from acidic (SR3), to neutral (SR1), and to alkaline (SR2). The SR1 and SR2 soils contained low organic carbon (<0.6 %) whereas the SR3 soil contained moderate amount at 1.3 %. The SR1 was sandy, containing 88 % sand. Although the SR2 soil was completely composed of fine particles (100%), no clay mineral peak ($2\theta<20$) was identified in the soil by XRD (Figure 1b). This implied that the SR2 soil was mainly composed of silt rather than clay fraction. On the other hand, the SR3 soil mainly consisted of clay (e.g. kaolinite) (Figure 1c). Less P was detected in the SR1 and SR2 soils, but SR3 was rich in P, containing 4.2 % P. This is because the SR3 range was built on a soil containing phosphatic clay (Cao et al. 2003a) In addition, the SR3 soil contained higher amounts (~6.5%) of Al, Ca, and Fe than those (2.8-5.4 %) in both the SR1 and SR2 soils (Table 1). Iron and Al (hydro) oxides, though do not form chemical compounds with Pb, provide large surfaces for sorption, resulting in reduced lead mobility and leachability in soils (Traina et al. 1996).

Lead concentrations in the SR1, SR2, and SR3 soils were as high as 57 and 36, and 13 g kg^{-1}, respectively (Table 1), far exceeding the USEPA soil screening level of 400 mg kg^{-1} (USEPA 1996). Due to the low levels of Fe and Al in the SR1 and SR2 soils, high concentrations of TCLP Pb, i.e. 1,940 and 466 mg L^{-1}, were obtained (Table 1). Correspondingly, as much as 68% and 26% of the total Pb were extracted with TCLP in the SR1 and SR2 soils. On the contrary, the SR3 soil had a low concentration of TCLP Pb (37.4 mg L^{-1}), with only 5.9 % of the total Pb being extracted by TCLP, indicating a low Pb leachability. This could be attributed to the high levels of Ca, Fe and Al and presence of Pb as insoluble forms (discussed below). Nevertheless, TCLP-Pb concentrations in all three soils were much higher than the USEPA non-hazardous regulatory limit of 5 mg L^{-1} (USEPA, 1992) and they would be considered as hazardous waste (USEPA, 2001).

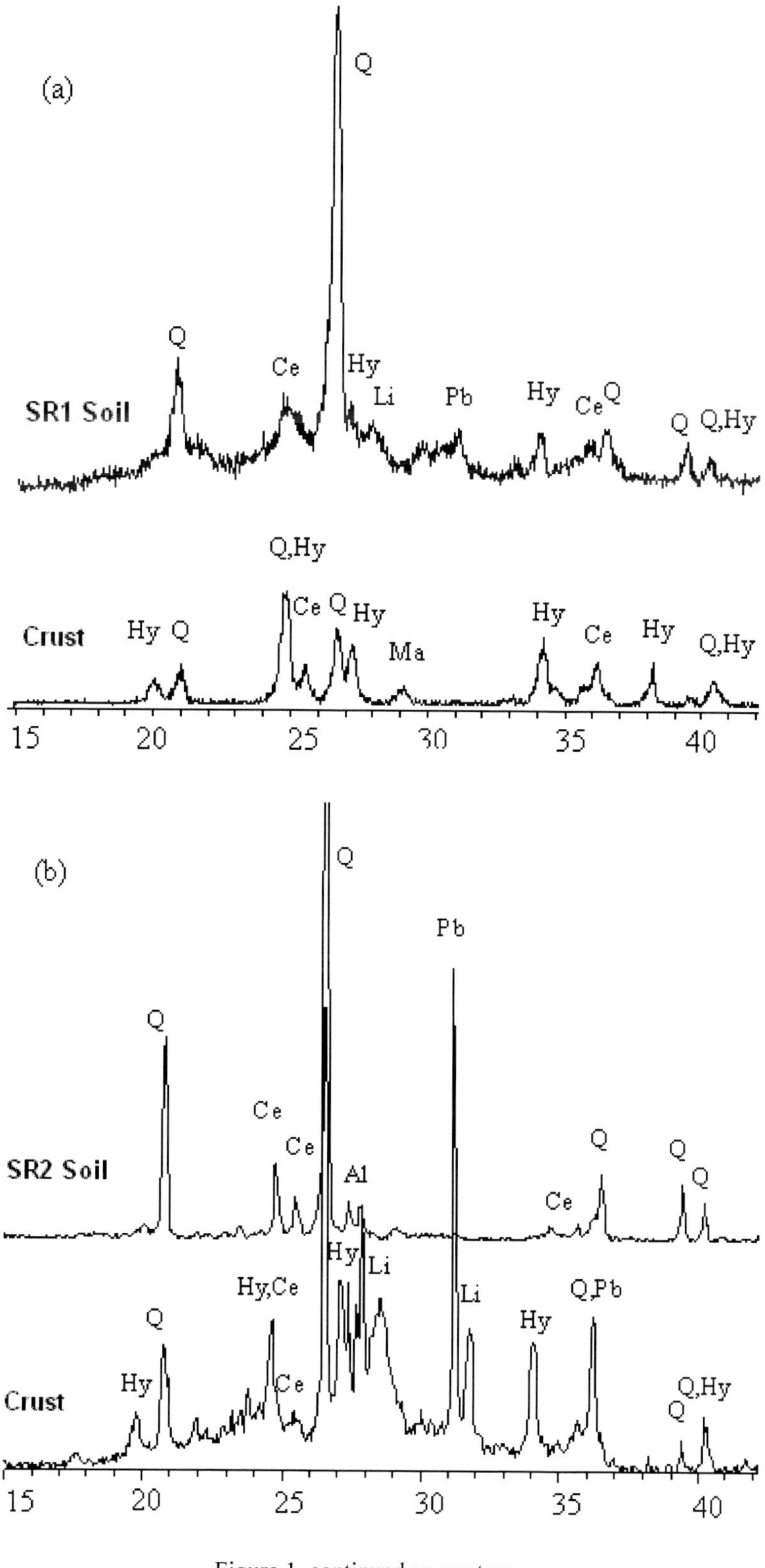

Figure 1. continued on next page.

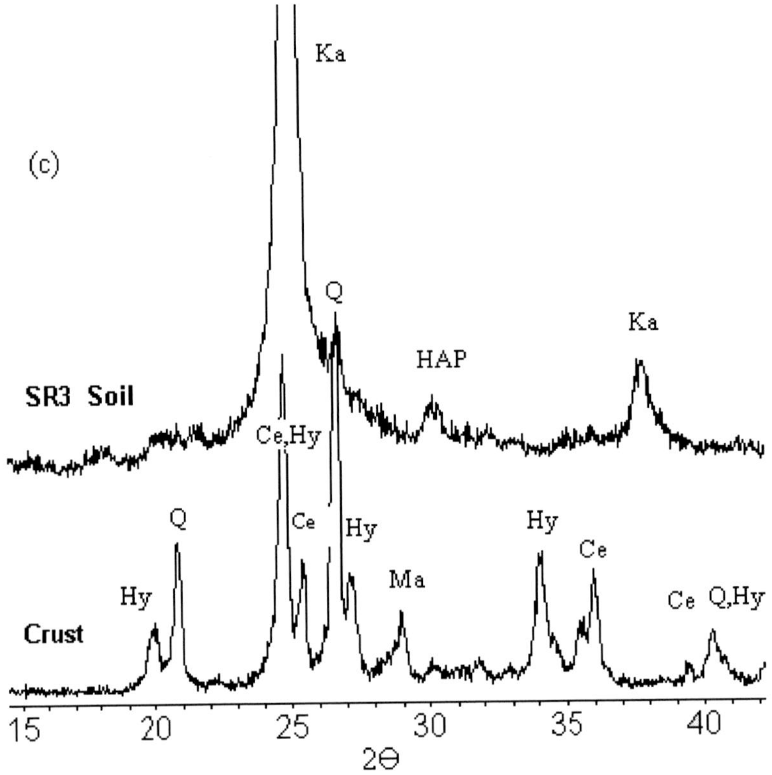

Figure 1. X-ray diffraction patterns of the soil and the crust formed on the surface of the Pb bullet fragments at (a) SR1, (b) SR2, and (c) SR3 ranges. Ce, cerussite; Hy, hydrocerussite, Q, qurtz; Ma, massicot; Li, litharge; Pb, metallic Pb; Al, albite; An, anorthoclase; Ka, kaolinite; and HyPy, Hydroxyppyromorphite

3.2. Transformation of Pb Bullets

Elevation of Pb concentrations in soils mainly resulted from the weathering of Pb bullets. This occurs when Pb bullets come in contact with a soil, exposing them to air and water (Lin, 1996; Jorgensen and Willems, 1987). While the lead species formed on the surface of Pb bullets depend on mostly environment conditions such as humidity and temperature, those in a soil are controlled by both soil properties and environment conditions (Lin, 1996; Cao et al., 2003a).

Figure 1 showed the XRD patterns for the three soils and the associated crusts formed on the weathered surface of Pb bullets. Same Pb minerals, i.e. Pb oxides (litharge PbO, and massicot PbO) and Pb carbonates (hydrocerussite $Pb_3(CO_3)_2(OH)_2$ and cerussite $PbCO_3$), were identified in the crust materials from the three ranges (Figure 1a, b, and c). This indicated that the weathering of Pb occurred along a similar path regardless of the soil properties in a range. Optical microscopy showed that the Pb fragments from bullets were visibly corroded, and encrusted with a yellowish gray colored material composed of Pb oxides and carbonates (data not shown). Using electron microprobe analysis, Lin (1996) showed that the crust on the surface of weathered Pb bullets consisted of two concentric rims: an outer rim of hydrocerussite and an inner rim of massicot. Similarly, Vantelon et al. (2005)

found that the crusts were composed of a mixture of litharge (α-PbO), hydrocerussite, and cerussite, with litharge dominating near the metallic Pb core and cerussite dominating in the outer crust, which is in contact with soils. These observations suggested that the Pb-oxides formed initially are being replaced by Pb-carbonates as the weathering of Pb bullets progress.

However, the weathering products from the crusts differed from those in the associated soil for a given range (Figure 1). As crust materials move into a soil as a result of natural disturbance such as rain and wind, they may undergo a series of geochemical processes that include oxidation/reduction, precipitation/dissolution, adsorption/desorption, and complexation/ chelation. Soil properties play an important role in these transformation processes, thus Pb minerals present in a soil may differ from those in bullet crusts (Cao et al. 2003a). The Pb species including cerussite, hydrocerussite, and metallic Pb were found in the SR1 soil, but cerussite, hydrocerussite, and massicot were found from the weathering products of lead bullets at the same site (Figure 1a). The presence of metallic Pb resulted from the fragmentation of Pb bullets. For the SR2 site, cerussite was the only Pb-based mineral detected in the soil, whereas the corresponding bullet crust material contained litharge, cerussite and hydrocerussite (Figure 1b).

Previous work has shown that transformation of Pb in a soil is site-specific, and soil pH is one of the most important factors (Cao et al., 2003a). A geochemical modeling shows that cerussite is more stable in a neutral pH condition (pH6-8), whereas hydrocerussite is predominant at higher pH range (8-10) (Cao et al., 2003a). However, both cerussite and hydrocerussite were found in the SR1 soil (pH 7.22), whereas the SR2 soil (pH 8.47) contained only cerussite. The stability ranges of cerussite and hydrocerussite in these two shooting ranges seemed to differ from the model-predicted results (Cao et al., 2003a). This discrepancy may be due to other controlling factors (e.g. soil carbonate and total soluble Pb concentrations) in addition to soil pH (Cao et al., 2003a). The presence of cerussite in the SR2 soil and crust on the surface of the weathered bullets was further evidenced by the SEM image (Figure 2a and b). In the SR3 soil, Pb carbonates were less likely to be formed because they would be dissolved at pH 5.8 even if the Pb carbonate formed. Instead, hydroxypyromorphite $Pb_5(PO_4)_3OH$ was formed in the soil due to its high concentration of P (Table 1 and Figure 2c). Formation of hydroxypyromorphite was further evidenced by the SEM elemental dot map where Pb was associated with P (Figure 2c). Thermodynamically, Pb phosphate minerals are the most stable among Pb compounds (Lindsay, 1979). Thus, they can be readily formed in the presence of adequate Pb and P. This is consistent with the study of Ruby et al. (1994) who demonstrated that weathering of galena ($PbSO_4$) to insoluble lead phosphates occurred in soils at a port facility historically used for shipment of phosphate ore.

Based on the observations discussed above and reported by others (Lin, 1996; Vantelon et al., 2005), we hypothesize the mechanisms of Pb bullet weathering consist of four processes: oxidation, followed by carbonation, dissolution and precipitation (Figure 3). The potential chemical reactions were detailed below:

I. Oxidation

$$2\ Pb(s) + O_2 = 2\ PbO(s)$$

II. Carbonation

$$PbO(s) + HCO_3^- = PbCO_3(s) + OH^-$$
$$3PbO(s) + H_2O + 2HCO_3^- = Pb_3(CO_3)_2(OH)_2\ (s) + 2OH^-$$

III. Dissolution

$$2PbCO_3(s) + 2H^+ = 2Pb^{2+} + 2HCO_3^-$$
$$Pb_3(CO_3)_2(OH)_2\ (s) + 4H^+ = 3Pb^{2+} + 2HCO_3^- + 2H_2O$$

IV. Precipitation

$$5Pb^{2+} + 3HPO_4^{2-} + H_2O = Pb_5(PO_4)_3(OH)\ (s) + 4H^+$$
$$Pb^{2+} + SO_4^{2-} = PbSO_4\ (s)$$

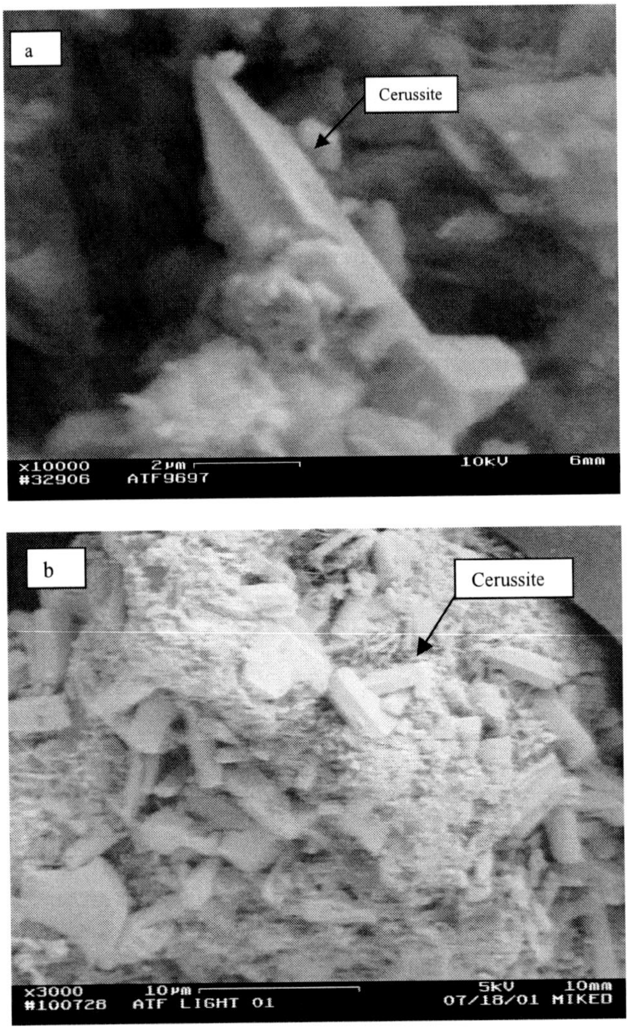

Figure 2. continued on next page.

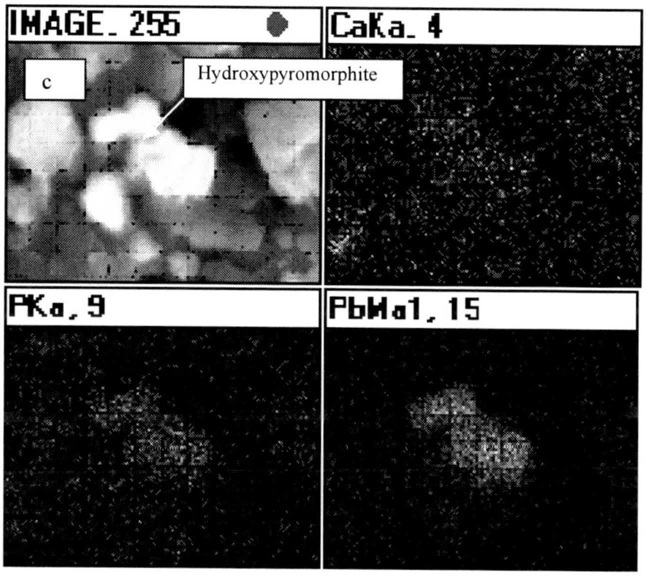

Figure 2. Scanning electron microscopy images showing the presence of (a) cerussite crystal in the SR2 soil and (b) cerussite crystal in the crust on the surface of the weathered Pb fragment, and (c) Scanning electron microscopy element dot map showing the association of Pb with P in the SR3 soil.

It is possible that when Pb bullets are exposed to air, atmospheric O_2 first induces the formation of Pb oxides (litharge or massicot) on the surfaces. Then CO_2 facilitates the transformation from Pb oxides to Pb carbonates (cerussite or hydrocerussite) (Lin et al. 1996, Vantelon et al. 2005). At the same time, dissolution of the secondary Pb minerals (Pb oxides and Pb carbonates) occurs due to the acidity in the soil solution (Essington et al. 2004). If there is sufficient amount of phosphate and sulfate present in the soil solution, then precipitation of $Pb_5(PO_4)_3(OH)$ and $PbSO_4$ are possible.

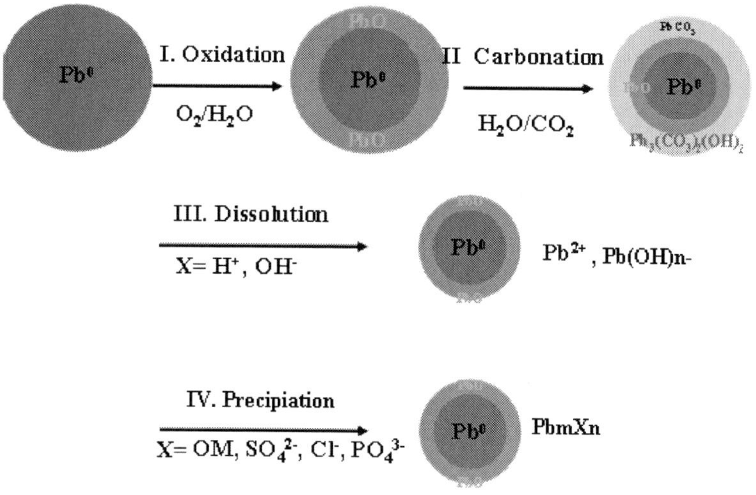

Figure 3. A proposed four-stage mechanism model for the weathering process of Pb bullets

3.2. Phosphate-Induced Pb Immobilization

Phosphate has been proven to effectively immobilize Pb in contaminated soil via formation of insoluble pyromorphite-like minerals (Ma et al. 1995). Efficiency of Pb immobilization by P is determined by the Pb minerals formed and the solubility of P sources. Figures 4 showed Pb concentrations and pH of TCLP leachates of the untreated and P-treated soils. Compared to the untreated SR1 soil, PR treatment reduced TCLP-Pb concentrations, but the TCLP-Pb concentrations were still well above the USEPA limit of 5 mg L^{-1} (Figure 4a). Generally, P-induced Pb immobilization is attributed to the formation of insoluble Pb phosphate minerals (Ma et al. 1995). However, the peaks (2θ=28°-32°) for Pb phosphate minerals, such as chloropyromorphite $Pb_5(PO_4)_3Cl$ and hydroxypyromorphite $Pb_5(PO_4)_3OH$ were absent in the PR-treated SR1 soil (Figure 5a). Instead, the PR was still present in the soil even after 28d of incubation mainly due to the low solubility of PR (Figure 5a). It is possible that the reduction of TCLP-Pb was due to the formation of Pb phosphate during the TCLP extraction (Scheckel et al. 2004), an artifact of low Pb concentrations in TCLP leachates. Increasing the incubation time from 1 day to 28 days did not significantly enhance TCLP-Pb reduction (Figure 4a)

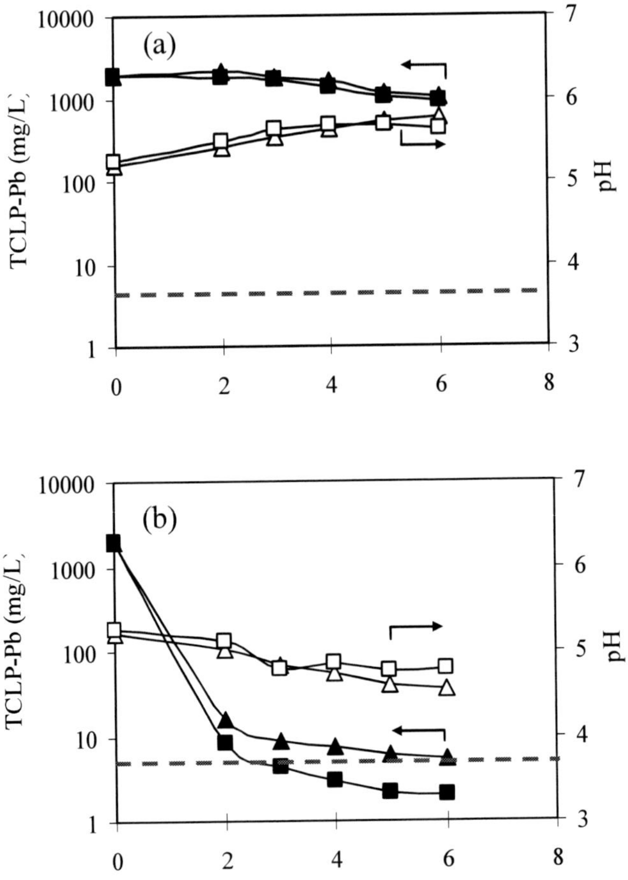

Figure 4. continued on next page

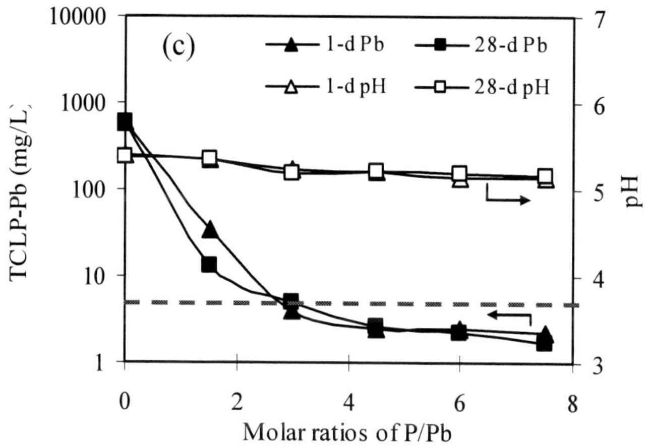

Figure 4. Lead concentrations and pH of the TCLP leachates from (a) PR-treated SR1 soil, (b) PR+PA-treated SR1soil, and (c) MCP-treated SR2 soil. PR, phosphate rock; PA, phosphoric acid; MCP, monocalcium phosphate Dashed horizontal lines indicate the U.S.EPA TCLP Pb regulatory limit of 5mg/L.

When PR was combined with phosphoric acid (PA), the treatment became more effective in immobilizing Pb. Application of PR+PA to the SR1 soil significantly reduced TCLP-Pb concentrations at all application rates, compared to the untreated (Figure 4b) or PR-treated SR1 soil (Figure 4a). The TCLP-Pb concentrations decreased with increasing P application in all soils. However, the treatment didn't reach equilibration after one day, since the TCLP-Pb concentrations further decreased with incubation time increasing to 28 days. The further reduction of TCLP-Pb with time in the PR+PA treated SR1 soil was partially due to the slow release of P from the PR. After 28 days of reaction, treatments with P application rates (P:Pb) $\geq$3.0 reduced the concentrations of TCLP-Pb to below the USEPA limit of 5 mg L^{-1} (Figure 4b). The reduction of TCLP-Pb resulted mainly from the formation of insoluble hydroxypyromorphite ($Pb_5(PO_4)_3OH$) at the expense of cerussite. This was evidenced by XRD analysis showing the presence of hydroxypyromorphite peaks ($2\theta=30°$) and disappearance of cerussite peaks ($2\theta=25°$) (Figure 5a). Although a part of PR was dissolved by PA, PR was still present in the soil as shown by XRD (Figure 5a). The formation of hydroxypyromorphite in the PR+PA treatment was further evidenced by SEM element maps (Figure 6). The maps indicated the presence of two particles, with particle 1 showing P associated with Ca (i.e., phosphate rock) and particle 2 showing Pb associated with P (i.e., hydroxypyromorphite). This was consistent with the XRD patterns (Figure 5a) that revealed the presence of both hydroxypyromorphite and undissolved rock phosphate.

Compared to PR, monocalcium phosphate (MCP) is much more soluble, and should be more effective in immobilizing Pb. Therefore, within 1 day of reaction and at P:Pb=3, the TCLP Pb was decreased in the SR2 soil to below the USEPA regulatory limit (Figure 4c), and the TCLP-Pb decreased with increasing MCP application. As a fact, all TCLP-Pb levels were below the limit when MCP was applied at P:Pb$\geq$3. XRD pattern demonstrated the formation of hydroxypyromorphite in the MCP-treated SR2 soil at the expense of cerussite (Figure 5b), which was responsible for the reduction of TCLP-Pb. Unlike PR and PR+PA treatmens, MCP didn't show time effect, i.e., TCLP-Pb was not further reduced as time increased from 1 day to 28 days (Figure 4c). It was possible that the formation of hydroxypyromorphite was mostly complete after one day of reaction, indicating that, in the presence of soluble phosphate, formation of hydroxypyromorphite was rapid.

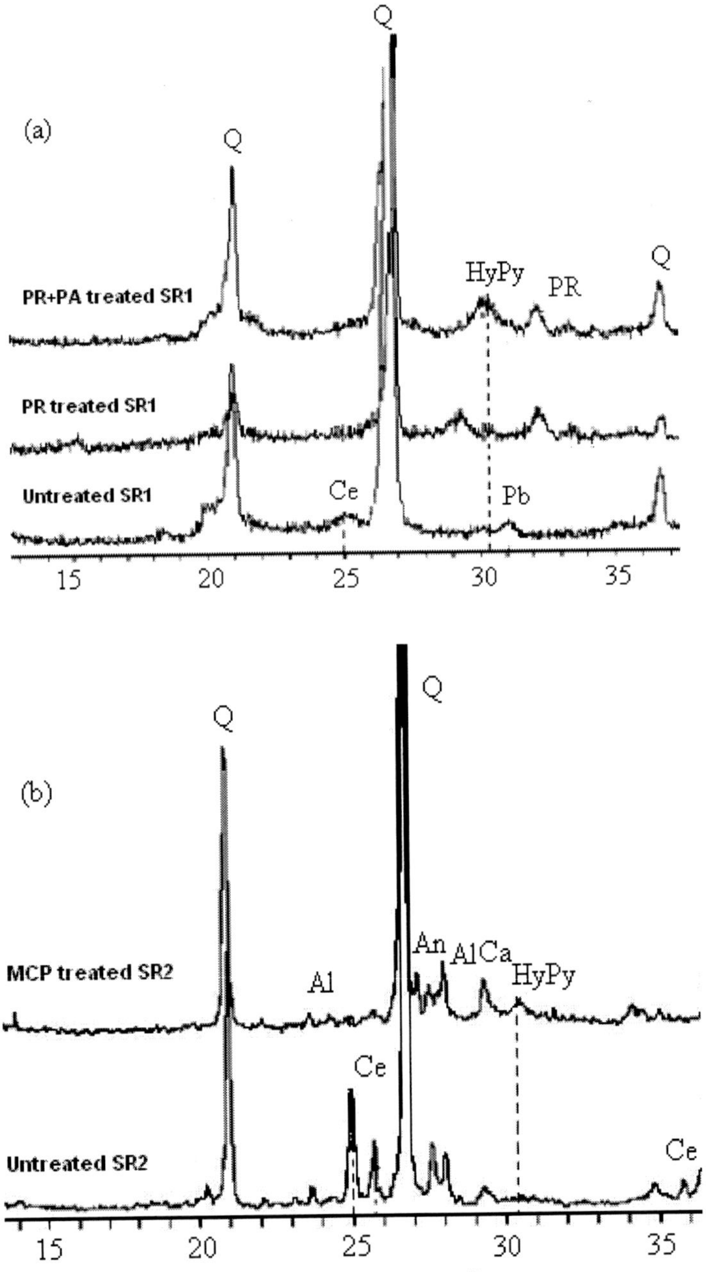

Figure 5. X-ray diffraction patterns of (a) SR1 and (b) SR2 soils treated with phosphate (P/Pb=3 and 28-d reaction). HyPy, hydroxypyromorphite $Pb_5(PO_4)_3(OH)$; PR, phosphate rock; PA, phosphoric acid; MCP, monocalcium phosphate.

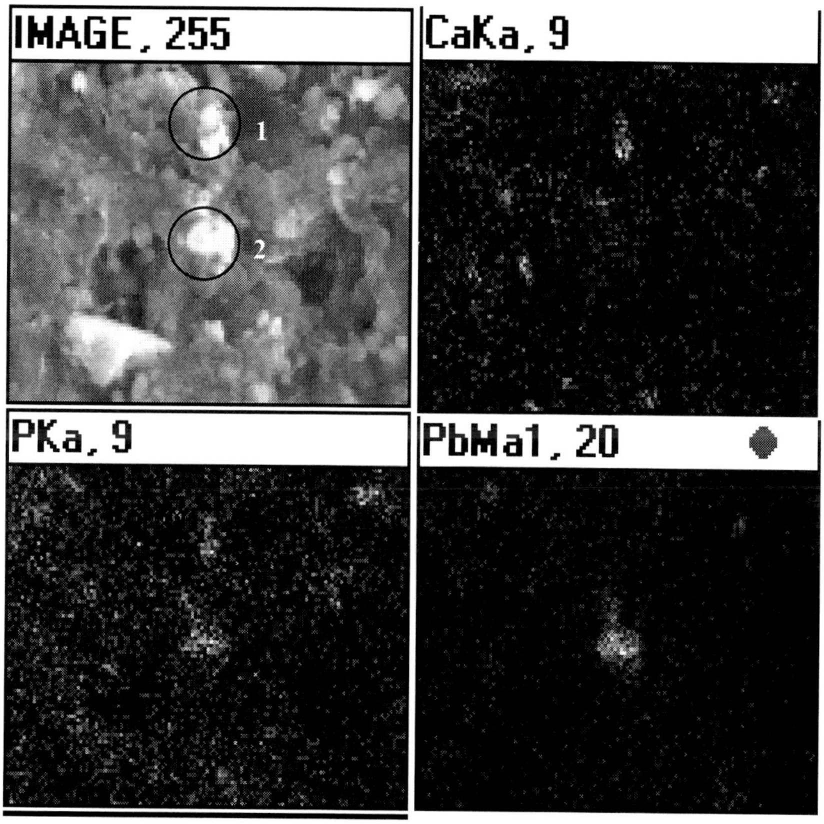

Figure 6. Scanning electron microscopy element dot map of the PR+PA-treated SR1 soil, showing presence of two particles, particle 1 is phosphate rock (Ca-P), and particle 2 is hydroxypyromorphite (Pb-P).

It has been demonstrated that pH of TCLP leachate affect Pb leachability (Dermatas and Meng 2003). Compared to pH 5.2 in the untreated SR1 soil, PR addition increased the pH to 5.7. As a result, the TCLP-Pb decreased from 1,940 to 920 mg L^{-1} (Figure 4a). It was consistent with the observation of Dermatas and Meng (2003) who showed that TCLP-Pb declined as pH increased in low pH range (pH<7). In this case, adsorption process may have controlled Pb leachability. This further confirmed the previous hypothesis that PR-induced immobilization of soil Pb was mainly attributable to surface adsorption by phosphate rock. By contrast, PR+PA decreased pH from 5.1 (the control) down to 4.6 (at the highest application rate of P:Pb=6) and the pH decreased with increasing RP application rate (Figure 4b). However, TCLP-Pb was reduced from 1,940 to 2.0 mg L^{-1}. This was contrary to the results of Dermatas and Meng (2003) where TCLP-Pb increased with pH decreasing in low pH range. Reduction of the TCLP-Pb with decreasing pH suggest that Pb immobilization would be attributed to the precipitation of insoluble Pb phosphate minerals (i.e., hydroxypyromorphite), rather than to Pb adsorption mechanism, consistent with the XRD results (Figure 5a). Similar results were observed in the MCP-treated SR2 soil where soil pH decreased with MCP addition while TCLP-Pb decreased, attributing to the formation of hydroxypyromorphite (Figure 5b).

Table 2. Solution pH and concentrations of P and Pb in the TCLP leachates of the SR1 and SR2 soils with and without phosphate treatment (P/Pb=3 and after 28-d of reaction)

Treatments	SR1			SR2		
	P(mg L^{-1})	pH	Pb(mg L^{-1})	P(mg L^{-1})	pH	Pb(mg L^{-1})
Untreated	0.16	5.15	1940	0.14	5.35	466
PR[a]	0.57	5.62	1654	-	-	-
PR+PA	88.2	4.78	4.26	-	-	-
MCP	-[b]	-	-	145	5.19	4.89

[a] PR, phosphate rock; PA, phosphoric acid; MCP, monocalcium phosphate.
[b] -, not available

3.3. Phosphorus Leachability

Along with the consideration of the effectiveness of phosphate treatment, a potential problem P leachability in the treatments of different P sources. This was important, especially for field application from the viewpoint of potential off-site P contamination and for evaluating P

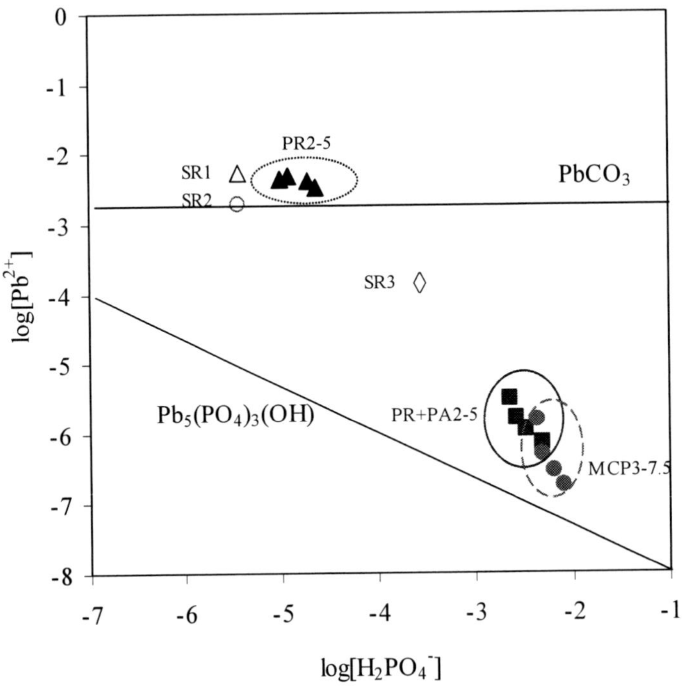

Figure 7. Lead activity-ratio diagram of cerussite ($PbCO_3$) and hydroxypyromorphite ($Pb_5(PO_4)_3OH$) with solution Pb activity in the TCLP leachates from the SR1, SR2, and SR3 soils. Plotted lines assumes the soil average condition with $[Cl^-] = 2\times10^{-4}$M, $[SO_4^{2-}]=1\times10^{-4}$M, $[PO_4^{3-}]=1\times10^{-4}$M, $[CO_2(g)]=3\times10^{-3}$M, and pH=6.5. SR1, SR2, and SR3 mean Pb concentrations in the TCLP leachates from the untreated SR1, SR2, and SR3, respectively; PR2-5 means P:Pb =2:1, 3:1, 4:1, and 5:1 as phosphate rock treatment for SR1 soil; PR+PA2-5 means P:Pb=2:1, 3:1, 4:1, and 5:1 as treatment of phosphate rock and phosphoric acid for SR1 soil; and MCP3-7.5 means P:Pb=3:1, 4.5:1, 6:1, and 7.5:1 as monocalcium phosphate treatment for SR2 soil.

of secondary P contamination should be taken into account. Unfortunately, the latter issue was not addressed adequately in the previous work (Cao et al., 2002). This study assessed the utilization efficiency. Table 2 presented P concentrations in the TCLP leachates from the SR1 and SR2 soil treated with P at P:Pb=3 after 28 days of reaction. As expected, P concentrations in the leachates of the control soil was low, i.e. 0.14-0.16 mg L^{-1} (Table 2). Phosphorus concentrations in the leachates from the treated soils were significantly elevated as aresult of added phosphate (Table 2). Phosphate rock is slightly soluble in water ($K_{sp}=1\times10^{-36}$ at 25°C), while MCP is more soluble (20 g L^{-1} at 25°C). Thus, MCP treatment resulted in the highest P leaching (145 mg L^{-1}), followed by PR+PA (88.2 mg L^{-1}) > PA (0.57 mg L^{-1}) (Table 2). In addition, application of PR should be more economic than MCP. Taking into consideration of the treatment effectiveness, economic efficiency, and P leaching, the PR+PA treatment seemed to be most promising, which was the most cost effective with the least risk of off-site P leaching among the three treatments. Although PR+PA treatment leached much less P than MCP, leachable-P in this treatment was still high (Table 2). Therefore, further reduction in P leaching in the PR+PA treatment may be achieved by reducing the ratio of PA in the PR+PA or by changing the sequence of PR and PA addition.

3.4. Modeling

The Visual MINTEQ model was applied to predict TCLP-Pb leachability in the three untreated soils and two phosphate-treated SR1 and SR2 soils. Figure 7 showed experimental results and modeled values of Pb^{2+} concentrations. In the untreated SR3 soil, Pb activity was supersaturated with regard to hydroxypyromorphite (Figure 7), agreeing with the XRD analysis where hydroxypyromorphite was present (Figure 1c). For the untreated SR1 and SR2 soils, Pb^{2+}concentrations were supersaturated with regard to cerussite (Figure 7), indicating that cerussite controlled Pb leaching, consistent with the XRD analysis (Figure 1a and b) showing the presence of Pb as cerussite in the SR1 and SR2 soils. Compared to the untreated SR1 soil, the treatment with PR alone decreased the Pb^{2+} concentrations. However, the Pb^{2+} concentrations were still high and were also supersaturated with cerussite (Figure 7), i.e. cerussite still controlled the Pb leaching in the PR-treated SR1 soil. When combined with PA, PR significantly reduced the Pb^{2+} concentrations, which were supersaturated with insoluble hydroxypyromorphite (Figure 7), indicating hydroxypyromorphite controlled the Pb leachability in the PR+PA treated SR1 soil. Similar results were observed in the SR2 soil, i.e. MCP significantly reduced the Pb^{2+} concentrations, which were supersaturated with insoluble hydroxypyromorphite (Figure 7).

4. Conclusions

Spent Pb bullets have been readily transformed into reactive Pb species in shooting range soils. The end products of the transformation depend on soil chemical conditions (e.g., pH, carbonate, and available anion ligands) as well as environment conditions. Lead in both SR1 and SR2 soils was converted into soluble cerussite ($PbCO_3$) or hydrocerussite ($Pb_3(CO_3)_2(OH)_2$), while Pb in the SR3 soil was present as insoluble hydroxypyromorphite ($Pb_5(PO_4)_3OH$) due to presence of high level of soil P. The Pb transformation followed a four-step process. i.e., Pb was first oxidized by O_2 to Pb-oxides (litharge, massicot), followed by carbonation of Pb–oxides through CO_2 to form cerussite and hydroxycerussite. The Pb carbonates was then dissolved into soluble Pb^{2+} by H^+ and the soluble Pb can be transformed into stable Pb phosphates by PO_4^{3-} present in soil solution. Such a Pb transformation resulted in contamination of Pb in soils, with total Pb concentrations as high as 5.7 %, and all these soils showed a high Pb leachability. In the SR1 and SR2 soils, as much as 68 % and 26 % Pb was extracted by TCLP. However, only 5.9 % of Pb was leached out in the SR3 soil due to the formation of insoluble hydroxypyromorphite. Upon application of P to SR1 and SR2 soils, Pb concentrations in TCLP leachates were significantly reduced, especially the treatments of phosphate rock + phosphoric acid or monocalcium phosphate, which reduced TCLP-Pb to be below USEPA limit of 5 mg L^{-1} at P:Pb molar ratios $\geq$3.0 after 28-d of reaction in both two soils. Lead immobilization in the phosphate-treated SR1 and SR2 soils was attributed to the formation of insoluble hydroxypyromorphite. Chemical analysis was consistent with modeling predictions, i.e., Pb leachability in the SR3 soil was controlled by insoluble hydroxypyromorphite, while in the SR1 and SR2 soils it was controlled by cerussite. When SR1 and SR2 soils were treated with P, Pb activity was controlled by hydroxypyromorphite. The results obtained from this study indicated that shooting ranges are significantly contaminated with Pb from the transformation of Pb bullets. Application of phosphate is an effective method for best management practices of Pb in shooting ranges through stabilizing Pb to minimize adverse impacts of Pb to the range environment.

Acknowledgments

This research was funded in part by Florida Center of Solid Waste and Hazardous Materials (FCSWHM). The authors would like to thank Dr. Willie Harris for assistance in mineralogical analysis.

References

Adriano, D.C. Trace Elements in Terrestrial Environments: Biogeochemistry, Bioavailability, and Risks of Metals. Springer: New York, 2001

ASTM. Annual book of ASTM standards. American Society for Testing and Materials. *Soil and Rock*. Vol. 4.08, 2000

Basta, N.T., R. Gradwohl, K.L. Snethen, and J.L. Schroder 2001 Chemical immobilization of lead, zinc, and cadmium in smelter-contaminated soils using biosolids and rock phosphate. *J. Environ. Qual.* **30**:1222-1230

Brown, S., R. Chaney, J. Hallfrisch, J. Ryan, and W. Berti 2004 In situ soil treatments to reduce the phyto- and bioavailability of lead, zinc, and cadmium. *J. Environ. Qual.* **33**:522-531

Cao, X.D., L.Q. Ma, M. Chen, S.P. Singh, and W.G. Harris 2002 Impacts of phosphate amendments on lead biogeochemistry at a contaminated site. *Environ. Sci. Technol.* **36**:5296-5304

Cao, X.D., L.Q. Ma, M. Chen, D.W. Hardison, and W.G. Harris 2003 Weathering of lead bullets and their environmental effects at outdoor shooting ranges. *J. Environ. Qual.* **32**:526-534

Cao, X.D., L.Q. Ma, D. Rhue, and C.S. Appel 2004 Mechanisms of lead, copper, and zinc immobilization by phosphate rock. *Environ. Pollut.* **131**:435-444

Craig, J.R., J.D. Rimstidt, C.A. Bonnaffon, T.K. Collins, and P.F. Scanlon 1999 Surface water transport of lead at a shooting range *Bull. Environ. Contam. and Toxicol.* **63**: 312-319.

Darling, C.T.R. and V.G. Thomas 2003 The distribution of outdoor shooting ranges in Ontario and the potential for lead pollution of soil and water. *Sci. Total Environ.* **313**:235-243

Davies, B.E Lead. In Heavy Metals in Soils (Alloway, B.J. Ed), Blackie Academic & Professional, London, 1995

Deiss, J., C. Byers, D. Clover, D. D'Amore, A. Love, M.A. Menzies, J. Powell, M.T. Walter 2004. Transport of lead and diesel fuel through a peat soil near Juneau, AK: A pilot study. *J. Contam. Hydrol.* **74**: 1-18

Dermatas, D. and X. Meng 2003 Utilization of fly ash for stabilization/solidification of heavy metal contaminated soils. *Engr Geol.* **2189**:1-18

Essington, M.E., J.E. Foss, and Y. Roh 2004 The soil mineralogy of lead at Horace's Villa. *Soil Sci. Soc. Am. J.* **68**: 979-993.

Gustafsson, J.P. Visual MINTEQ (version 2.32); Department of Land and Water Resources Engineering, The Royal Institute of Technology: Stockholm, Sweden, 2005; http://www.lwr.kth.se/english/OurSoftWare/Vminteq/.

Hardison Jr, D.W., L.Q. Ma, T. Luongo, and W.G. Harris 2004 Lead contamination in shooting range soils from abrasion of lead bullets and subsequent weathering. *Sci. Total Environ.* **328**: 175-183

He, Z.L., A.K. Alva, Y.C. Li, D.V. Calvert, and D.J. Banks 1999 Sorption-desorption and solution concentrations of phosphorus in a fertilized sandy soil. *J. Environ. Qual.* **28**:1804-1810.

Houlihan, J. and R. Wiles 2002 Lead pollution at outdoor firing ranges. Environmental Working Group, Washington, DC. Available at: www.ewg.org/sports/ posionouspastime/ leadpoll.pdf

Interstate Technology and Regulatory Council (ITRC) 2003 Characterization and remediation of soils at closed small arms firing ranges. Interstate Technology and Regulatory Council, Small Arms Firing Range Team, Blacksburg, VA. Pp204. Available on line at: www.itrcweb.org/SMART-1.pdf

Jorgensen, S.S. and M. Willems 1987 The fate of lead in soils: The transformation of lead pellets in shooting range soils. *Ambio* **16**: 11-15

Laperche, V., T.J. Logan, P. Gaddam, and S.J. Traina 1997 Effect of apatite amendments on plant uptake of lead from contaminated soil. *Environ. Sci. Technol.* **31**: 2745-2753.

Lin, Z., B. Comet, U. Qvarfort, R. Herbert 1995 The chemical and mineralogical behavior of Pb in shooting range soils from Central Sweden. *Environ. Pollut.* **89**:303-309.

Lin, Z 1996 Secondary mineral phases of metallic lead in soils of shooting ranges from Orebro County, Sweden. *Environ. Geol.* **27**: 370-375

Ma, L.Q., T.J. Logan, and S.J. Traina 1995 Lead immobilization from aqueous solutions and contaminated soils using phosphate rocks. *Environ. Sci. Technol.* **29**:1118-1126

Manceau, A., M.C. Boisset, G. Sarret, R.L. Hazemann, M. Mench, P. Cambier, R. Prost 1996 Direct determination of lead speciation in contaminated soils by EXAFS spectroscopy. Environm. *Sci. Technol.* **30**: 1540-1552.

Mellor, A., and C. McCartney 1994 The effects of lead shot deposition on soils and crops at a clay pigeon shooting site in northern England. *Soil Use and Manag.* **10**: 124-129.

Migliorini, M., G. Pigino, N. Bianchi, F. Bernini, and C. Leonzio 2004 The effects of heavy metal contamination on the soil arthropod community of a shooting range. *Environ. Pollut.* **129**:331-340

Nelson, D.W., and L.E. Sommers Total carbon, organic carbon, and organic matter, p. 539-577. In D.R. Keeney et al. (ed.) Methods of Soil Analysis. Part 2: Chemical and Microbiological Properties, Vol. 9, ASA, Madison, Wisconsin, 1982

Olsen, S.R., and L.E. Sommers Phosphorus, p. 403-427. In D.R. Keeney et al. (ed.) Methods of Soil Analysis. Part 2: Chemical and Microbiological Properties Vol. 9, ASA, Madison, Wisconsin, 1982

Richards, B.K., T.S. Steenhuis, J.H. Peverly, M.B. Mcbride 2000 Effect of sludge processing mode, soil texture, and soil pH on metal mobility in undisturbed soil columns under accelerated loading. *Environ. Pollut.* **109**: 327-346

Ruby, M.V., A. Davis, A. Nicholson 1994 In situ formation of lead phosphates in soils as a method to immobilize lead *Environ. Sci. Technol.* **28**: 646-654

Scheckel, K.G., C.A. Impellitteri, J.A. Ryan, and T. Mcevoy 2004 Assessment of a sequential extraction procedure for perturbed lead-contaminated samples with and without phosphorus amendments *Environ. Sci. Technol.* **37**:5296-5304

Scheuhammer, A.M., and S.L. Norris 1996 The ecotoxicology of lead shot and lead fishing weights. *Ecotoxicology* **5**: 279-295

Singh, S.P., L.Q. Ma, and W.G. Harris 2001 Heavy metal interactions with phosphatic clay: Sorption and desorption. *J. Environ. Qual.* **30**:1961-1968.

Traina, S.J. and V. Laperche 1999 Contaminant bioavailability in soils, sediments, and aquatic environments. Proc. Natl. Acad. Sci. USA. 96:3365-3373

U.S. Environmental Protection Agency (USEPA) 2001 Best management practices for lead at outdoor shooting ranges. EPA-902-B01-001. United States Environmental Protection Agency, *Regions* **2**

USEPA 1992 Solid Waste Leaching Procedure Manual. SW-924. United States Environmental Protection Agency, Cincinnati, OH

USEPA 1996 Soil Screening Guidance: User's Guidance. EPA 540/R-60-018, Office of Solid and Emergency Response, Washington, DC.

USEPA 2003 America's Children and the Environment: Measures of Contaminants, Body Burdens, and Illnesses (2nd Ed.), EPA /240-R-03001, Office of Children's Health Protection, Washington, DC.

USGS 2002 Lead in December 2001 Mineral industry surveys. U.S. Geological Survey Minerals Information Publications Service 984 National Center. Reston, VA 20192

VanBon, J., and J. J. Boersema. 1988 Sources, effects and management of metallic lead pollution: the contribution of hunting, shooting and angling. In "Second international TNO/BMFT conference on contaminated sites". Academic Publishers, Hamburg, Kluwer.

Vantelon, D., A. Lanzirotti, A.C. Scherinost, and R. Kretzschmar 2005 Spatial distribution and speciation of lead around corroding bullets in a shooting range soil studied by micro-X-ray fluorescence and absorption spectroscopy *Environ. Sci. Technol.* **38**:5296-5304

Yang, J., D.E. Mosby, S.W. Casteel, and R.W. Blancher 2001 Lead immobilization using phosphoric acid in a smelter-contaminated urban soil. *Environ. Sci. Technol.* **35**:3553-3559.

Zhang, P., and J.A. Ryan 1999 Transformation of Pb(II) from cerrusite to chloropyromophite in the presence of hydroxyapatite under varying condition of pH. *Environ. Sci. Technol.* **33**: 625-630.

In: Environmental Pollution: New Research
Editor: Rachel H. Plattenberg, pp. 121-139
ISBN: 1-60021-285-9
© 2007 Nova Science Publishers, Inc.

Chapter 4

MANAGEMENT OF IRRIGATED AGRICULTURE TO INCREASE ORGANIC CARBON STORAGE IN SOILS

James A. Entry[1*], R.E. Sojka[1**] and Glenn E. Shewmaker[2]***
[1] USDA Agricultural Research Service, Northwest Irrigation and Soils Research Laboratory, 3793 North, 3600 East, Kimberly, Idaho 83341
[3] University of Idaho, Research and Extension Center, Twin Falls, Idaho 83303-1827.

Abstract

Increasing the amount of C in soils may be one method to reduce the concentration of CO_2 in the atmosphere. We measured organic C stored in southern Idaho soils having long term cropping histories that supported native sagebrush vegetation (NSB), irrigated moldboard plowed crops (IMP), irrigated conservation -chisel- tilled crops (ICT) and irrigated pasture systems (IP). The CO_2 emitted as a result of fertilizer production, farm operations and CO_2 lost via dissolved carbonate in irrigation water, over a 30 year period, was included. Net organic C in ecosystems decreased in the order IP>ICT>NSB>IMP. In this study, if NSB were converted to IMP, 0.15 g C m^{-2} would be emitted to the atmosphere, but if converted to IP 3.56 g C m^{-2} could be sequestered. If IMP land were converted to ICT, 0.95 g C m^{-2} could be sequestered in soil and if converted to IP 3.71 g C m^{-2} could be sequestered. There are $2.6x10^8$ ha of land worldwide presently irrigated. If irrigated agriculture were expanded 10% and the same amount of rainfed land were converted back to native grassland, an increase of $3.4x10^9$ Mg C (5.9% of the total C emitted in the next 30 yr) could potentially be sequestered. The total projected release of CO_2 is $5.7x10^{10}$ Mg C worldwide during the next 30 years. Converting rainfed agriculture back to native vegetation while modestly increasing areas in irrigated agriculture could have a significant impact on CO_2 atmospheric concentrations while maintaining or increasing food production.

* E-mail address: jentry@kimberly.ars.pn.usbr.gov; Phone (208) 423 - 6553, Fax (208) 423-6555. Corresponding author.
** E-mail address: sojka@kimberly.ars.pn.usbr.gov; Phone (208) 423 - 6562, Fax (208) 423-6555
*** E-mail address: gshew@uidaho.edu; Phone (208) 736-3600, Fax (208) 736-0843

Abbreviations

C	carbon;
NSB	native sagebrush;
IMP	irrigated moldboard plowed crop land;
ICT	crop land managed with conservation (chisel) tillage;
IP	irrigated pasture;
C_S	carbon projected to sequestered in soils worldwide over the next 30 years;
C_{EW}	carbon projected to be emitted worldwide over the next 30 years.

Introduction

In 1992, nearly all countries of the world signed the Framework Convention on Climate change. Its long term goal is to stabilize the concentration of greenhouse gases in the atmosphere at concentrations that should prevent dangerous anthropogenic interference with the climate system. To stabilize or reduce CO_2 concentrations, the gas must be transferred from the atmosphere to marine or terrestrial ecosystems. Processes or activities that remove greenhouse gases from the atmosphere are defined as sinks in the Framework Convention. In the 1997 Kyoto Protocol, agricultural soils are specifically recognized in the list of potential sinks of greenhouse gases.

Soils are the largest pool of C in the terrestrial environment (Jobbagy and Jackson, 2000; Schlesinger, 1990; 1995). The amount of C stored in soils is twice the amount of C in the atmosphere and three times the amount of C stored in living plants (Schlesinger, 1990; 1995; Kimble and Stewart, 1995), therefore, a change in the size of the soil C pool could significantly alter the atmospheric CO_2 concentration (Wang et al., 1999). The current concentration of C in soils reflects the balance between past C accumulation and loss. Accumulation of C in soils is derived from litter and root input, while losses result from microbial degradation of organic matter, eluviation and erosion (Entry and Emmingham, 1998). At equilibrium, the rate and amount of C added to the soil via vegetation are equal to the rate and amount of C lost through organic matter degradation and other losses (Henderson, 1995). Within limits, soil C increases with increasing soil water and decreasing temperature (Hontoria et al., 1999; Wang et al., 1999; Burke et al., 1989). The effect of soil water is much greater than the effect of soil temperature (Birch and Friend, 1956; Hontoria et al., 1999; Liski et al., 1999). Increasing water within temperature zones can increase plant production and, thus, C input to soils via increased plant litter and root production (Liski et al., 1999).

Land use changes can impact the amount of C stored in the soil by altering C inputs and losses. In forest, grassland and wetland ecosystems, conversion of native vegetation to agricultural cropping has resulted in substantial C transfer to the atmosphere as a result of loss of climax vegetation to the lower equilibrium C concentration in soil (Lal et al., 1999; Wang et al., 1999; Cambardella and Elliot, 1992; Johnson, 1992). In arid and semi-arid environments plant survival and growth is limited by available water and irrigation is required to increase plant production to the point where crops become economically viable. Irrigation also increases C input to soils via increased litter and root production.

When assessing the potential of irrigation of arid or semiarid land to increase C storage in soils, one needs to assess C loss from CO_2 emitted to the atmosphere as a result of 1) fertilizer manufacture, storage, transport and application, 2) fossil- fuel CO_2 emitted from pumping irrigation water, 3) farm operations, such as tillage and planting, and 4) CO_2 lost via dissolved carbonate in irrigation water (West and Marland, 2001; Schlesinger, 1999). Schlesinger (1999) used a fertilization value of 336 kg N ha^{-1} yr^{-1}, which is an unusually high fertilization rate in U.S. farms. The CO_2 released during fertilizer production of 336 kg N ha^{-1} yr^{-1} is approximately 16.7 g C m^{-2} yr^{-1} (Schlesinger, 1999). It has been noted that a more realistic rate is 100-150 kg N ha^{-1} yr^{-1} (West and Marland, 2001). Carbon dioxide released from pumping irrigation water in the United States ranges from 126 kg C ha^{-1} yr^{-1}, when using gasoline, to 266 kg C ha^{-1} yr^{-1} when using electricity (West and Marland, 2001). In addition, C may be lost as CO_2 from the irrigation water itself. Irrigation water in arid and semiarid regions often contains as much as 1% dissolved CO_2. When water is applied to the soil, $CaCO_3$ can precipitate, releasing CO_2 into the atmosphere. If irrigation water containing 0.05 g L^{-3} dissolved Ca is used to irrigate crops in semi-arid climates, the calculated increase in plant C is 2000 g C m^{-2} per year over C contained in native soils and vegetation. The net CO_2 released via irrigation water is calculated to be 8.4 g C m^{-2} yr^{-1} (Schlesinger,1999).

Farm management practices, including conservation tillage and erosion control, have reduced the amount of CO_2 emitted to the atmosphere in both Canada and the United States (West and Marland, 2001; Janzen et al., 1997; Paustian et al., 1997; Rasmussen and Collins, 1991). Intensively managed crop or pasture lands has potential for C gain through the use of improved grazing regimes, improved fertilization practices and irrigation management (Follett, 2001; Bruce et al., 1999). We hypothize that increasing plant growth on arid and semiarid lands by conversion to irrigated agriculture is one method that may increase C storage in soils. The objective of this research was to determine if land managed as irrigated moldboard plowed crops converted to irrigated conservation tillage or irrigated pasture could sequester additional C. We use our findings to pose several possible scenarios of altered land management polices that could favor global C sequestration based on the C budgets that we have estimated.

Materials and Methods

Site Descriptions

The study area is located on the Snake River Plain, between 42° 30' 00" and 43° 30' 00" N. and 114° 20' 00" and 116° 30' 00" W. The sites occur across an elevational gradient ranging from 860 m to 1300 m. The area is classified as a temperate semi-desert ecosystem (Bailey, 1998). The climate is typified by cool, moist winters and hot, dry summers with annual precipitation ranging from 175 to 305 mm, two-thirds of which occurs during October through March (Collett, 1982). Average annual temperature ranges from 9 to 10 °C. Soils are typically well-drained loams and silt loams derived from loess deposits overlying basalt. Vegetation throughout the general area was historically dominated by basin big sagebrush (*Artemisia tridentata* var. *tridentata* Nutt.), Wyoming big sagebrush (*Artemisia tridentata* var. *wyomingensis* Nutt.), and perennial bunch grasses, including Sandberg bluegrass (*Poa secunda* J. Presl), bottlebrush squirreltail (*Elymus elymoides* Raf. Swezy.), bluebunch wheatgrass (*Pseudoroegneria spicata* Pursh. A. Love), and Thurber's needlegrass (*Achnatherum thurberianum* (Piper) Barkworth).

Experimental Design

The experiment was arranged in a completely randomized design (Kirk, 1982). Soil samples were taken from: 1) three sites supporting native sagebrush vegetation (NSB) located near agricultural land in Southern Idaho (each site supported a basin big sage and a Wyoming big sage vegetation type) ; 2) three sites that were formerly crop land and converted to and maintained as irrigated pasture (IP) for the past 30 years; 3) three sites that were irrigated crop land and have been managed with conservation tillage (ICT) for the past 8 years; and 4) three irrigated agricultural crop lands in moldboard plowing systems (IMP) that were each growing a) alfalfa, b) wheat, c) potato and d) beans. There were four treatments (NSB, IMP, ICT, and IP) x three sites for each treatment x five cores taken within each treatment at each site (replications) x four soil depths (0-5 cm, 5-15 cm, 15-30 and 30-100 cm) . We took a total of 240 samples.

Native Vegetation Sagebrush Sites

Native sagebrush sites were vegetated with native steppe vegetation and a low composition of exotic annual grasses. Sites were chosen for this study based on a history of no livestock grazing (BLM, Bruneau Resource Area, unpublished data). All study sites had 5-10 % slope and were on areas that supported basin big sagebrush or Wyoming big sagebrush or communities (Table 1). Soil was classified as a fine, montmorillonitic, mesic Xerollic Haplargid on the Brown's Creek site, a coarse-loamy, mixed non-acid, mesic Xeric Torriorthents on the Simco site and a loamy, mixed, mesic lithic Xerollic Camborthids on the Kuna Butte site (Collett, 1982).

Irrigated Pasture Sites

Three irrigated pastures were selected that were formerly crop land and converted to and maintained as irrigated pasture (IP) for the past 30 years. The Buhl site was vegetated with Kentucky bluegrass (*Poa pratensis* L.)–orchardgrass.(*Dactylis glomerata* L.) on a Rakane-Blacknest soil complex, fine-loamy, mixed, mesic Xerollic Durargids soil. The Gooding site was vegetated with smooth brome (*Bromus inermis* Leyss.)-orchardgrass on a Paulville-Idow soil complex, fine-loamy, mixed, mesic Xerollic Haplargid soil. The Kimberly site was vegetated with smooth brome-orchardgrass pasture on a Portneuf soil, coarse-silty, mixed, superactive, mesic Durinodic Xeric Haplocalcid soil. Grazing rates on these pastures were 10-12 animal unit months yr^{-1}.

Table 1. Location and characteristics of study sites.

Site Name	Latitude Longitude (10' block)	Treatment	Elevation (m)	Site Description	Last Known Fire(s)	Current Dominant Vegetation
Bowns Creek	43° 20' 00'' 115° 56' 00"	Native	1010	*Artemisia tridentata* Basin /*Pseudoroegneria* (Pursh) A. Love. *Artemisia tridentata* Wyoming /*Pseudoroegneria* Soil Type: Xeroic Haplargid	unknown (>50 years)	*Artemisia tridentata* Basin, Wyoming / *Pseudoroegneria spicata / Poa secunda*
Simco	43° 22' 00'' 115° 51 '00"	Native	1095	*Artemisia tridentata* Basin /*Pseudoroegneria* *Artemisia tridentata* Wyoming /*Pseudoroegneria* Soil Type: Xeric Torriorthent	unknown (>50 years)	*Artemisia tridentata* Basin, Wyoming / *Pseudoroegneria spicata / Poa secunda*
Kuna Butte West	43° 27' 00'' 116° 27' 00"	Native	950	*Artemisia tridentata* Basin /*Pseudoroegneria* *Artemisia tridentata* Wyoming /*Pseudoroegneria* Soil Type: Xerollic Camborthid	unknown (>50 years)	*Artemisia tridentata* Basin, Wyoming / *Poa secunda*
Kimberly Pasture	42° 00' 00'' 114° 30' 00"	Irrigated Pasture	1210	*Dactylis glomerata L. / Bromus inermis* Lyess. Soil Type: Xeric Haplocalcid	(>100 years)	50 % *Dactylis glomerata.* 50 % *Bromus inermis*
Gooding Pasture	42° 46' 00'' 114 ' 37' 00"	Irrigated Pasture	1070	*Bromus inermis / Dactylis glomerata* Soil Type: Xerollic Haploargid	(>100 years)	50 % *Dactylis glomerata* 50 % *Bromus inermis*
Buhl Pasture	42° 30' 00'' 114° 30' 00"	Irrigated Pasture	1073	*Poa pratensis L. / Dactylis glomerata.* Soil Type: Xerollic Durargids	(>100 years)	50 % *Poa pratensis* 50 % *Dactylis glomerata*
Kimberly	42° 00' 00'' 114° 30' 00"	Irrigated moldboard plow	1260	Rotations of : *Phaseolus vulgaris L / Triticum aestivum L. / Solanum tuberosum L / Medicago sativa L.* Durinodic Xeric Haplocamberid	(>100 years)	*Phaseolus vulgaris L / Triticum aestivum L. / Solanum tuberosum L / Medicago sativa L.*
University of Idaho	42° 46' 00'' 114 ' 37' 00"	Irrigated moldboard plow	1220	Rotations of : *Phaseolus vulgaris L / Triticum aestivum L. / Solanum tuberosum L / Mmedicago savtia L.* Durinodic Xeric Haplocalcid	(>100 years)	*Phaseolus vulgaris L / Triticum aestivum L. / Solanum tuberosum L / Mmedicago savtia L.*
South Farm	42° 30' 00'' 114° 30' 00"	Irrigated moldboard plow	1200	Rotations of : *Phaseolus vulgaris L / Triticum aestivum L. / Solanum tuberosum L / Medicago sativa L.* Durinodic Xeric Haplocalcid	(>100 years)	*Phaseolus vulgaris L / Triticum aestivum L. / Solanum tuberosum L / Medicago sativa L.*

Table 1. Continued

Site Name	Latitude Longitude (10' block)	Treatment	Elevation (m)	Site Description	Last Known Fire(s)	Current Dominant Vegetation
Kimberly	42° 00' 00'' 114° 30' 00"	Irrigated conservation tillage	1260	Rotations of : *Phaseolus vulgaris L / Triticum aestivum L. / Solanum tuberosum L / Medicago sativa L.* Durinodic Xeric Haplocamberid	(>100 years)	*Phaseolus vulgaris L / Triticum aestivum L. / Solanum tuberosum L / Medicago sativa L.*
University of Idaho	42° 46' 00'' 114 ' 37' 00"	Irrigated conservation tillage	1220	Rotations of : *Phaseolus vulgaris L / Triticum aestivum L. / Solanum tuberosum L / Medicago sativa L.* Durinodic Xeric Haplocalcid	(>100 years)	*Phaseolus vulgaris L / Triticum aestivum L. / Solanum tuberosum L / Medicago sativa L.*
South Farm	42° 30' 00'' 114° 30' 00"	Irrigated conservation tillage	1200	Rotations of : *Phaseolus vulgaris L / Triticum aestivum L. / Solanum tuberosum L / Medicago sativa L.* Durinodic Xeric Haplocalcid	(>100 years)	*Phaseolus vulgaris L / Triticum aestivum L. / Solanum tuberosum L / Medicago sativa L.*

Irrigated Conservation Tillage and Crop Sites

Three sites with fields rotating among alfalfa (*Medicago sativa* L.), wheat (*Triticum aestivum* L.), potato (*Solanum tuberosum* L.) and beans (*Phaseolus vulgaris* L.) were sampled. All sites were located on fields managed by USDA Agricultural Research Service's Northwest Irrigation and Soils Research Laboratory or the University of Idaho, Research and Extension Center. Soil on all sites was classified as a coarse-silty, mixed, superactive, mesic Durinodic Xeric Haplocalcid, with 0.1-0.21 g/g clay and 0.6-0.75 g/g silt, and organic matter of approximately 13 g kg^{-1}. The soil has a pH of 7.6-8.0. Slope on these sites ranges from 1.0 to 3.0% (Table 1).

Sampling Procedures

Soil cores were taken from each site during winter (January), spring (April), summer (August), autumn (November) in 1999. We sampled the top 1 m of soil each season (winter, spring, summer and autumn) to determine if the amount of C in soil would be affected by irrigation, tillage and vegetation. Sampling locations were randomly chosen at each site or field. Separate 10 cm diameter replicate cores were randomly taken and partitioned into 0-5 cm, 5-15 cm, 15-30 and 30-100 cm depths. Roots greater than 1.0 cm diameter were measured separately. Carbon in above-ground vegetation was estimated by measuring the amount of material in 10 separate 1.0 m^2 areas in each site or field (Entry and Emmingham, 1998).

Carbon in Soil and Above-ground Vegetation

Concentration of organic C in each sample of mineral soil was determined by the Walkley-Black procedure and loss on ignition (Nelson and Sommers, 1996). The amount of C per ha^{-1} of the 0-100 cm of mineral soil was calculated assuming 0.44 g C g^{-1} organic matter with correction for soil bulk density. Ten separate 10 cm diameter soil cores were taken to a 1.0 m depth, divided into 0-5 cm, 5-15 cm, 15-30 and 30-100 cm depths to determine bulk density. Bulk density was measured by dividing by the oven dry weight after drying at 105 °C for 48 hours by the volume of the sample (Blake and Hartage, 1982). Above-ground vegetation was collected and separated into sage, grass, forbs, herbs and duff. Above-ground material was dried at 80 °C for 48 h, weighed and ground to pass a 1 mm opening. Carbon in above-ground vegetation was determined by loss on ignition (Nelson and Sommers, 1996). The amount of C in the above ground material was assumed to contain 0.44 g C g^{-1} organic matter on an ash free basis (Nelson and Sommers, 1996). Calculations estimating C stored in soils in the Western United States and worldwide are based on the Walkley Black procedure.

Calculations

Concentration of organic C determined by the Walkley-Black procedure was converted, using bulk density measurements, to a meter square basis to a depth of 1 meter. Organic C in kg m^{-2} was converted to Mg C ha^{-1} by multiplying by 10,000 (land area) and dividing by 1,000 (C weight), which is a 1: 10 conversion. We divided the resulting number by 10 to account for a 10 % conversion of one treatment (land area). The amount of C sequestered in the Pacific Northwestern United States, the 11 western states in the United States and worldwide was estimated by multiplying Mg C ha^{-1} by the number of hectares of irrigated land in each area. There are 9,055, 979 ha of land in irrigated crop land in the Pacific Northwest, 24,322,029 ha in the Western U.S. and 260, 000,000 ha worldwide (Bucks et al., 1990; Tribe, 1994; Howell, 2000). The carbon sequestered worldwide (C_S) relative to the amount of C projected to be emitted worldwide during the next 30 yr (C_{EW}) was calculated by dividing the Mg C sequestered in each treatment x area by the total projected worldwide release of CO_2 - C during the next 30 years (5.7×10^{10} Mg C) multiplied by 100.

Statistical Analysis

All data were subjected to a one way vegetation type analysis of variance (ANOVA) for a completely randomized design (Snedecor and Cochran, 1980; Kirk, 1982). Residuals were normally distributed with constant variance. Statistical Analysis Software programs (SAS Institute Inc. 1996) were used to conduct the analysis of variance. Significance of treatment means were determined at $P < 0.05$ with the Least Square Means test.

Results and Discussion

Site Specific Findings

Statistical comparisons in the ANOVA showed that soil bulk density, soil C, site C, net C in soil, net site C in site x vegetation interactions are not significant at $P \leq 0.05$. Therefore, results are discussed with respect to vegetation differences (Snedecor and Cochran 1980; Kirk 1982). Bulk density was greater in soils in IP, ICT and IMP than in NSB soils. Bulk density was less in the NSB 0-5 cm soil depth than the 5-15, 15-30 and 30-100 cm depths and all other soils (Table 2). Soil C was greater in the NSB 0-5 cm soil depth than the 5-15, 15-30 and 30-100 cm depths and all other soils (Table 2). Soil was greater in the in the 0-5 cm, 5-15 cm and 15-30 cm depths in the IP than the IMP or ICT treatments.

Organic C contained in above-ground vegetation was greater on NSB sites than IP; however, IP biomass was removed by grazing. Crops were not considered as permanent vegetation. Prior to adjustment for agricultural CO_2 emissions, (total), soil C and C on site was greatest to least in the order IP>ICT>IMP>NSB (Table 3). Conversion of land to IP resulted in less C emitted to the atmosphere than IMP- or ICT- managed crops because less fertilizer and farm operations were necessary. After adjustment for agricultural CO_2 emissions, (net) C in soils was greatest to least in the order IP>ICT>NSB>IMP.

We estimated that if NSB sites were converted to IMP a net loss of 0.15 kg C m^{-2} over 30 years would occur (Table 3). We estimated a net gain of 0.80 kg C m^{-2} over 30 years If NSB sites were converted to ICT , and if converted to IP one could expect net gain of 3.56 kg C m^{-2} over a 30 year period. We estimated that if IMP was converted to ICT a net gain of 0.95 kg C m^{-2} over 30 years would occur. If IMP land was converted to IP an estimated net gain of 3.71 kg C m^{-2} over a 30 year period would occur.

Table 2. Bulk density, Walkley Black C and loss on ignition C (LAI-C) in soils growing native sagebrush, irrigated moldboard plowed cropland, irrigated conservation tilled cropland and irrigated pastures in Southern Idaho.

Treatment	Soil Depth	Bulk Density	WBC†	LAI-C
		-- Mg m^{-3} --	---------- g C kg^{-1} soil----------	
Native sagebrush	0-5	0.97 b	12.7 a	10.1 a
	5-15	1.28 ab	4.7 c	4.1 c
	15-30	1.34 a	5.4 c	5.1 c
	30-100	1.36 a	4.0 c	3.5 c
Irrigated moldboard plow crop	0-30 ‡	1.28 ab	7.8 b	6.9 b
	30-100	1.37 a	6.0 bc	5.1 bc
Irrigated conservation tilled crop	0-15 ‡	1.38 a	8.9 b	7.6 b
	15-30 ‡	1.38 a	6.9 bc	6.8 bc
	30-100	1.37 a	3.7 c	3.1 c
Irrigated pasture	0-30 ‡	1.33 a	8.5 b	7.9 b
	30-100	1.40 a	4.3 c	3.8 c

† In each column, values followed by the same letter are not significantly different as determined by the Least Square Means Test (P # 0.05), n = 30.

‡ Statistical comparisons in the ANOVA showed that soil bulk density with respect to soil depth were not significant at $P \leq 0.05$. Therefore, data were combined (Snedecor and Cochran 1980; Kirk 1982).

Regional and Global Implications

Changes in agricultural practices have great potential to sequester C. In most cases converting selected land managed as IMP to ICT or IP can be implemented with modest economic impact to landowners and pose relatively few socioeconomic issues. If agricultural land is managed properly, these practice shifts would also potentially reduce erosion and water or air pollution. Estimating the potential for C sequestration in terrestrial ecosystems is difficult because the dynamics that control C flow among plants, soils and the atmosphere are poorly understood. Storage of C in below ground systems is the best long-term option in terrestrial ecosystems because C in soils has a longer residence time than most plant biomass. Using the values obtained in southern Idaho, we estimated C storage in soils locally, regionally and globally in soils if: 1) 10% of irrigated land now in IMP agriculture was converted back to NSB, 2) all land presently in IMP was converted to ICT and 3) 10 % of land in irrigated IMP was converted to IP. Since increased agricultural production will be necessary to feed an increasing population, it is impractical to suggest that a large portion of land in IMP can be converted to IP

The reported amounts of C stored in native sagebrush vegetation and irrigated agricultural systems are similar throughout the United States and worldwide (Bowman et al., 1999; Collins et al., 1999; Amthor and Huston, 1998; Potter et al., 1998; Rasmussen and Parton, 1994; Schlesinger, 1977;). These data were used to calculate potential C storage for irrigated agriculture in the Pacific Northwestern U.S., the Western U.S., and worldwide over a 30 year period. If land currently in IMP is converted to NSB we estimated a gain of 1.5 Mg C ha^{-1} (Table 4). Little of this land is managed with conservation tillage. We estimate an increase of 9.5 Mg C ha^{-1} over 30 years if the land presently managed with IMP were converted to ICT. Using this value we calculated that 8.6 x 10^7 Mg C (0.15% of the total C emitted in the next 30 yr) could potentially be sequestered in irrigated soils in the Pacific Northwestern U.S (Table 4). Using these values to represent C gains for all irrigated crop land in the western U.S. and if land in IMP were converted to ICT, a possible 2.3 x 10^8 Mg C (0.40% of the total C emitted in the next 30 yr) could be sequestered in irrigated agricultural soils in the next 30 years. If the world's IMP land were converted to ICT, 2.5x10^9 Mg C (4.38% of the total C emitted in the next 30 yr) could be sequestered in the next 30 years. A shift of 10 % of current IMP land to ICT is a reasonable conservation practice goal. Similarly, we can not expect all land presently in IMP to be converted to IP, but a 10 % conversion is feasible. If we predict a storage increase of 37.1 Mg C ha^{-1} for IMP converted to IP and assume 10 % of IMP land is converted to IP, an estimated 3.4x10^7 Mg C (0.05% of the total C emitted in the next 30 yr) could be sequestered in Pacific Northwestern U.S. soils and 9.0x10^7 Mg C (0.16% of the total C emitted in the next 30 yr) in the Western U.S. soils in the next 30 years. If our study values are used to represent C gains for irrigated crop land worldwide, an estimated 9.6x10^8 Mg C (1.68% of the total C emitted in the next 30 yr) could be sequestered if irrigated land presently managed as IMP were converted to ICT (Table 4). Conversion of crop land to irrigated pasture could also relieve grazing pressure on public rangelands, an issue of heated debate between environmentalists and ranchers.

If irrigated agriculture is expanded due to increase in water use efficiency, one could expect a gain in C sequestration. If NSB were converted to ICT, 8.0 g C ha^{-1} could be sequestered (Table 4). Predicting a storage increase of 8.0 Mg C ha^{-1} for NSB converted to ICT and assuming 10 % expansion of irrigated agriculture, an estimated 7.2x10^6 Mg C (0.01% of the total C emitted in the next 30 yr) could be sequestered in Pacific Northwestern U.S. soils and 1.9x10^7 Mg C (0.033% of the total C emitted in the next 30 yr) in the Western U.S. soils in the next 30 years. If our study values are used to represent C gains for irrigated crop land worldwide, an estimated 2.1x10^8 Mg C (0.37% of the total C emitted in the next 30 yr) could be sequestered (Table 4). If NSB were converted to IP, 35.6 g C ha^{-1} could be sequestered. Predicting a storage increase of 35.6 Mg C ha^{-1} for NSB converted to IP and assuming 10% expansion of irrigated agriculture, an estimated 3.2x10^8 Mg C (0.5% of the total C emitted in the next 30 yr) could be sequestered in Pacific Northwestern U.S. soils and 8.7x10^8 Mg C (1.53% of the total C emitted in the next 30 yr) in the Western U.S. soils in the next 30 years. If our study values are used to represent C gains for irrigated crop land worldwide, an estimated 9.3x10^9 Mg C (16.3% of the total C emitted in the next 30 yr) could be sequestered

Table 3. Organic carbon in: soils[1], above ground biomass[2] and on sites at present[3], carbon emitted during agricultural operations[4], net organic carbon in soils[5] and net carbon gain on sites[6].†

Vegetation	---------- Carbon Present ----------				------ Net Carbon Gain ------	
	Soil[1] ‡	Above ground[2] ¶	Site[3]	Carbon emitted [4] §	Soil[5] ‡	Site[6]
	kg C m^{-2}					
Native sagebrush	5.91 c	0.42 a	6.34 c	0.00 d	5.91 c	6.34 c
Irrigated moldboard plow crops	7.29 b	0.00 c	7.29 b	1.10 a	6.19 b	6.19 c
Irrigated conservation till crops	8.01 b	0.00 c	8.01 b	0.87 b	7.14 b	7.14 b
Irrigated pasture	10.14 a	0.05 b	10.19 a	0.29 b	9.85 a	9.90 a

† In each column, values followed by the same letter are not significantly different as determined by the Least Square Means Test (*P* # 0.05), n = 30.

‡ Values of organic C stored in soils are based on the Walkley Black procedure.

¶ Carbon in soils, above ground vegetation and on the sites at the present time.

§ Estimated carbon emitted in production of fertilizer, fuel consumption in farm operations and via irrigation water over a 30-year period.

Table 4. Potential organic carbon gain by conversion of irrigated lands currently in moldboard plowing systems to conservation tillage, conversion of native sagebrush to irrigated conservation tillage, conversion of native sagebrush to irrigated pasture and conversion of 10% of irrigated lands currently in moldboard plowing systems to irrigated pasture over the next 30 years.

Vegetation Conversion	C gained from a 10 % conversion	Pacific Northwest United States ‡		Western United States ‡		Worldwide ‡	
	--Mg C ha^{-1}--	Mg C	% C_S/C_{EW} §	Mg C	% C_S/C_{EW} §	Mg C	%C_S/C_{EW} §
Irrigated moldboard plow to irrigated conservation tillage †	9.5	8.6×10^7	0.15	2.3×10^8	0.40	2.5×10^9	4.38
Native sagebrush to irrigated conservation tillage	8.0	7.2×10^6	0.01	1.9×10^7	0.03	2.1×10^8	0.37
Native sagebrush to irrigated pasture	3.56	3.2×10^8	0.56	8.7×10^8	1.53	9.3×10^9	16.32
10% of irrigated moldboard plow to irrigated pasture †	37.1	3.4×10^7	0.06	9.0×10^7	0.16	9.6×10^8	1.68

† Estimated C gain from 100% conversation of moldboard plow to conservation tillage and 10% conversion of moldboard plow agriculture to irrigated pasture.

‡ Land area in irrigated cropland in Pacific Northwest - 9, 055,979 ha, Western United States 24,322,029 ha, worldwide 260,000,000 ha.

§ %C_S/C_{EW} = carbon sequestered (C_S) divided by the amount of C projected to be emitted worldwide during the next 30 yr, which is 5.7×10^{10} Mg C (C_{EW}) multiplied by 100.

Table 5. Potential carbon transfer by converting an equal amount (10%) of rainfed moldboard plow land back to native forest or grassland on a basis of 1 unit of irrigated rainfed agricultural land to 1 unit of native forest or grassland and conversion of equal amount (10%) of rainfed moldboard plow land back to native forest or grassland on the basis of 1 unit of irrigated rainfed agricultural land to 2 units of native forest or grassland

Conversion of Vegetation	C Stored From Conversion	Pacific Northwest United States		Western United States		Worldwide	
	Mg C ha^{-1}	Mg C	$\%C_S/C_{EW}$ †	Mg C	$\%C_S/C_{EW}$ †	Mg C	$\%C_S/C_{EW}$ †
Rainfed moldboard plow to native forest on a 1 unit:1 unit basis	5.6	5.1×10^7	0.09	1.4×10^8	0.24	1.5×10^9	2.63
Rainfed moldboard plow to native grassland on a 1 unit:1 unit basis	13.0	1.2×10^8	0.21	3.2×10^8	0.56	3.4×10^9	5.96
Rainfed moldboard plow to native forest on a 2 unit:1 unit basis	5.6	1.1×10^8	0.18	2.8×10^8	0.49	3.0×10^9	5.26
Rainfed moldboard plow to native grassland on a 2 unit:1 unit basis	13.0	2.4×10^8	0.42	6.4×10^8	1.20	6.8×10^9	11.93

† $\%C_S/C_{EW}$ = carbon sequestered (C_S) divided by the amount of C projected to be emitted worldwide during the next 30 yr, which is 5.7×10^{10} Mg C (C_{EW}) multiplied by 100.

Since the earth releases 1.9 x 10^9 Mg C yr^{-1} (Schlesinger, 1995; Amthor and Huston, 1998), the conversion of 10% IMP to IP, resulting in a possible 9.6x10^8 Mg C sequestered over 30 yrs, may be insignificant (Table 4). However, if crops were produced via high output irrigated agriculture while less productive rainfed agricultural land were returned to temperate forest or native grassland, an increase of 5.6 and 13 Mg C ha^{-1}, respectively, could be gained over 30 years for each unit of rainfed land converted to native vegetation (Table 5). The amount of irrigated agriculture can likely be increased at least 10% solely through increases in irrigation efficiency and waste water reuse (Howell, 2000). Using a conversion basis of 1 unit of irrigated agriculture to return 1 unit of rainfed agricultural land to native forest, if irrigated agriculture were expanded 10% (meaning that an additional 2.6 x 10^7 ha of arid or semiarid land were irrigated) and the equal amount of land being managed as rainfed agricultural land were converted to native forest, there is potential to sequester 5.1x10^7 Mg C (0.09% of the total C emitted in the next 30 yr) in the Pacific North West, 1.41x10^8 Mg C (0.24% of the total C emitted in the next 30 yr) in the western U.S. and 1.5x10^9 Mg C (2.6% of the total C emitted in the next 30 yr) worldwide (Table 5). If the rainfed agricultural land were converted to native grassland, there is a potential to sequester 1.2x10^8 Mg C (0.2% of the total C emitted in the next 30 yr) in the PNW, 3.2x10^8 Mg C (0.6% of the total C emitted in the next 30 yr) in the western U.S. and 3.4x10^9 Mg C (5.9% of the total C emitted in the next 30 yr) worldwide (Table 5).

However, irrigated agricultural land typically produces twice the crop yield of rainfed agricultural land (Bucks et al., 1990; Howell, 2000). If irrigated agriculture were expanded 10%, each hectare of new irrigated land could produce the same crop yield as 2 ha of rainfed land (Bucks et al., 1990; Tribe, 1994; Howell, 2000). Under this scenario, the conversion of irrigated land to native forest could potentially sequester 1.0x10^8 Mg C (0.2% of the total C emitted in the next 30 yr) in the PNW, 2.8x10^8 Mg C (0.5% of the total C emitted in the next 30 yr) in the western U.S. and 3.0x10^9 Mg C (5.26% of the total C emitted in the next 30 yr) worldwide (Table 5). If converted to native grassland in this 2:1 conversion scenario, there is a potential to sequester 2.4x10^8 Mg C (0.4% of the total C emitted in the next 30 yr) in the PNW, 6.4x10^8 Mg C (1.2% of the total C emitted in the next 30 yr) in the western U.S. and 6.8x10^9 Mg C (11.9% of the total C emitted in the next 30 yr) worldwide. If highly erosive rainfed lands were selected or if rainfed lands urgently needed for habitat restoration were targeted for such a conversion, significant additional erosion, water quality and habitat benefits could also result.

Since native desert or semi-desert has relatively little ecosystem C compared to forest, grassland or wetland ecosystems (Houghton et al., 1999; Amthor and Huston, 1998; Schlesinger, 1977), converting irrigated moldboard plow crop land back to desert or semi-desert could result in a soil C gain of 0.15 kg C m^{-2}. A sequestration of only 1.90 x 10^6 Mg C 30 yr^{-1} might be expected in the Pacific Northwestern U.S., 5.10 x 10^6 Mg C 30 yr^{-1} in the western U.S. and 5.46 x 10^7 Mg C 30 yr^{-1} worldwide. This modest C accumulation in soil would be tempered by the fact that the conversion would require substantial policy incentives and decades to implement. Substantially more C may be sequestered by selectively returning rainfed agricultural land derived from forest, grassland or wetlands back to native vegetation. Tropical and temperate forests typically contain from 10-12 kg C m^{-2}, grasslands contain from 18-20 kg C m^{-2}, and wetland ecosystems contain from 60-70 kg C m^{-2}, whereas arid and semi-arid lands contain 5-7 kg C m^{-2} (Houghton et al., 1999; Ross et al., 1999; Schimel et al., 2000). Since nearly a third of the yield and nearly half of the value of crops in the U.S. are produced on irrigated lands predominantly in arid or semi-arid climatic zones (Bucks et al., 1990; Tribe, 1994; Howell, 2000), a strong strategic rationale can be made for expanding irrigated agriculture in these areas for both crop production and C sequestration, if accompanied by selective return of rainfed agricultural land derived from forest, grassland or wetlands back to native vegetation.

Factors Affecting Interpretation

Grazing affects the quantity and chemical composition of soil organic matter and the distribution of C in the soil profile (Schuman et al., 1999; Frank et al., 1995; Dommar and Williams, 1990). Schuman et al. (1999), Frank et al. (1995) and Dommar and Williams (1990) found that grazing often increases the concentration of soil C. Ecosystems co-evolved with herbivores. The fact that C storage in range and grassland ecosystems may be unaffected, but usually is increased with light grazing, suggests that grazing is an important part of long-term sustainability of these ecosystems (Schuman et al., 1999).

We recognize that the values for potential C gain in our study are estimates. To obtain a more precise estimate of potential C sequestration from management conversions on a worldwide basis it would be necessary to investigate the potential C accumulated in soils in many different vegetation types. Use of these data from Idaho provide an indication of the potential for these kinds of management shifts on a larger scale. Our estimated values for C gain may actually be conservative due to improving land management methods and improving irrigation technology. The C trends that we monitored were the end result of management that predated new technology now available that would have prevented much of the erosion and loss of soil C on our monitored irrigation sites. Most irrigated cropping world-wide uses surface irrigation, with substantial runoff resulting in some transport offsite of C via erosion with sediment and dissolved C in the water.

Flood and furrow irrigation also transport nutrients, pesticides and enteric microorganisms offsite and ultimately to surface and ground water (Sojka and Entry, 2000; Sojka et al., 1998a; 1998b). Conversion of furrow irrigation to sprinkler irrigation reduces C transport off site via sediment and water because of dramatic reductions of offsite flow and leaching (Aase et al., 1998). The use of conservation tillage and improved sprinkler irrigation systems to reduce erosion, especially in combination with new technologies such as the use of polyacrylamide, has potential to further reduce C transport and degradation (Aase et al., 1998). Additional C that may be sequestered resulting from improved long-term inputs of technology needs to be determined to more accurately predict potential C gains by irrigated agriculture in the future. Our estimates made no attempt to adjust C budgets for loss of C due to erosion. Because great improvements in controlling irrigation-induced erosion have occurred in recent years, it is likely that our C storage estimates for irrigated agriculture are conservative.

Conclusions

As ecosystems mature, they accumulate soil C to a maximum carrying potential, which is controlled by climate, topography, soil type and vegetation (Van Cleve et al., 1993; Dewar,

1991; Harmon et al., 1990). Therefore, at equilibrium, the amount of C added to the soil via vegetation is equal to the amount of C lost through organic matter degradation and other losses (Henderson, 1995) and eventually a maximum limit will be reached. Although irrigated agricultural systems can not accrue C indefinitely, with improved management they can potentially remove substantial amounts of C from the atmosphere for the next 30-50 years. The potential C sequestered on site by conversion of native vegetation to irrigated agriculture is above the steady state equilibrium of native vegetation. This is in contrast to rainfed agricultural systems which are currently attempting modest C gains to attain near- baseline C concentrations by implementing reduced or no-tillage practices. Rainfed agricultural lands with or without no-till practices have soil C values far below those of native vegetation. A third of the yield and nearly half of the value of crops in the U.S. and worldwide are produced on the irrigated 15-17% of arable lands that are largely in arid or semi-arid climatic zones (Tribe, 1994; Bucks et al., 1990). Since the earth releases 1.9 x 10^9 Mg C yr^{-1} (Amthor and Huston, 1998; Schlesinger, 1995), the conversion of 10 % IMP to ICT resulting in a possible 2.5 x 10^9 Mg C sequestered over 30 years is a modest 4.47% compared to the total C released into the atmosphere over that time period. However, if crops were produced via high output irrigated agriculture, while selected less-productive rainfed agricultural land were returned to temperate forest or native grassland on a world-wide basis, there could be substantial reductions in atmospheric CO_2. Policy makers and agricultural research infrastructure should recognize the enormous potential benefit of land and water management strategies, polices and incentives that could expand arid zone irrigated agriculture as a means for efficient food and fiber production along with substantial C sequestration potential. This potential would be enhanced if coupled with selective return of less efficient rainfed agricultural lands derived from forest, grassland or wetlands back to native vegetation. We recognize that such an expansion would have to be accompanied by renewed efforts of water development. While water resource development has been occurring at a modest pace worldwide since 1990, Howell (2000) indicated the potential for increased extent of irrigation via efficiency improvements and waste water use. Recognition of these potential C benefits should provide an incentive to fund research and pursue management strategies that are possible without sacrificing production and which could increase restoration of native ecosystems, reduce erosion and improve water quality through appropriate targeting of the strategy.

References

Aase, J. K., D. L. Bjorneberg, and R.E. Sojka, 1998. Sprinkler irrigation runoff and erosion control with polyacrylamide - laboratory tests. *Soil Sci. Soc. Am. J.* **62**:1681-1687.

Amthor, J.S. and M.A. Huston 1998. Terrestrial ecosystem responses to global change: a research strategy, ORNL/TM-1998/27, Oak Ridge National Laboratory. 157 pp.

Bailey, R.G. 1998. Ecoregions of North America. U.S. Department of Agriculture. Forest Service, US Government Printing Office,Washington D.C.

Birch, H.F. and M.T. Friend. 1956. The organic matter and nitrogen status of east Africa soils. *J. Soil. Sci.* **7**:156-167.

Blake, G. R. and K.H. Hartage. 1982. Bulk density. In p. 363-375 A.L. Page, R.H. Miller, and D.R. Keeney. (eds). Methods of soil analysis. Part 2. Chemical and microbiological properties. *Am. Soc. Agron.* Madison, WI.

Bowman, R.A., M.F. Vigil, D.C. Nielsen, and R.L. Anderson. 1999. Soil organic matter changes in intensively cropped dryland systems. *Soil Sci. Soc. Am. J.* **63**:186-191.

Bruce, J.P., M. Frome, E. Haites, H. Janzen, R. Lal and K. Paustian. 1999. Carbon sequestration in soils. *J. Soil Water Conserv.* **59**:382-389.

Bucks, D.A., T.W. Sammis, and G.L. Dickey. 1990. Irrigation for arid areas. pp 449-548. In: Management of Farm Irrigation Systems. G.J. Hoffman, T.A. Howell and K.H. Solomon (eds).*Am. Soc. Ag. Eng.*, St. Joseph, MI.

Burke, J.C., C.M. Yonker, W.J. Parton. C.V. Cole, K. Flach, and D.S. Schimel. 1989. Texture, climate and cultivation effects on soils organic matter content in U.S. grassland soils. *Soil Sci. Soc. Am. J.* **53**:800-805.

Cambardella, C.A., and E.T. Elliott. 1992. Particulate soil organic matter changes across a grassland cultivation sequence. *Soil Sci. Soc. Am. J.* **56**:777-783.

Collett, R.A. 1982. Soil Survey of Ada County area. U.S. Department of Agriculture, Natural Resources Conservation Service., US Government Printing Office, Washington D.C. 182 pp.

Collins, H.P., R.L. Blevins, L.G. Bundy, D.R. Christenson, W.A. Dick, D.R. Huggins and E.A. Paul. 1999. Soil carbon dynamics in corn-based agroecosystems: *Soil Sci. Soc. Am. J.* **63**:584-591.

Dewar, R.C., 1991. Analytical model of carbon storage in the trees, soils, and wood products of managed forests. *Tree Physiol.* **8**: 239–258.

Dommar, J.F., and W.D. Williams. 1990. Effect of grazing and cultivation on some chemical properties of soils in the mixed prairie. *J. Range. Manage.* **43**:456-460.

Entry, J. A., and R. E. Sojka, 2000. The efficacy of polyacryalmide and related compounds to remove microorganisms and nutrients from animal wastewater. *J. Environ. Qual.***29**: 1905-1914.

Entry, J. A., and W.H. Emmingham. 1998. Influence of forest age on forms of carbon in Douglas-fir soils in the Oregon Coast Range. *Can. J. For. Res.* **28**: 390-395.

Follett, R.F. 2001. Soil management concepts and carbon sequestration in cropland soils. *Soil Tillage & Research.* **61**:77-92.

Frank, A.B., D.L. Tanaka, L. Hofmann and R.F. Follett. 1995. Soil carbon and nitrogen of Northern Great Plains grasslands as influenced by long term grazing. *J. Range. Manage.* **48**:470-474.

Harmon, M.E., W. Ferrell and J.F. Franklin. 1990. Effects on carbon storage of conversion of old growth forests to young growth forests. *Science* **247**: 699–702.

Henderson, G.S. 1995. Soil organic matter: a link between forest management and productivity. In p 419-436. W.F. McFee and J. M. Kelley (eds.) Carbon forms and functions in soils. Soil Science Society of America Press., Madison, WI.

Hontoria, C., J.C. Rodriguez-Murillo and A. Saa. 1999. Relationships between soil organic carbon and site characteristics in peninsular Spain. *Soil Sci. Soc. Am. J.* **63**:614-621.

Howell, T.A. 2000. Irrigations role in enhancing water use efficiency. In: p 67-80.R.G. Evans, B.L. Benham and T.P. Trooien (eds). National Irrigation Symposium. American Society of Engineers, St. Joseph, MI.

Houghton, R.A., J.L. Hackler and K.T. Lawrence. 1999. The U.S. carbon budget: contributions from land use change. *Science* **285**:574-578.

Janzen, H.H., C.A. Campbell, E.G. Gregorich and B.H. Ellert. 1997. Soil carbon dynamics in Canadian ecosystems. p 57-80. In: R. Lal, J. Kimble, E. Levine and B.A. Stewart (eds.) Soils and global change. CRC Press, Boca Raton, FL.

Jobbagy, E.G., and R.B. Jackson. 2000. The vertical distribution of organic carbon and its relation to climate and vegetation. *Ecol. Appl.* **10**: 423-436.

Johnson, D. W. 1992. Effects of forest management on soil carbon storage. *Water, Air, Soil Pollut.* **64**: 83-120.

Kimble, L., and B.A. Stewart. 1995. World soils as a source or sink for radiatively- active gasses. In: p 1-7 R. Lal, J. Kimble, E. Levine and B.A. Stewart R. (eds.) *Soils and global change.* CRC Press, Boca Raton, FL.

Kirk, R.E., 1982. Experimental Design: Procedures for the Behavioral Sciences. 2nd ed. Brooks/Cole Publishing, Monterey, CA, 911 pp.

Lal, R., R.F. Follett, J. Kimble and C.V. Cole. 1999. Managing U.S. cropland to sequester carbon in soil. J. Soil Water Conserv. **59**:374-381.

Liski, J., H. Iivesniemi, A. Makela and C.J. Westman. 1999. CO_2 emissions from soil in response to climatic warming are overestimated - the decomposition of old soil organic matter is tolerant of temperature. *Ambio* **28**:171-174.

Nelson, D.W., and L.E. Sommers. 1996. Total carbon, organic carbon and organic matter. *In* p 961-1010. Methods of soil analysis. Part 3, Chemical and microbiological properties. J.M. Bigham (ed.). ASA, CSSA, SAAJ, Madison, WI.

Paustian, K., O. Anderon, H. Janzen, R. Lal, P. Smith, G. Tian, H. Tiessen, M. van Noordwijk and P. Woomer. 1997. Agricultural soil as a C sink to offset CO_2 emissions. *Soil Use and Management* **13**:230-244.

Potter, K.N., H.A. Torbert, O.R. Jones, J.E. Matocha, J.E. Morrison. Jr., and P.W. Unger. 1998. Distribution and amount of soil organic C in long term management systems in Texas. *Soil & Tillage Research.* **47**: 309-321.

Rasmussen, P.E., and H.P. Collins. 1991. Long-term impacts of tillage, fertilizer and crop residue on soil organic matter in temperate semi-arid regions. *Adv Agron.* **45**:93-134.

Rasmussen, P.E., and W.J. Parton. 1994. Long-term effects of residue management in wheat-fallow: I. Inputs, yield, and soil organic matter. *Soil Sci. Soc. Am. J.* **58**: 523-530.

Ross, D.J., K.R Tate, N.A. Scott and C.W. Feltham. 1999. Land-use change: effects on soil carbon, nitrogen and phosphorus pools and fluxes in three adjacent ecosystems. *Soil Biology and Biochemistry* **31**: 803-813.

SAS Institute Inc. 1996. SAS User's Guide: Statistics - Version 6.03 Edition. Statistical Analysis System (SAS) Institute Inc., Cary, NC. 584 pp.

Schimel, D., J. Melillo, H. Tian, A.D. McGuire, D. Kicklighter, T. Kittel, N. Rosenbloom, S. Running, P. Thornton, D. Ojima, W. Parton, R. Kelly, M. Sykes, R. Neilson and B. Rizzzo. 2000. Contribution of increasing CO_2 and climate to carbon storage by ecosystems in the United States. *Science* **287**:2004-2006.

Schlesinger, W.M. 1977. Carbon balance in terrestrial detritus. *Ann. Rev. Ecol.* **8**:51-81.

Schlesinger, W.M. 1990. Evidence from chronosequence studies for a low carbon-storage potential of soils. *Nature* **348**:232-234.

Schlesinger, W.M. 1995. An overview of the C cycle. p 9-26. *In* R. Lal, J. Kimble, E. Levine and B.A. Stewart. (eds.) Soils and global change. CRC Press, Boca Raton, FL

Schlesinger, W.M. 1999. Carbon sequestration in soils. Science 284:2095.

Schuman, G.E., J.D. Reeder, J.T. Manley, R.J. Hart, and W.A. Manley. 1999. Impact of grazing management on the carbon and nitrogen balance of a mixed-grass rangeland. *Ecological Applications*. **9**:65-71.

Snedecor, W.G. and W.G. Cochran. 1980. *Statistical methods*. 7th ed. Iowa State University Press, Ames, Iowa. 354 p.

Sojka, R.E., R.D. Lentz, C.W. Ross, T.J. Trout, D.L. Bjorneberg, and J.K. Aase. 1998a. Polyacrylamide effects on infiltration in irrigated agriculture. *J. Soil. Water. Cons.* **54**:325-331.

Sojka, R.E., R.D. Lentz, and D.T. Westermann, 1998b. Water and erosion management with multiple applications of polyacrylamide in furrow irrigation. *Soil Sci. Soc. Am. J.* **62**:1672-1680.

Sojka, R. E. and J.A. Entry. 2000. Influence of polyacrylamide application to soil on movement of microorganisms in runoff water. *Environ. Pollut.* **108**:405-412.

Tribe, D. 1994. Feeding and Greening the World, the Role of Agricultural Research. CAB International. Wallingford, United Kingdom. 274pp.

Van Cleve, K., C.T. Dryness, G.M. Marion, and R. Erickson. 1993. Control of soil development on the Tanana River floodplain, interior, *Alaska.Can. J. For. Res.* **23**: 941-955.

Wang, Y., R. Amundson and S. Trumbore. 1999. The impact of land use change on C turnover in soils. *Gobal Biogeochemical Cycles*.**13**:47-57.

West, T.O. and G. Marland. 2001. A synthesis of carbon sequestration, carbon emissions, and net carbon flux in agriculture: comparing tillage practices in the United States. Agric. Ecosyst. Environ. **91**: 217-232.

In: Environmental Pollution: New Research
Editor: Rachel H. Plattenberg, pp. 141-184
ISBN: 1-60021-285-9
© 2007 Nova Science Publishers, Inc.

Chapter 5

Predicting the Uptake and Bioaccumulation of Organic Pollutants from Natural Sediments

Noemí R. Verrengia Guerrero*
Toxicology and Legal Chemistry, Dept. of Biological Chemistry, Faculty of Exact and Natural Sciences, University of Buenos Aires. 4° Piso, Pabellón II, Ciudad Universitaria Buenos Aires - Argentina. C1428EHA

Abstract

Sediments provide habitat for biota, substrate for aquatic plants and hunting ground for predators. They are sinks for nutrients but at the same time they can interact with the many organic and inorganic pollutants that are released into the aquatic systems. As a consequence, sediments are able to concentrate a wide range of chemicals at levels that largely exceed those present in the water column posing a risk to the whole aquatic ecosystem. However, the complex and variable composition of the natural particles makes it very difficult to predict the potential bioaccumulation and harmful effects to biota. It is widely accepted that the bioavailability of sediment-bound contaminants depends on three main factors: the composition and characteristics of the natural particles, the physical and chemical properties of the contaminants, and the behavior and physiological characteristics of the organisms. All these aspects will be reviewed in this chapter.

As concerns for sediment quality criteria increase, several approaches have been proposed in literature to understand and clarify the problem, such as the use of several types of formulated control sediments and also the use of artificial particles. For this chapter the usefulness of a negatively charged artificial particle (with defined surface properties) was investigated in the presence and absence of commercially available humic acids, as an experimental model for natural sediments. Acute bioassays (48 h) were conducted to determine the bioaccumulation of several organic contaminants by the freshwater fingernail clam *Sphaerium corneum* exposed in systems containing either water only, a sample of a natural sediment, or the artificial particles with and without humic acids. Radiolabelled molecules were used to analyze nmol L^{-1} concentrations, octanol-water partition coefficients (Kow) were applied to investigate the extent of uptake, particle-water partition coefficients

* E-mail address: noev@qb.fcen.uba.ar, Tel/Fax: 54 11 4576 3342

(Kd) were calculated to describe aqueous and particulate sources, and dissociation constants were applied to distinguish the contributions of ionized and undissociated forms of the molecules. The results are analyzed to identify the major factors involved in bioaccumulation studies and they show that the negatively charged particles coated with humic acids may be considered as a very good experimental model for the natural particles. The significance of this work is in applying novel features to a better understanding of the role of sediment in the uptake and accumulation of organic contaminants by benthic organisms.

1. Introduction

Aquatic systems comprise about 75% of the planet (Menzer, 1991). Basically they can be classified as marine and freshwater systems, and estuaries. They have long been suffering from direct and undirect discharges of contaminants from multiple anthropogenic activities. In particular, estuaries and coastal areas are the most severely affected (Elbaz-Poulichet et al., 1984; GESAMP, 1991; Menzer, 1991). Simultaneously, these areas present a high biological productivity since they are the preferred sites for the development and growth of larvae of fishes, mollusks and crustacean (Ketchum, 1983). In addition, a great number of these biological species have relatively short life cycles and frequent reproductive cycles that they can be easily disturbeded by changes in the environmental conditions (Ford, 1989).

As concerns for environmental pollution increase, efforts to reduce and to control the levels of discharges have been also implemented. However, these efforts are not always conducted with the same intensity all over the world. On the other hand, there are still a great number of non-point sources of contamination that reach the aquatic systems (Baudo and Muntau, 1990).

Historically, water quality receives almost all the attention. A great number of efforts were and still are dedicated to assess and to control the levels of both organic and inorganic contaminants present in natural waters. However, it soon became evident that reduction of pollutants in the aqueous phase was necessary but not enough to protect the aquatic communities properly. It is widely recognized that most contaminants concentrate in sediment particles at levels that greatly exceed those present in the water column.

When contaminants are bound to sediments they may remain undetected in monitoring programs that rely on analyses based solely on water samples (Neal et al., 1998). But even when these programs include the analysis of total concentrations in sediment samples they are not able to assess the bioavailability of sediment-bound contaminants (Samoiloff, 1989; Arjonilla et al., 1994). In fact, the complex and variable composition of the natural particles makes it very difficult to predict the potential bioaccumulation and harmful effects to biota. It is widely accepted that the bioavailability of sediment-bound contaminants depends on three main factors (Landrum and Robbins, 1990; Landrum and Faust, 1991):

- the physical and chemical properties of the contaminants
- the composition and characteristics of the natural particles
- the biological characteristics of the organisms, such as habitat, behavior, and physiology

Consequently, during the past decade, sediment quality criteria have received increasing attention from regulatory agencies.

This chapter will describe how and why the contaminants distribute in the sediment particles, the main characteristics and composition of the particles, and the resulting interactions between particles and contaminants. These issues will allow understanding the potential factors that may be involved in the bioavailability of chemicals. Much attention will be paid to hydrophobic organic contaminants. Different approaches to assess bioavailability will be also discussed. Among them, a novel experimental model, based on an artificial particle coated with humic acids, will be presented as analogue for natural sediments.

1.2. Distribution of Contaminants

When contaminants are released to the aquatic systems they can be distributed in any of the following compartments (Figure 1):

- the soluble aqueous phase
- colloids and micelles
- suspended particulate matter
- bottom sediments

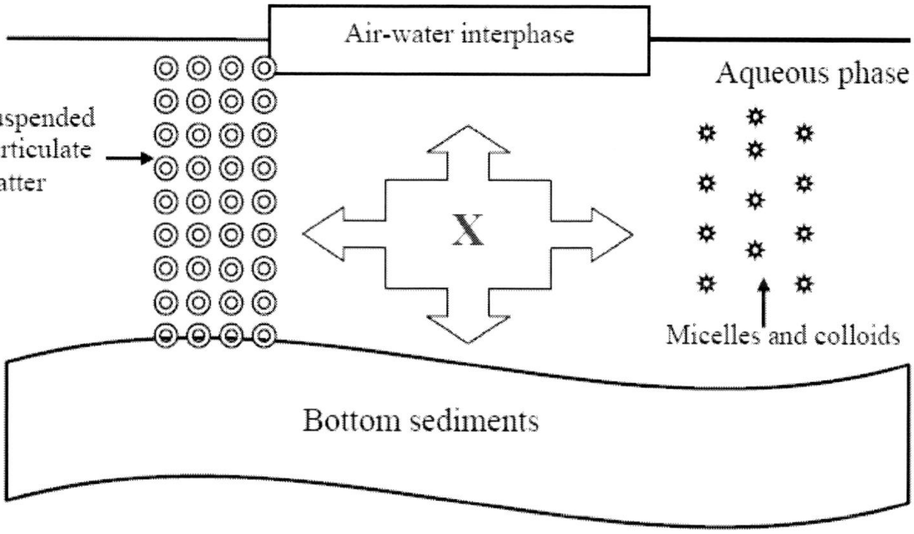

Figure 1. Distribution of a given contaminant X in the aquatic environment

The soluble aqueous phase constitutes a purely arbitrary fraction that it is solely defined by the fractionating procedure used. The most common procedure involves the filtration through a filter of 0.45 μm of pore size (APHA-AWWA-WPCF, 1980; Ashton y Chan, 1987; Turner and Millward, 2002). Therefore, the soluble aqueous phase is that that passes the filter, while the fraction remaining after filtration is called the particulate suspended matter.

Different substances may remain in the soluble aqueous phase, some of them are inorganic ions such as Cl^-, HCO_3^-, CO_3^{-2}, F^-, NO_3^-, NH_4^+, OH^-, SO_4^{2-}, SH^-, S^{2-}, CN^- (Brezonik et al., 1991; Florence et al., 1992). Chloride ions are responsible for the salinity of the water, in turn carbonate and bicarbonate ions are responsible for the alkalinity. Most of the inorganic compounds that contain atoms of carbon, hydrogen, oxygen, nitrogen and phosphor are essential nutrients for the primary production processes (Lerma et al., 1989). In addition, the aqueous phase may contain a certain proportion of dissolved organic matter from natural origin, mainly constituted by humic acids.

Most of the metal and hydrophobic organic contaminants that reach the aquatic systems have very low water solubility and consequently they become associated to the particles of both the suspended and bottom sediments (Baudo and Muntau, 1990).

Sediments can be defined as a heterogeneous mixture of particles of different composition. The particles may be transported in the aqueous phase as suspended particulate matter, or be settled on the bed of the aquatic systems. For these reasons some authors prefer the word *sediment* to refer to both kinds of particles, since they have basically the same composition, but differ from each other by their size (Horowitz, 1988; Förstner, 1990). However, some times it is necessary to limit the word sediment to the settled material, due to the different transformation and transport processes that may be involved in each material.

Colloids and micelles are formed by discrete particles whose sizes vary between nm and μm. Therefore, according to their size some proportion of the colloids will cross the filter, while another will be retained. Part of the colloidal material derives from a natural origin and it is mainly constituted by dead biological tissues from microorganisms, bacteria, protozoa, phytoplankton and zooplankton (Means and Wijaratne, 1982; Witkowski et al., 1987). Another part may be formed by compounds from anthropogenic origin like anionic surfactants such as detergents (Quiroga et al., 1989). The colloidal material presents a non-polar inner core, an external ionic phase, and an intermediate region of variable polarity (Valsaraj and Thibodeaux, 1989). Therefore, metal and also some polar organic compounds may interact with the polar groups of the external phase, while hydrophobic compounds may become associated to the non-polar groups of the inner core and the intermediate region (Means and Wijaratne, 1982; Valsaraj and Thibodeaux, 1989). It is worth noting that the chemical and biological properties and also the dynamic of colloids in the aquatic environment are closely related to the suspended particulate matter. In addition, particles of sediments together with colloids and micelles share a common role in the aquatic system that is their ability to interact with the different anthropogenic contaminants. For this reason, they are referred as geosorbents.

1.3. Sediment Components

Schematically, as shown in Figure 2, five principal components may be distinguished in sediment particles: the matrix, inorganic coatings, non-viable organic matter, biological components, and entrapped substances. In general, the inorganic or biological matrix is surrounded by a wide variety of inorganic and organic compounds that are distributed as coatings on the mineral matrix or as discrete phases (Leenheer, 1988; Turner and Millward, 2002).

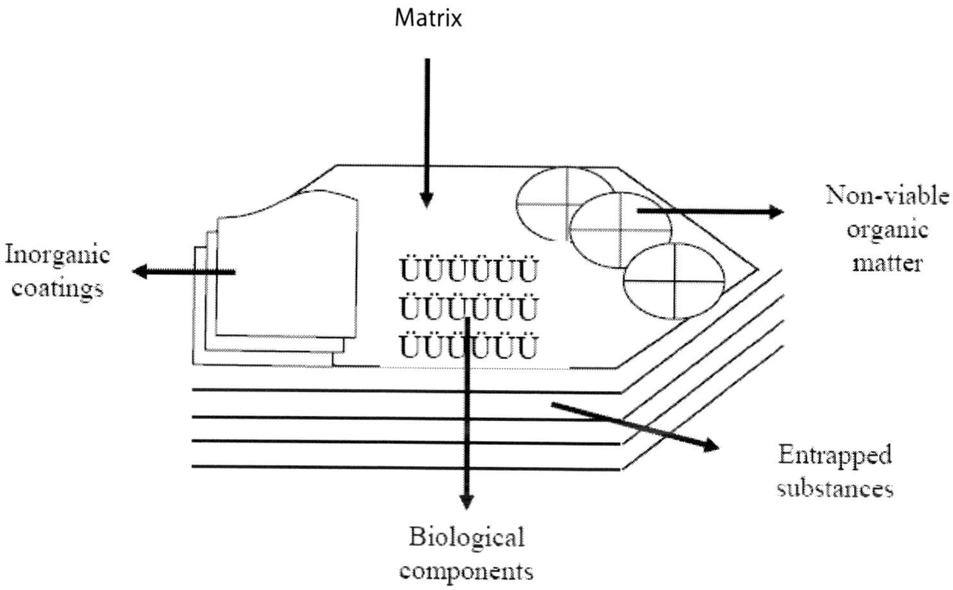

Figure 2. Principal components of natural sediment particles

Additionally, in bottom sediments, there is a water phase located between the particles usually termed as *pore* or *interstitial water* (Ingersoll, 1995; Rand et al., 1995). Typically it represents 50% by volume of surface sediment particles (Ingersoll, 1995).

For particles having a mineral matrix, the main components are quartz and other primary silicates, such as feldspar, and secondary silicates like clays (Witkowsky et al., 1987; Horowitz, 1988; Turner and Millward, 2002). The particles may be separated on a size basis by sieving or centrifugation methods. In this way sand (50-2000 μm), silt (2-<50 μm) and clay (<2 μm) fractions may be obtained (Ingersoll, 1995). However, this classification is partially correct. When sieving a natural sediment sample, other components may be also found in the fractions such as shell, seeds and detritus (Witkowsky et al., 1987). Other minor mineral components include materials in different states of mineralization. The calcareous exoskeletons of dead organisms may be present mixed within the principal mineral components (Witkowsky et al., 1987).

1.3.1. Inorganic Components

Oxides and hydroxides of iron, aluminum and manganese, carbonates and sulfides are the most frequent inorganic compounds present in the particles (Horowitz, 1988; Lai, 1988; Förstner, 1990; Turner and Millward, 2002). They can be distributed on the external surfaces, within the interlayers, and associated to the pores (Luthy et al. 1997). The metal compounds may act as weak bases conferring some positive charges to the sediment surfaces (Witkosky et al., 1987; Horowitz, 1988; Ingersoll, 1995).

1.3.2. Non-viable Organic Matter

The non-viable organic matter present in the natural particles can be classified in three main classes: a heterogeneous mixture of biopolymers; humic substances, and kerogen and black carbon (Figure 3).

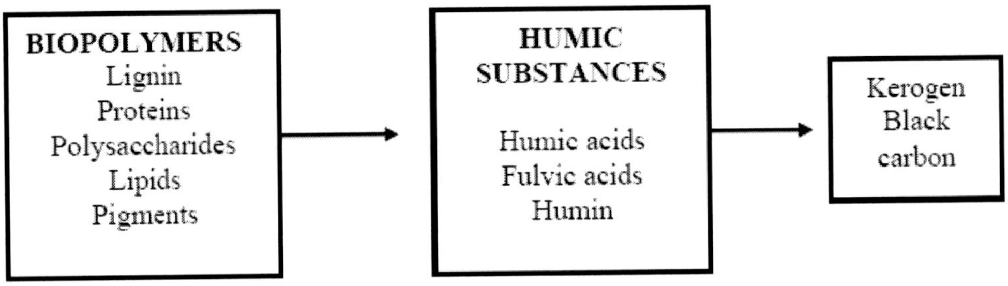

Figure 3. Principal classes of non-viable organic matter

The main constituents of biopolymers are lignin, proteins, polysaccharides, lipids and pigments (Rand et al., 1995; Weber et al., 2001, Huang et al., 2003; Ehlers and Loibner, 2006). It is considered that humic substances are derived from biopolymers by degradation of dead vegetable and animal tissues (Hayes, 1997, Rand et al., 1995). During diagenesis, biopolymers are degraded and cross-linked to form humic materials (Ehlers and Loibner, 2006).

Humic substances are by far the principal component of the total organic matter present in soils (reaching to 70-80%) and sediments (Hayes, 1997). Aiken et al. (1985) defined humic substances as *"a general category of naturally occurring heterogeneous organic substances that can be generally characterized as being yellow to black color, of high molecular weight, and refractory"*.

According to their solubility, three fractions may be distinguished: a) *fulvic acids*, soluble both in acid and alkaline solutions; b) *humic acids*, soluble in alkaline but insoluble in acid conditions, and c) *humin*, found only in soils and sediments since they are insoluble in both acid and alkaline conditions (Aiken et al., 1985; Rand et al., 1995; Weber et al., 2001). The chemical structures of humic acids are not well defined; they have a great number of carboxylic acids, phenol and alcoholic hydroxyls, and carbonyls as principal functional groups (Schulten et al., 1991). These groups confer a net negative charge to the particles (Wang et al, 1997; Turner and Millward, 2002).

The fraction of humic acids is relatively less polar and more condensed that the fulvic acids, presenting also a lower content of carboxylic and phenolic groups (Kenworthy and Hayes, 1997; Perminova et al., 1999). The chemical properties of humin are less known, due to experimental difficulties to isolate this fraction. Some authors have suggested that this fraction may be constituted by unchanged or transformed lignins and polysaccharides, humic acid-like material, and black carbon (Song et al., 2002, Weber et al., 2001; Ehlers and Loibner, 2006).

The organic matter content of sediments does not present a conservative pattern. On the contrary, it may be transformed by both biotic and abiotic processes. The principal abiotic processes are mediated by photolysis and hydrolysis reactions, and changes in salinity and pH (Leenheer, 1991). The biotic processes are mediated by the microorganisms of the biological component. In this way, humic materials can be further modified to render more refractory macromolecules, such as kerogen and black carbon (Leenheer, 1991; Luthy et al., 1997). Kerogen has a three dimensional structure where the aromatic moieties are disposed in sheets and cross-linked by aliphatic chain-like bridges (Ehlers and Loibner, 2006). Black carbon may be formed either by weathering of graphitic carbon in rocks, or by an incomplete combustion of fossil fuels and vegetation (Koelmans et al., 2006). Two types of black carbon can be distinguished. Soot is the product obtained from the combustion of liquid fuels and is characterized by highly aromatic globular structures in grape-like aggregates (Huang et al., 2003; Koelmans et al., 2006). Instead char is the product of an incomplete combustion containing less structured carbonized residues (Schmith et al., 2000; Koelmans et al., 2006).

The modifications in organic matter composition induce changes in polarity, aromatic carbon content, and composition of the functional groups as a result of changes in the relative amounts of H/O and O/C atoms (Luthy et al., 1997). It has been observed a decrease in the content of biomolecules for deeper particles of bottom sediments (Leenheer, 1991)

The natural organic components may also interact with the inorganic components of the matrix. Usually these interactions may occur through electrostatic or covalent binding, hydrogen binding, and van der Waals sorption between non-polar substrates. The final result is the displacement of water molecules from the mineral particle as a consequence of the adsorption of such large macromolecules with multifunctional groups. Organic coatings on sediments have been considered similar as a bilayer membrane model, where the polar groups, mostly solvated, are found on the external surface, and the non-polar compounds are internally located (Leenheer, 1991).

1.3.3. Biological Components

Among the biological components there are several species of microorganisms, like bacteria, fungi and protozoan, plankton. Death biological tissues, plant debris and fecal matter can be also included in this category while they have not undergone extensive transformation processes. Microorganisms play an important role in the biotic transformation of many contaminants, in the breakdown of the non-viable organic matter, and in their ability to catalyze redox reactions (Heip et al., 1995; Moffett and Ho, 1996; Turner and Millward, 2002).

1.3.4. Entrapped Substances

Expansive clays, like momtmorillonite, allow the entrapment of both natural and anthropogenic substances (Witkowsky et al., 1987). Most of the entrapped substances are from anthropogenic origin, like sewage solids, plastics, tar, solvents, surfactants, mine tailing, coal dust and fly ash (Turner and Millward, 2002).

1.4. Interactions with Contaminants

Both organic and inorganic contaminants may associate to the sediment particles through different ways. They can interact with the external surfaces of the mineral matrix; by swelling within the clay interlayer surfaces, and also with the internal surfaces or pores that present varying range diameters from <20 to 500 Å (Horowitz, 1988; Farrington, 1989; Luthy et al., 1997). Interactions with the mineral components of the particles will depend on their origin and geological history. Although a large quantity of the entrapped substances derives from an anthropogenic origin, so they can be considered as contaminants per se, they are reactive molecules with possibilities for additional interactions with other contaminants (Luthy et al., 1997).

The physical and chemical properties of the particles are determined by the mineral and chemical characteristics of the different components. According to these properties, a wide range of interactions may occur when chemical contaminants are released to the aquatic environment. Among them, the following processes are the most important (Lai, 1988; Rand et al., 1995):

- Precipitation, coprecipitation or dissolution
- Redox reactions
- Sorption, adsorption, and absorption processes
- Partitioning
- Desorption processes
- Ionic exchange
- Complexation

Some of these interactions need to be clearly defined. *Sorption* is perhaps the most general term since it is used to describe any type of association of a contaminant with a solid phase, without any reference to a specific physical or chemical process (Lyman, 1995; Pignatello and Xing, 1996). The term *adsorption* is usually limited to denote the interactions of the contaminants by physical mechanisms to a two-dimensional surface (Lyman, 1995; Ehlers and Loibner, 2006). Instead, *absorption* is used to define the penetration of contaminants into the three-dimensional matrix (Ehlers and Loibner, 2006). *Partitioning* denotes the distribution of a contaminant between the particles (solid phase) and the aqueous phase once the equilibrium state has been reached (Ehlers and Loibner, 2006).

Several factors are able to affect the sorption and desorption processes, such as the water solubility of the contaminants, their relative affinity for sediment organic matter and dissolved organic matter; the pH; redox potential; proportion of acid-volatile sulfides; grain size, and the mineralogy of the particles (Ingersoll, 1995). Those processes are generally controlled by the size of the particles, their superficial area, density, porosity and superficial charges (Lai, 1988).

The size of the particles is perhaps the principal factor in determining the transport of the sediment particles, adsorption-desorption processes, and the transference of chemicals between the water column and the particulate matter (Lai, 1988; Förstner, 1990). Particle size is also related to the superficial area that is higher when the particle size decreases. The superficial area reflects the number of actives sites available for adsorption of contaminants (Lai, 1988). The surface of the particles presents superficial charges due to the presence of the different compounds that coat the particles. Superficial charges govern the process of ionic exchange, ion adsorption and precipitation of sediments (Lai, 1988).

As a consequence of the interactions with the particles, an elevated concentration of contaminants, usually 100-1000 times higher, is found in the sediments in comparison with the levels present in the aqueous phase (Horowitz, 1988). For these reasons, the distribution of contaminants in the particulate matter and bed sediments constitutes a process of great concern for the proper assessment of water quality.

The process of distribution of sediment-bound contaminants depends not only on the properties of the particles and the chemical substance, but it is also influenced by the physical, chemical, geological and hydrological characteristics of the basin (Bordalo Costa and Peneda, 1989; Marcus, 1989; Pardo et al., 1990). Consequently, changes in environmental conditions, such as modifications in pH, salinity and redox potential, or in the case of discharges of effluents containing a high proportion of complexing agents, a redisolution of the chemicals initially bound to the particles may occur enhancing the proportion of contaminants in the aqueous phase (Horowitz, 1988; Bordalo Costa and Peneda, 1989; Pardo, et al., 1990; Eggleton and Thomas, 2004).

Therefore, particles of the suspended matter and sediments may play a double role in the aquatic system: a) binding chemical contaminants and acting as a reservoir, b) as a source of potential release of contaminants to the aqueous phase (Baudo and Muntau, 1990; Landrum and Robbins, 1990). In this way, two opposites behaviors may be expected, the challenge is how to evaluate and predict them.

1.4.1. Association to the Particles

As a first approach, it may be considered that the sorption processes occur rapidly reaching the chemical equilibrium at a relative short time (Jepsen and Lick, 1999). Under these conditions, the sorption phenomenon may be described by the following equation:

$$[X]s = Kp\,[X]w \quad (1)$$

where $[X]s$ is the concentration of the contaminant X in the sediment particles (in mg kg^{-1}), $[X]w$ is the concentration in the dissolved aqueous phase (in mg L^{-1}), both at equilibrium conditions, and Kp is the water-particle partition coefficient expressed in units of L kg^{-1}.

By plotting $[X]s$ versus $[X]w$ at a constant temperature the adsorption isotherm is obtained. According to equation (1), Kp is a constant equal to the slope of the isotherm. Linear adsorption isotherms are commonly found for hydrophobic organic chemicals having a low relative molecular mass and water solubility, present at very low concentrations, and with low affinity towards the mineral components of the matrix (Podoll and Mabey, 1987; Luthy et al., 1997). The linear model assumes that the sorption process depends basically on a partitioning process of the contaminants with the gel-like humic material components (Luthy et al., 1997). These interactions are not very strong and most of them are nonspecific (Podoll and Mabey, 1987).

However, given the heterogeneity in composition of the particles and the variations in the environmental conditions, it is not surprising to observe departures from this simple lineal model in field studies. There could be many reasons for this. Firstly, because the chemical equilibrium may not always reach even after a long time (Power et al., 1991; Jepsen and Lick, 1999). On the other hand, as a consequence of the different components present in the particles, differences in affinity and binding constants should be expected. For instances, it has been established that the preferential sites involved in the interactions of inorganic contaminants are, in decreasing order, the manganese oxides, humic substances, hydrous iron oxides, and finally clay minerals (Förstner, 1982; Horowitz, 1988).

In the case of hydrophobic organic contaminants, it is generally accepted that the organic matter content is the predominant sorption component present in the particles (Ingersoll, 1995; Eggleton and Thomas, 2004; Koelmans et al., 2006). This fact is particularly true when the organic material is in the range of >0.01 - 2% (Ehlers and Loibner, 2006). In this case, sorption to the mineral components is low. Therefore, to describe properly the sorption process the following equation has been proposed:

$$[X]oc = Koc\ [X]w \qquad (2)$$

where $[X]oc$ is the concentration of the contaminant X in the organic matter content of the particles (in mg kg^{-1}), $[X]w$ is the concentration in the dissolved aqueous phase (in mg L^{-1}), both at equilibrium conditions, and K*oc* is the normalized partition coefficient expressed in units of L kg^{-1} (Ingersoll, 1995; Park and Erstfeld, 1999).

The normalized K*oc* is related to the physicochemical properties of the contaminants, such as water solubility and the octanol-water partition coefficient (Kow), being practically independent of the sediment type (Mackay et al., 1991; Park and Erstfeld, 1999). However, K*oc* values are closely related to the characteristics of the organic matter content of the particles, especially in relation to the composition of humic substances and aging processes, which will be described in the next section (Ehlers and Loibner, 2006).

It may be considered that at low Kow values the transport of the organic non-polar compounds occurs principally by the aqueous solution, since they have little affinity for the particles. On the contrary, for hydrophobic non-polar chemicals with higher Kow values their fate is governed by the transport associated with the particles.

Nevertheless, for certain sediment particles the linear model cannot be applied and many reports of non-linear isotherms for describing the sorption processes have been published (Baker, 1991; Young and Weber, 1995; Pignatello and Xing, 1996; Jepsen and Lick, 1999). In some of these cases the Freundlich isotherm may be useful to describe the interactions:

$$[X]s = Kf\,[X]^{n}w \qquad (3)$$

where K*f* is the Freundlich constant.

Obviously, n = 1 for linear isotherms. Freundlich isotherms assume that the contaminants are sorbed into specific sites on the particles and are commonly found for polar or ionizable organic chemicals (Podoll and Mabey, 1987).

Other alternative it is the use of the Langmiur isotherm that considers that there are a finite number of functional groups available for the sorption process. In this case, the initial part of the isotherm is linear reaching a plateau at higher $[X]w$.

Some studies have shown that the non-linearity of the adsorption isotherms could be due to differences in the composition of the organic matter, in particular at the actives sites available for the interactions (LeBoeuf and Weber, 1999; Koelmans et al., 2006). Taking into account their structure, humic and fulvic acids may be considered as expanded three-dimensional polymers coating the particle surfaces (Pignatello, 1998). This fraction (often called soft, rubbery or amorphous fraction) resembles a gel-like rubbery portion, where partitioning processes are highly likely to occur (Huang et al., 2003; Koelmans et al., 2006). Instead, when the humic substances have been transformed, rendering kerogen, soot and char materials, they may be regarded as a harder, condensed, or less-flexible glassy fraction (Huang et al., 2003; Koelmans et al., 2006). Interactions with this second fraction, collectively termed black carbon, are principally governed by adsorption processes and may lead to non-linear isotherms (Ehlers and Loibner, 2006).

The black carbon fraction frequently represents 1-20% of the total organic carbon in aquatic sediments (Accardi-Dei and Geshwend, 2003; Cornelissen and Gustafsson, 2006). For certain hydrophobic organic contaminants, like polycyclic aromatic hydrocarbons (PAHs), polychlorinated byphenyls (PCBs) polychlorinated dibenzo-p-dioxins and -furans, sorption to black carbon may be 10-1000 times stronger than to the gel-rubbery portion (Jonker and Koelmans, 2001, 2002; Accardi-Dei and Geshwend, 2002; Cornelissen and Gustafsson, 2006). However, its sorption ability may become attenuated and decrease on a temporal basis, as a consequence of the previous presence of the same substances already sorbed. This fact shows that in this fraction there would be a limited number of sorption sites available for interactions (Cornelissen and Gustafsson, 2004, 2006).

In summary, there are many active sites available for interactions, each of them with different capacities and affinities for sorbates. Therefore, for natural particles the equilibrium constants should consider the total contribution of these different "domains", some of which will exhibit linear behaviors while others non-linear behaviors (Luthy et al., 1997; Huang et al., 2003; Koelmans et al., 2006). It is considered that for hydrophobic organic contaminants there are three principal types of domains (Weber et al., 2001):

- Domain I: consists in mineral sites, which exhibit almost a linear behavior for sorption.
- Domain II: consists in expanded, highly amorphous and swollen organic matter, exhibiting also linear behaviors.
- Domain III: formed by the relatively condensed organic matter, where interactions generally follow non-linear behaviors.

1.4.2. Desorption Processes

It is generally accepted that only a certain proportion of sediment-bound contaminants are able to desorb from the particles. Some authors used the term *hysterisis* to describe this phenomenon (Stapleton et al., 1994; Huang et al., 2003). It means that not all a sorbed solute can be released during desorption as predicted by the equilibrium partition (equation 1). This justifies the view of sediments as large reservoirs for anthropogenic chemicals. Initially this may be regarded as some kind of a natural remediation process. Moreover, when contaminants are bound to the particles they may undergo both chemical and microbial transformations that render often less toxic compounds.

It is a well known fact that many hydrophobic organic chemicals become progressively less susceptible to desorption while increasing the contact time to the particles (Alexander, 2000; Reid et al., 2000; Weber et al., 2001; Huang et al., 2003). This phenomenon is termed *aging* or *weathering*, and may occur by different causes. The contaminant molecules may move into certain sites of the particle matrix where they become inaccessible by microorganisms and even their extracellular enzymes. They may be entrapped by the organic matter component, or move to the pores located in the surface area of the particles. Some of these pores have a diameter of about 0.3-1.0 mm, quite similar to the size ranges of many hydrophobic organic contaminants. The chemicals are virtually sequestered by the particles and desorption is unlike due to the formation of bound residues. Similarly, dissolved metal contaminants associate firstly to relatively weak binding sites on the particle surfaces, where desorption may occur, but then they migrate to higher energy binding sites, where desorption is unlike (Millward and Liu, 2003).

Unfortunately, the dark side of that view is evident in a number of situations. Desorption processes may occur as a consequence of changes either in environmental conditions or sorbent properties (Ingersoll, 1995; Eggleton and Thomas, 2004; Ehlers and Loibner, 2006). In other cases, transformation reactions may lead to even more toxic compounds, such as the formation of methylmercury from inorganic mercury (Walker et al., 2001). However, in comparison there is considerable much more literature on the kinetics and mechanisms involved in sorption than in desorption processes.

There are many situations where either natural or anthropogenic environmental changes have shown to play a crucial role in desorption of chemicals from sediments. An increase in dissolved oxygen may induce a positive change in the redox potential, therefore anoxic sediments become oxidized, leading to an increase in microbial activity and a decrease in sediment pH. Consequently, a redissolution of metals such as Cd, Cu, Pb, Ni, Hg and Zn, may occur, principally from sulfide bound complexes, to the water column (Caille et al., 2003; Eggleton and Thomas, 2004). On the contrary, it has also been reported that oxygen depletion, induced for eutrophication, increases the desorption rates of mercury (Samoiloff, 1989). These apparently controversial results reflect that desorption processes are very difficult to predict in field conditions. In fact, the amount of both metal and hydrophobic organic substances desorbed from the particles will depend on many factors, such as the competitive interactions of different contaminants for adsorption sites; the variable binding strength, number and affinity of the active sites for sorption; the composition and concentration of dissolved ligands present in the aqueous phase (Yang et al., 2001; Eggleton and Thomas, 2004). Therefore, desorption processes for both metal and organic substances are highly conditioned by the characteristics of the sediment particles, the particular contaminant, and also by the particular aquatic system (Reeves et al., 2004).

Changes in sorbent properties can release the contaminants by transporting them directly to the water column or to more accessible sites. As it was explained in the previous section, there are different domains, exhibiting different affinities, to interact with contaminants. Therefore, multiple desorption rates are expected operating over different time scales, ranging from hours, days, through years (Weber et al., 2001). Nevertheless, as a first approach it is generally accepted that hydrophobic organic chemicals desorb from sediment particles in two phases (Reid et al., 2000; van Noort et al., 2003; Griffiths, 2004; Dunnivant et al., 2005). The first one is relatively fast, due to labile /reversible associations, and in general sorption of contaminants follows linear isotherms. It is usually termed as the *free* fraction and it would represent the fraction more readily available for uptake by biota. On the contrary, the second one is slow, due to non-labile/irreversible associations (bound-residues). In these cases, the sorption of contaminants is usually described by Langmuir isotherms. In agreement with this, Talley et al. (2002) have found that the proportion of PAHs associated to the clay/silt fraction of the particles is subjected to a significant transformation by both abiotic and biotic processes, relatively faster release rate and lower desorption activation energy, in comparison with the proportion that is associated to the organic matter and coal fractions.

However, it has been recently showed that the second phase could be comprised of two fractions: the slow and the very slow one (van Noort et al., 2003). The compounds associated to these binding sites would follow Langmuir isotherms, which will differ with respect to both sorption affinity and sorption capacity. By stereochemical factors, planar PAHs and PCBs would be more strongly sorbed to the condensed fraction of organic matter since desorption rates are lower than for non-planar congeners (van Noort et al., 2003).

In summary, desorption processes are crucial factors to understand and to predict the fate and risk of contaminants in aquatic systems. They are extremely relevant for water and sediment quality, remediation purposes, and also they are intimately related to the bioavailability of sediment-bound contaminants.

1.5. Bioavailability of Sediment-Bound Contaminants

Several definitions have been proposed in the literature to define and clarify the term bioavailability (Spacie et al., 1995; Alexander, 2000; Ehlers and Loibner, 2006). Probably the most intuitive is the one that describes bioavailability as *"the fraction of the total chemical in the surrounding environment which is available for uptake"* (Ruus et al., 2005).

In aquatic systems, the *surrounding environment* may be the soluble aqueous phase, the particulate suspended matter and bottom sediments. But aquatic organisms are exposed not only to all these possible routes of uptake, also to a variety of dietary sources. Consequently, the relevance of each route is highly conditioned by the habitat and biological characteristics of the living organism under consideration. Organisms can, therefore, be exposed to sediments simply because they lay on the surface of the substrates, because they are embedded in them, they feed on them, or even respire through them.

Benthic organisms that ingest sediment particles as food, or incorporate them accidentally while feeding, are particularly exposed to sediment-bound contaminants (Luoma and Ho, 1995). Those species that selectively ingest sediment fractions are likely to receive a larger dose than non-selective feeders (Kukkonen and Landrum, 1995). In addition, organisms including marine mammals, shore birds and even human consumers, may be exposed to contaminants that derive from sediments as they pass up the food chain (Lee, 1992; Luoma and Ho, 1995).

When an organism ingests sediment particles, a certain proportion of the contaminants that were initially bound to the particles may be released in the digestive tract and consequently be absorbed. Organisms may develop different strategies for dealing with these absorbed substances. Contaminants may be stored in certain tissues or subcellular components as a defense mechanism to avoid the onset of toxic effects. Chlorinated aromatic compounds, like the pesticide DDT and PCBs, are usually accumulated in lipid-rich tissues (Timbrell, 2000; Walker et al., 2001). Certain metals are efficiently removed from circulation by sulfide-rich proteins (metallothioneines) or by inorganic granules (Simkiss and Taylor, 1981, 1994; Rainbow, 2002; Masala et al., 2004). In these cases, contaminants become bioaccumulated and act as a source of contaminants for biomagnification in the food chain (Ingersoll, 1995). In other cases, the organism may be able to biotransform the chemicals promoting their excretion. However, the organism may also suffer some kind of toxic effect when the chemical binds to subcellular components and interferes with crucial physiological processes and metabolic pathways.

The extent of the toxicokinetics processes (absorption, distribution, biotransformation and excretion) and the characteristics of the toxicodynamics processes (action upon the target) are widely variable within the species of the biological communities. In addition, there are a large number of aspects of the normal physiology and metabolic pathways that still remain unknown in many aquatic invertebrates, even for those species that have been selected as standard organisms for sediment toxicity tests. Some aspects related to differences in the toxicokinetics processes will be further emphasized in the Discussion section.

Thus, the concept of bioavailability cannot be considered as a simple assessment of the interactions between contaminants and sediment particles. It also implies a link to biological aspects. Therefore, in a broad sense, the bioavailability of sediment-bound contaminants is not only related to the fraction actually available for uptake but includes the quantity that can be absorbed by the aquatic organisms and induce toxic effects.

When assessing bioavailability it is critically important to provide information about the relationship between accumulated body burdens and their subsequent impact on changes in physiological conditions (Power et al., 1991). In addition, changes at a given population level should be related to their negative impact on the community level and ecological significance (Chapman et al., 2002).

1.5.1. Assessing Bioavailability

Undoubtedly, one of the classical approaches for assessing bioavailability of sediment-bound contaminants is the Equilibrium Partitioning Model (Landrum and Robbins, 1990; Di Toro et al., 1991; Lee, 1992). Basically, this model assumes that the uptake and bioaccumulation of contaminants is a consequence of a thermodynamic equilibrium between the contaminants present in pore water and in the soft tissue of plants or animals (Figure 4). The concentration in pore water may be determined by the particle-water partition coefficients and in the case of hydrophobic organic chemicals, the normalized K*oc* is the best approach. Contaminant body burdens can be estimated by the bioaccumulation factor. In particular, for hydrophobic organic chemicals, the concentration of the chemical in the organism is normalized by the lipid content. Therefore, biota sediment bioaccumulation factors (BSAF) are defined by the following equation (Spacie et al., 1995):

$$(\text{BSAF}) = \frac{[\text{X}]\text{b} / f_{\text{lipid}}}{[\text{X}]\text{sed} / f_{\text{oc}}}$$

where [X]b = concentration of a given chemical X in the plant/organism expressed in mass units by wet weight; f_{lipid} is the concentration of lipids expressed as a fraction of mass (g lipid/g tissue); [X]sed is the concentration of X in sediment, expressed on a dry weight basis; f_{oc} total organic carbon in sediment expressed as a fraction mass (g organic carbon/g sediment).

According to this model, BSAF should be constant. Therefore by knowing the concentration of chemicals in the sediment organic carbon and the lipid content of a given organism, it should be possible to calculate the chemical body burden. Calculations of BSAF require relatively simple analytical determinations but great care must be taken to ensure the equilibrium conditions. Unfortunately, it is not always clear whether all natural aquatic systems are actually at equilibrium conditions (Landrum and Robbins, 1990). Thus even if they are, intermittent or unexpected discharges of effluents, and several environmental factors (e.g. changes in nutrients loads, amount of suspended matter, salinity, etc.) will induce departures from the steady state.

Experimentally, BSAF may be determined by field or laboratory studies (bioassays). In field studies, analyses of contaminants are performed using bioindicator organisms collected from polluted sites, assuming that the equilibrium conditions have been reached. On the other hand, these factors may be determined by exposing laboratory-reared standard organisms to samples collected from the environment (suspended particulate matter, bottom sediments, pore water, elutriates, etc.). Sediment bioassays will be described in the next section.

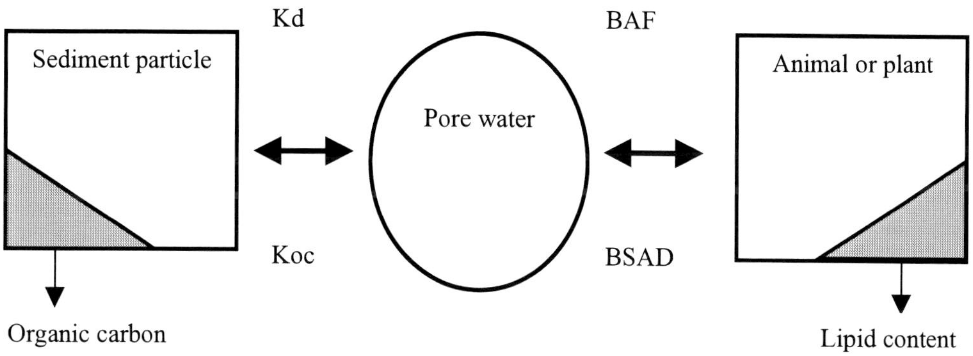

Figure 4. Schematic illustration of the Equilibrium Partition model

The Equilibrium Partitioning Model has been widely applied but it soon became clear that departures could be attributed to a variety of different causes. The most frequent are the heterogeneous composition of the organic matter present in the sediment particles, the particular lipid content structure of the organisms, and their biological and physiological characteristics, such as trophic level, habitat, and feeding behavior (Landrum and Robbins, 1990; Tracey and Hansen, 1996; Wong et al., 2001).

According to the model, pore water is the main route of uptake of chemicals. This is true for several benthic organisms. However, contributions from food and overlying water may be equal or even more significant routes (Landrum and Robbins, 1990). Some organisms, like bivalves, ingest also particles from both suspended matter and bottom sediments and they differ in their efficiency for digesting the particles (Peijnemburg and Jager, 2003). However, in fishes the aqueous phase is usually the predominant route of uptake. Therefore departures from the model may be found in every case where routes of uptake other than pore water are involved.

It is also implicit in the model that the uptake occurs by passive diffusion (Peijnemburg and Jager, 2003). Therefore if there is any process that involves active transport and/or regulation of contaminants it will induce deviations (Simkiss, 1996; Davies et al., 1997; Peijnemburg and Jager, 2003).

In favor of the model, it is worth noting that the lack of reliable methods for measuring freely dissolved concentrations in pore water has been an important disadvantage to a proper validation. Recently, however, Kraaij et al. (2003) have reported that the bioaccumulation of hydrophobic organic chemicals having a log Kow $\leq$ 7.5 in *Tubificidae* (oligochaete) can be described by the equilibrium partitioning model. The authors had previously developed a technique, using matrix solid phase microextraction, to accurately determine levels of contaminants in pore water (Mayer et al., 2000).

During the last years, different experimental approaches have been developed in order to assess properly the bioavailable fraction from sediment-bound contaminants. Biological methods, generally termed *biomimetic* methods, consist in using particular laboratory extraction techniques to investigate the extent of contaminant desorption from the sediment particles (Parkerton et al., 2000, Ehlers and Loibner, 2006). The extraction techniques are carefully selected so as to mimetize as much as possible the amount of contaminants that could be available for uptake by biota under field conditions. Scavenging resins like Tenax®

TA and XAD-4 are generally the most used (Lamoureux and Browanwell, 1999; De Lange et al., 2005; Leppanen and Kukkonen, 2006). Basically, water suspensions of sediment particles are mixed with the resin for a given time and then analyses of chemicals associated to the resin are performed. Some workers have reported that this approach may be particularly useful to estimate the bioavailability of hydrophobic organic chemicals with low water solubility (Reid et al., 2000). However, although some encouraging results have been obtained so far, more studies are necessary to validate the use of these resins.

A relatively recent and novel approach is based on isolating gut digestive fluids to perform some kind of *in vitro* extraction of sediment particles (Weston and Mayer, 1998a,b; Voparil et al., 2004). Digestive fluids from polychaetes (*Arenicola brasiliensis* and *A. marina*) are removed either with a pipette inserted in the organism's midgut or by dissection. They are centrifuged or filtrated to remove any particulate matter, mixed with the particles, and shaken for 4 h. It is considered that surfactant micelles present in the digestive fluid are responsible for the solubilization of the chemicals. The model is able to predict with great accuracy the bioavailability of highly hydrophobic organic chemicals when the Equilibrium Partitioning model often fails.

In addition, some workers have proposed a numerical model, principally based on chemical properties of the substances and the sediment organic carbon content to predict the levels of hydrophobic organic chemicals in aquatic organisms (Erstfeld and Park, 1997; Park and Erstfeld, 1997, 1999).

1.5.2. Sediment Bioassays

Historically, the first sediment toxicity tests were performed to evaluate the potential risk of disposal of dredge material into the oceans. Recommended bioassays were conducted for 10 days using clams, polychaetes, shrimps and fishes, selecting survival and bioaccumulation as endpoints (Ingersoll, 1995).

Since then, many species of organisms have been used for sediment toxicity tests as is shown in Table 1. Although the list does not pretend to be complete, it contains the most frequent cited species, including those recommended as standard bioindicator organisms.

Ideally, animals should meet several criteria to be selected as useful test organisms. The species should have broad geographical distribution, exist in direct contact with sediment particles, be easily to identify and maintain in the laboratory, and be ecologically relevant (ASTM, 1995). Such organisms should have demonstrated a relative sensitivity to the contaminants of interest, be tolerant to a wide range of sediment physicochemical characteristics, and be compatible with selected exposure methods and endpoints (ASTM, 1995). The organisms can be collected from the natural environment or be reared in the laboratory (Lee, 1992; Jager et al., 2005). It should be remembered than when animals are collected from highly contaminated aquatic systems, they could have developed tolerance and even resistance responses to some substances (Timbrell, 2000; Chapman et al., 2002; Rainbow, 2002). Alternatively, animals either originally reared in the laboratory or collected from a clean area can be transplanted to a natural contaminated site (Riget et al., 1997).

According to their habitat, benthic invertebrates are considered as ideal bioindicator organisms, especially to test whole sediment samples. Many species of chironomids and oligochaetes ingest sediment particles as food (Reynoldson and Day, 1995), whereas algae, bacteria, and fish from different trophic levels are often more appropriate to test elutriates, sediment extracts and pore water.

Table 1. Species of organisms that have been proposed for sediment toxicity tests

Organism	Species	Observations
Algae	*Selenastrum capricornutum* *Chlorella vulgaris*	Not widely used (1)
Bacteria	*Photobacterium phosphoreum* (Microtox assay) *Spirillum volutans* *Escherichia coli* *Salmonella typhimurium*	Used for sediment elutriates, extracts, and pore water (1,2,3)
Benthic invertebrates		
Amphipods	*Gammarus pulex* *Gammarus lacustris* *Diporeia sppi* *Hyalella azteca* *Rhepoxynius ebronius*	Widely used and recommended (1,4,5,6)
Annelids 1) Oligochaetes 2) Polichaetes	 *Lumbriculus variegatus* *Tubifex tubifex* *Neanthes arenaceodentata* *Neanthes sp* *Capitella capitata* *Nereis virens*	 Widely used and recommended (4,6) (7)
Cladoceran	*Daphnia magna* *Ceriodaphnia dubia.*	Widely used (1,4,6,8)
Insects	*Chironomus riparius* *Chironomus tentans* *Hexagenia limbata* *Hexagenia spp.*	Larvae stage are used, they ingest sediment particles. Widely used and recommended (1,4,6,9)
Mollusks (1) bivalves (2) gastropods	 *Mytilus edulis* *Crassostrea virginica* *Lymmacea acuminata*	 Field studies principally (4)
Nematodes	*Panegrellus redivirus*	Not widely used (1)
Fish	*Pimephales promelas*	No contact with sediments Used only for sediment elutriates, extracts, and pore water (1)

(1) Reynoldson and Day, 1995; (2) Luoma and Ho, 1995; (3) NRWI, 1991; (4) Ingersoll, 1995; (5) ASTM, 1992; (6) ASTM, 1995; (7) ASTM, 1994a; (8) Environment Canada, 1992; (9) OECD, 2000.

Both freshwater and marine amphipods are the species most frequently recommended and used for toxicity testing (Reynoldson and Day, 1995; ASTM, 1995; US EPA, 2000; Chapman et al., 2002). Among them, *Hyalella azteca* is considering the most promising and several standardized methods have been developed (Environment Canada, 1997; US EPA, 2000; ASTM, 2002). This species is widely distributed in nature, relatively easy to culture under laboratory conditions, has a short life cycle (2-3 months) and a wide tolerance to sediment grain size (Reynoldson and Day, 1995). However, recently some criticisms have arisen about its use. In sediment laboratory tests, the animals tend to burrow into the sediment particles and ingest them, just because they are deprived of their natural food (Wang et al., 2004). Consequently, it is assumed that the animals are sediment deposit feeders. But in its natural habitat, *H. azteca* responds mainly to contaminants in the overlaying water column and feeds mainly on algal communities, being actually an epibenthic grazer (Wang et al., 2004). Therefore, further research is needed about the possibility that animals selected for bioassays may display different feeding strategies in the field than when they are constrained to laboratory conditions. In fact, the search for standard organisms has not ended. Recently, the cladocerean *Chydorus sphaericus* has been suggested as an alternative to the commonly used daphnids. This species is actually a benthic organism which differs from daphnids in that it can be used in whole sediments as well as in pore water samples (Dekker et al., 2006).

When using some organism species it is necessary to feed the animals during the treatments. This is the case for instance with the midges *C. riparius*, *C. tentants*, the cladoceran *Daphnia magna*, and the oligochaetes *Tubifex tubifex* (ASTM, 1995). Otherwise, animals may be negatively affected not only by sediment toxicity but also by food deficiency (De Lange et al., 2005). However, feeding organisms may seriously affect the bioavailability of contaminants in a way that is not always predictable (Harkey et al., 1994; Taylor et al., 1998; De Haas et al., 2004). In nature, chironomid larvae are tube dwelling deposit feeders, feeding on detritus and organic matter present in the sediment particles (De Lange et al., 2005) whereas in standard bioassays, the animals selectively prefer the added food and avoid the ingestion of the particles (Åkerblom and Goedkoop, 2003). Even in nature, the availability of nutrients and trophic status has a direct influence on the distribution of contaminants between sediments, the water column and biota (Berglund et al., 2001).

For laboratory bioassays, sediment samples are usually collected from the natural environment. During this process, and also during handling and storage, the samples are subjected to changes due to alterations in sediment structure, disruption of chemical equilibrium, chemical speciation, and redox potential (Lee, 1992; Ingersoll, 1995; Luoma and Ho, 1995; Geffard et al., 2004). Sunlight should be avoided to prevent photolysis reactions of susceptible contaminants (Ingersoll, 1995). Also, for anoxic samples it is necessary to limit the contact with air to prevent oxidation processes (Ingersoll, 1995). It is highly advised to keep the collected samples in dark, at 4ºC, and be tested within short periods. Normalized protocols have been proposed in the literature for the best procedures (ASTM, 1994b; Luoma and Ho, 1995; US EPA, 2000; ASTM, 2002).

A preliminary characterization of the sediment samples should include at least the following parameters: pH, total organic and inorganic carbon, acid volatile sulfides, percentage of water, grain size, as well as percentages of sand, silt and clay materials (Ingersoll, 1995).

Sediment toxicity can be also evaluated by means of elutriates, sediment extracts, and pore or interstitial water. However, it is worth noting that for any of these liquid phases some benthic species are not suitable organisms for conducting toxicity tests. In the absence of particulate matter, the animals are unable to burrow and are likely to be much more stressed and mobile (Ingersoll, 1995; Davies et al., 1999a; Miño et al., 2006). Instead, organisms commonly used for water toxicity testing are preferred.

Elutriates are commonly obtained by vigorous shaking (usually for 15-30 min) of one part of a sediment sample with four parts of water. The mixture is allowed to settle and the liquid phase is isolated by centrifugation and eventually filtered (Burton and MacPherson, 1995). It is considered that this fraction may provide information and prediction about the potential release of contaminants from the sediment to the water column (ASTM, 1994b). In addition, different sediment and water ratios have been proposed, as well as times, temperatures and shaking techniques (see Arizzi Novelli et al., *in press*).

Extractions of sediment samples using different organic solvents have been conducted in order to assess the bioavailability of organic chemicals with low water solubility (Luoma and Ho, 1995; Reid et al., 2000). The sediment extracts have been tested principally with bacteria. However, several authors have found that the sediment extracts usually overestimated the bioavailability and toxicity as compared with intact sediments (Luoma and Ho, 1995; Eriksson Wiklund and Dag Broman, 2005).

Pore or interstitial water from sediments can be obtained by several methods including centrifugation, suction, squeezing, and dialysis (Ingersoll, 1995).

Sediment tests are usually performed for 7, 10, 14, 21 and 28 days of exposure, although longer periods may be used for chronic tests or those that use marine species (US EPA, 2000; ASTM, 2002).

The most common endpoints for contaminant exposure are survival, bioaccumulation, growth and/or reproduction (Leppanen and Kukkonen, 1998). However, other endpoints have also been selected such as burrowing behavior, larval settlement, emergence, sexual maturation, fecundity, embryo aberrations, and avoidance reactions (ASTM, 1995; Luoma and Ho, 1995; Eriksson Wiklund and Dag Broman, 2005). In addition subcellular biomarkers have also being recommended (Huggett et al., 1992; Peakall, 1992; Martín-Díaz et al., 2004).

The extent of these responses depends basically on the toxicokinetic and toxicodynamic processes that occur once the chemicals have been incorporated into the organism. These factors are especially relevant when bioaccumulation is the endpoint of the bioassay. Most bioaccumulation studies include the investigation of the rate of uptake and elimination of chemicals by the animals (Kukkonen and Landrum, 1994; Basack et al., 1997; Belfroid and Sijm, 1998; Fisk et al., 1998). However the effect of the possible biotransformation of contaminants has received less attention and many of the metabolic processes that might be involved in detoxification are still not well known.

In environmental studies most bioaccumulation tests are based on assessing the contaminants present in the natural ecosystem, but for laboratory studies it is a common practice to use ^{14}C or ^{3}H radiolabelled compounds. These materials allow the use of low levels of contaminants during the exposure so as to reproduce environmental conditions and the subsequent analyses are then performed by liquid scintillation counting. This method is suitable for substances that are not readily metabolized, but it may lead to an overestimation of the bioaccumulation burden if metabolic transformations occur (Spacie and Hamelink, 1995). In addition, when assessing biological responses, correlations between dose-effects of chemicals may be difficult to establish.

It has been reported that amphipods and polychaetes readily metabolized PAHs, while bivalves have a limited ability (Lee, 1992). Evidence of significant differences in the biotransformation of several organic compounds has been found among three freshwater species namely the larvae of the midge *Chironomus riparius*, the oligochaete *Lumbriculus variegatus* and the fingernail clam *Sphaerium corneum* (Borchert et al., 1997; Verrengia Guerrero et al., 2002).

Finally, after standard bioaccumulation tests with sediment, depuration treatments are normally required to avoid an overestimation of the chemical body burden. Although a 24 h depuration period is usually recommended and adopted in most cases (Chapman, 1985; Ankley et al., 1991; Phipps et al., 1993), some studies have suggested that in several bivalve species a period of up to 3-4 days may be required for the complete assimilation and digestion of ingested particulate material (Reinfelder et al., 1997). Similarly, in the cladocerean *Daphnia magna* exposed to metal-contaminated sediment, there was no significant decrease in gut contents after 48 of purging in clean water (Gillis et al., 2005). On the other hand, if depuration is too long, bioaccumulation of chemicals may be underestimated due to elimination processes of the organisms (Kukkonen and Landrum, 1994; Sheedy et al., 1998; Mount et al, 1999; Dawson et al., 2003).

1.6. The Approach

As it was described previously, there are different approaches to estimate the bioavailability of sediment-bound contaminants. However, the use of toxicity tests selecting bioaccumulation as the endpoint is perhaps one of the most direct and powerful tools.

For this chapter two sets of experimental results are presented. Firstly, acute bioassays were performed to investigate the uptake and accumulation of a diverse selection of organic contaminants by a bivalve mollusk. The freshwater fingernail clam *Sphaerium corneum* was selected as bioindicator organism. The clams were exposed to the contaminants either in aqueous solution or in solutions containing a sample of a natural sediment. The sample had been collected from a relatively "non-contaminated" site. Results were compared to verify if the uptake was actually higher when the animals were exposed to the chemicals in the presence of this sediment. Radiolabelled molecules were used to analyze nmol L^{-1} concentrations, so as to mimic possible environmental concentration levels. Octanol-water partition coefficients (Kow), obtained from the literature, were applied to investigate the extent of uptake of the different compounds. The water-particle partition coefficients (Kd) were calculated in order to establish the possible correlationship between them and the resulting uptake.

Then, an artificial particle was tested as an experimental model for the natural sediment. The use of control sediments has become an important tool to assess and predict the bioavailability of particle-bound contaminants under laboratory conditions (Ingersoll, 1995). Several types of formulated sediments have been proposed, consisting of different kinds of sand, silt and clay (Suedel and Rogers, 1994; Naylor and Rodrigues, 1995; Kemble et al., 1999). In addition to these natural materials, the use of artificial particles as analogues for aquatic sediments has also been suggested (Simkiss, 1995; Fleming et al., 1998). In particular, commercial materials commonly used as chromatographical resins have been selected because their size, backbone structure, and active groups on their surfaces are well characterized and standardized by the manufacturers (Davies et al., 1999 a, b; Simkiss et al., 2000, 2001). However, this model may be considered too simple, since it may not reflect fully the complexity of natural sediments. Thus, artificial particles will clearly not show the full range of ligands that are present in natural organic matter (Fleming et al., 1998; Kemble et al.,

1999). For these reasons a variety of formulated sediments have been proposed containing organic matter such as composted cow manure, peat moss, maple and cereal leaves, α-cellulose, and humic acids (US EPA, 1994; Kemble et al., 1999). Among these materials, humic acids constitute the most important and natural source of organic matter in sediments from aquatic systems (Rand et al., 1995) and they have the added advantage that they can be obtained in a consistent form as a commercial product.

Therefore, the selection of the artificial particles was based on the following two main aspects. Firstly, that the natural particles have a net negative charge. Therefore a cationic exchange resin was selected. Secondly, that the non-viable organic matter plays an important role in determining the interactions between particles and organic chemicals. Therefore the particles were pretreated with a solution of commercially available humic acids. This pretreatment was performed to render a similar organic matter content than the value measured for the natural sediment. Bioassays were finally performed to determine the resulting uptake in clams exposed to the artificial particles with and without humic acids, and to compare these values with those obtained in animals that had been exposed to the natural sediment.

2. Materials and Methods

2.1. Materials

The organic chemicals used for this work represent a diverse selection of molecular structures and environmental contaminants that may be present in aquatic systems. Chlorophenols, in particular 2,4-dichlorophenol (2,4-DCP), 2,4,5-trichlorophenol (2,4,5-TCP), pentachlorophenol (PCP), and polycyclic aromatic hydrocarbons, like pyrene, are ubiquitous and well known contaminants (Walker et al., 2001). Cyproconazole, [2-(4-chlorophenyl) -3-cyclopropyil – 1(1*H*- 1,2,4-triazol- 1-yl) butan-2-ol], is a systemic fungicide, effective against a broad range of plant diseases and acts as a steroid demethylation inhibitor (Tomlin, 1997). Trifluralin, [trifluoro-2,6- dinitro-N,N-dipropyl –*p* - toluidine], is a non-ionic herbicide, belonging to the family of 2,6-dinitroanilines (Edwards and McMinn, 1985) and acting as a cell division inhibitor (Tomlin, 1997). It is used in the pre-emergence control of many annual grasses (Tomlin, 1997). Cyfluthrin, [α-cyano-4-fluoro-3-phenoxybenzyl – 3(2,2-dichlorovinyl) – 2,2-dimethylcyclopropanecarboxylate], is a non-systemic insecticide belonging to the pyrethroid family (Tomlin, 1997). Their chemical structures are presented in Figure 5.

All the contaminants used were [^{14}C] radiolabelled. Data on specific activity, purity, source, and the values of water solubility, pKa and octanol-water partition coefficients (log Kow) are tabulated in Table 2.

The sample of natural sediment was collected from a non-contaminated site at the ARC Study Centre, Milton Keynes, United Kingdom by representatives of the Water Research Centre, Medmenham, United Kingdom. After collection the sample was sieved through a 1 mm filter to remove large objects and any endogenous organisms. The sediment was homogenized in a cement mixer before being stored at 4 °C. The organic matter content was 1.7% with 70% of the particles in the size range <63 μm. Dry weight was 50% of the wet weight.

Toyopearl SP 650M, a cationic exchanger, was selected as artificial particle and obtained from Fisher, Loughborough, United Kingdom. The backbone of the particles consists of polymers of ethylene glycol and methyl methacrylate, containing n-propylsulfonyl groups attached through oxygen atoms as functional groups. The particle size is 40 μm.

Commercially available humic acids (HA) were obtained from Fluka Chemie AG, Switzerland, (product number 53680). The composition reported by the manufacturer was 46.63% carbon, 4.30% hydrogen and 0.72% nitrogen. A solution containing 20 mg HA L-1 was prepared by dissolving the humic material in dechlorinated tap water that had been previously filtered through activated charcoal to remove any possible endogenous organic matter. All solutions were filtered through a 0.45 μm membrane filter to remove any particulates prior to use (Verrengia Guerrero et al., 2000). The final concentration of the solutions was checked by measuring their absorbance at $\lambda = 430$ nm (Benés et al., 1976). It is considered that a level of approximately 6 mg L-1 is the average concentration of humic acids dissolved in natural freshwaters (Benson and Long, 1991) so the selected concentration of 20 mg HA L-1 was rather high, but was more closely related to the organic matter content usually present in natural sediments (Ingersoll, 1995).

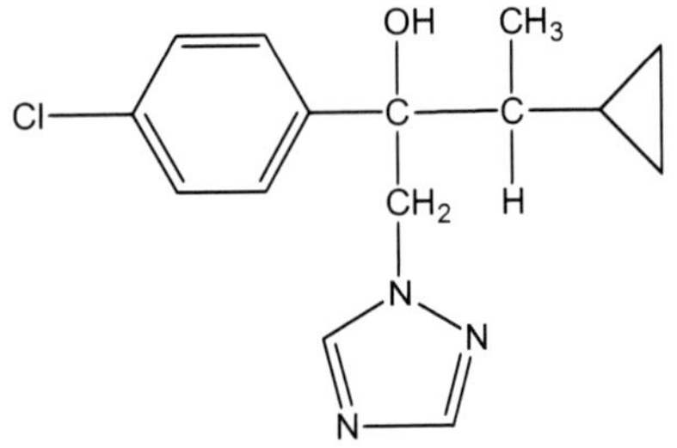

Cyproconazole

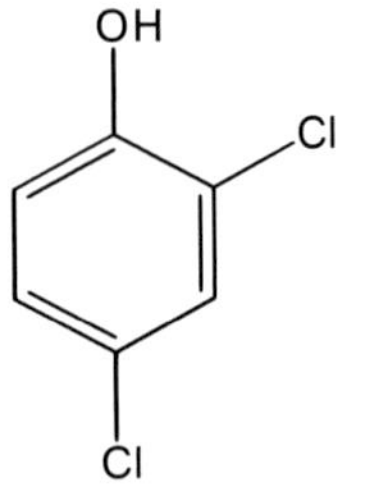

2,4-dichlorophenol

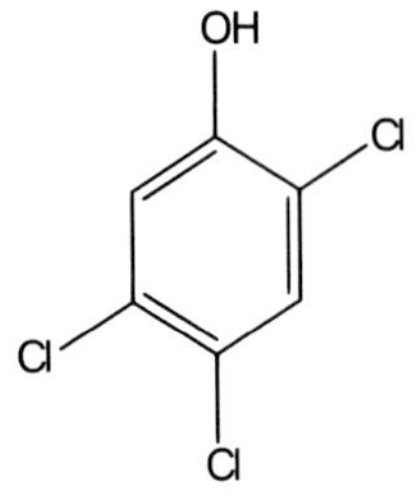

2,4,5-trichlorophenol

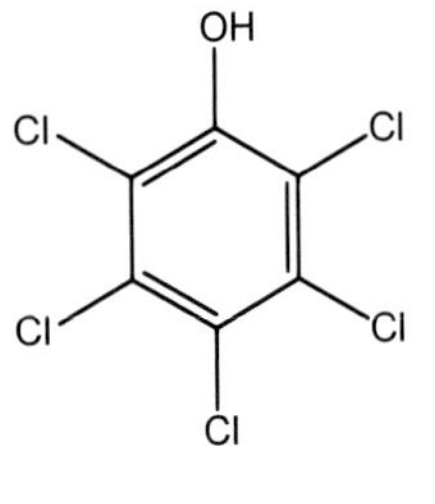

Pentachlorophenol

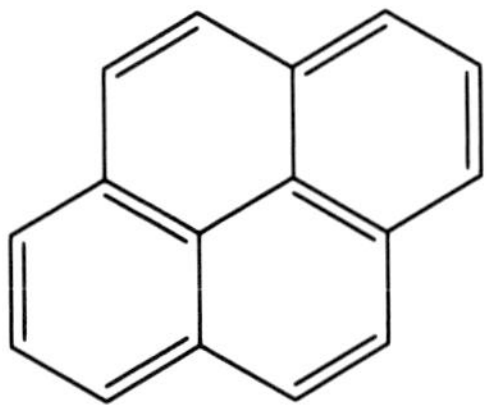

Pyrene

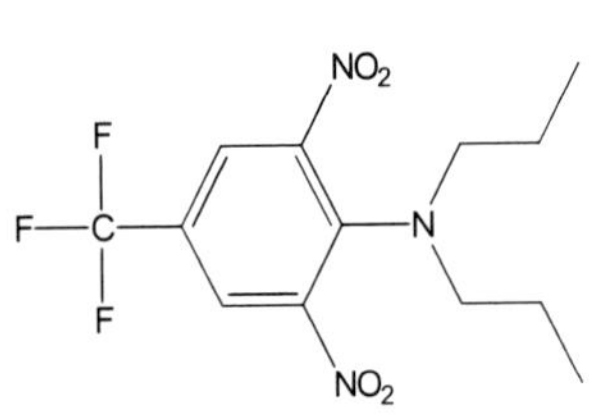

Trifluralin

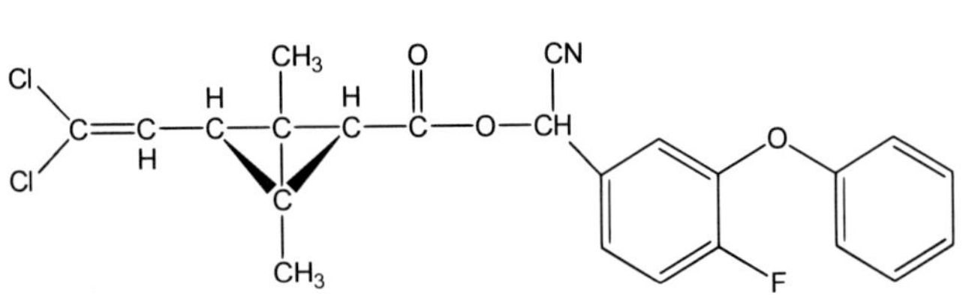

Cyfluthrin

Figure 5. Chemical structures of the selected compounds

Table 2. Some characteristics and properties of the selected compounds

Substance	Specific activity Bq x mmol	Purity %	Source	Water solubility	pKa	Log Kow
Cyproconazole	7.3 x 10^8	97.9	Novartis International Inc	140 mg L^{-1} (a)	- -	2.91 (a)
2,4-dichlorophenol	5.77 x 10^8	>98	Sigma-Aldrich Company	4.4 g L^{-1} (b)	7.85 (c)	3.23 (c)
2,4,5-trichlorophenol	2.70 x 10^8	97-98	Sigma-Aldrich Company	1.2 g L^{-1} (b)	6.94 (c)	4.19 (c)
Pentachlorophenol	3.85 x 10^8	98-99	Sigma-Aldrich Company	11 mg L^{-1} (b)	4.75 (c)	5.24 (c)
Trifluralin	25.9 x 10^8	99	Dow Agro Science (gift)	0.22 mg L^{-1} (a)	- -	4.83 (a)
Pyrene	21.72 x 10^8	>99	Sigma-Aldrich Company	Insoluble (d)	- -	5.13 (d)
Cyfluthrin	4.83 x 10^8	>99	Bayer (gift)	2.3 μg L^{-1} (a)	- -	5.95 (a)

(a) Tomlin, 1997; (b) Steinberg et al., 1992; (c) Escher and Schazenbach, 1996; (d) Xia and Ball, 1999.

2.2. Organism Selected

Freshly collected clams, *Sphaerium corneum*, were obtained from Sciento, Manchester, United Kingdom. Once in the laboratory, they were maintained in aerated plastic containers (18.0 x 13.5 x 6.0 cm), containing approximately 700 ml of dechlorinated tap water (pH = 8.3). The clams were placed in an incubator at 8.5 ± 0.5 °C, and fed with a commercial product consisting of a suspension of dead algae (Liquifry N° 1, Interpet, United Kingdom) every 48 h. The water medium was changed once a week. Bivalves were allowed to acclimate to the laboratory conditions for at least 7 days before conducting the bioassays (FAO, 1987). Only healthy animals of similar size (0.5 ± 0.1 cm shell length) were used in all experiments. The ability of clams to tightly close their valves in response to an external stimulus was selected as the criterion for identifying healthy organisms (FAO, 1987; Verrengia Guerrero, 1995).

2.3. Bioassays

All the bioassays were performed in triplicate for 48 h, using unfed animals at 8.5 ± 0.5 °C, in static conditions (without renewal of the water medium). Six animals were placed in 250 ml borosilicate glass bakers, containing 100 ml of each contaminant solution dissolved either in dechlorinated tap water, in the presence of the sample of natural sediment, in the 20 mg L^{-1} humic acid solution, or in the presence of the Toyopearl SP resin with and without the humic acid solution. Animals were exposed separately to the different test systems that contained 50 nmol L^{-1} of each compound. The experiments were performed in parallel allowing direct comparisons of the results.

The test system containing the sample of natural sediment consisted of 100 mg of dry material suspended in 100 ml of dechlorinated tap water plus 50 nmol L-1 of each contaminant, added in this order. The system was equilibrated by stirring for 48 h prior to the experiments (Simkiss et al., 2000; Verrengia Guerrero et al., 2003). In the case of the Toyopearl SP resin a similar procedure was followed but the system was equilibrated for 4 h. For the systems containing Toyopearl SP resin and humic material, the artificial particles were suspended in 100 ml of the 20 mg L-1 humic acid solution and equilibrated by stirring for 20 h. Then, an aliquot of each contaminant stock solution was added to obtain a final concentration of 50 nmol L-1 and the systems were allowed to equilibrate for another 4 h.

During the bioassays, the clams were placed in beakers on a stainless steel mesh with continuous gentle stirring to ensure a homogeneous suspension of the particles during the 48 h exposure period. After the treatments, the animals were allowed to depurate for 24 h to purge their gut contents. During this process the clams were transferred to clean 250 ml glass beakers, containing 100 ml of dechlorinated tap water and a drop of Liquifry N° 1 as food. Approximately 2.5 hr before the end of that period, the water medium was renewed to avoid reabsorption of any depurated chemicals.

Organisms that had not been exposed to the radiolabelled substances were employed as blank controls. There was no mortality during the experiments.

2.4. Contaminant Analyses

Animals were immobilized by cooling them on ice for 4-6 min and then sacrificed by opening the valves. The whole body soft tissue was carefully removed, weighed, and placed in a glass vial (20 ml). The tissues were solubilized by adding 1 ml of Soluene 350 (Packard, Pangbourne, Berkshire, United Kingdom) and heated in an oven at 50 °C overnight. After the digestion procedure, 10 ml of Hionic Fluor (Packard, Pangbourne, Berkshire, United Kingdom) were added to each vial. The concentration of each contaminant was determined from the specific activity of each stock solution after counting the samples in a Packard 2250 CA Tricarb Liquid Scintillation Analyser (LSA) (Packard Instruments, Meridan, CT, USA) and correcting for chemical and color quenching. Accumulation data were expressed as pmol g^{-1} of wet tissue.

2.5. Particle-Water Partition Coefficients (Kd)

Batch adsorption experiments were used to determine the particle-water partition coefficients (Kd), and were expressed as mol kg^{-1} solid/mol kg^{-1} solution. Eight different concentrations of each radiolabelled compound were prepared, using dechlorinated tap water from the stock solution. Experiments were performed in duplicate at a solid concentration of 1 g L^{-1} in conical flasks that were stirred continually. The samples of natural sediment and Toyopearl SP resin, with or without humic acids, were shaken over the same periods used for the bioassays prior to analyses. Previous experiments had shown that those periods were well in excess of the equilibrium times (Simkiss et al., 2000; Verrengia Guerrero et al., 2003).

The quantity of each contaminant adsorbed onto each kind of particle was determined as the difference between the initial and the equilibrium concentration. The equilibrium concentrations were determined in triplicate samples of 3 ml that had been centrifuged at 3000 rpm x 15 min. Then, aliquots of 0.1 ml of the supernatant were treated with 5.0 ml of Insta Gel (Packard, Pangbourne, Berkshire, United Kingdom) and measured by LSA.

2.6. Statistical Analyses

Regression analyses were carried out using Excel® software package (Microsoft, USA).

The data were evaluated statistically applying the test of Student (Sokal and Rohlf, 1997) to study the influence of the natural sediment on the bioaccumulation. Results were considered to be significant at $\alpha = 0.05$.

3. Results

3.1. Uptake of Contaminants from Water and from the WR Natural Sediment

Table 3 shows the concentrations of chemicals accumulated by *S. corneum* exposed to contaminated water in the absence or presence of a natural sediment. Ratios between the bioaccumulation from the systems containing the sample of natural sediment and from water only are also included.

No significant differences between the two treatments were observed for Cyproconazole and 2,4-DCP ($p > 0.05$). However, the presence of the natural sediment significantly increased the uptake of 2,4,5-TCP, PCP, Trifluralin, pyrene and Cyfluthrin ($p < 0.05$). As the particle size was below 63 μm in 70% of the sediment sample, a great proportion of the particles could be ingested by *S. corneum*. The results showed that when the animals were exposed to the more hydrophobic contaminants (log Kow > 3.3) they absorbed more of these chemicals presumably because of an increased uptake and accumulation from ingested particles.

Table 3: Levels of organic compounds bioaccumulated by *S. corneum* exposed to systems containing either water only or the sample of natural sediment (WR sed) (mean ± SD)

Substance	Accumulation from water only (pmol g^{-1})	Accumulation from WR sed (pmol g^{-1})	Ratio accumulation from WR sed vs accumulation from water
Cyproconazole	87.2 ± 44.6	111.0 ± 43.9	1.27
2,4-dichlorophenol	342.8 ± 98.1	399.9 ± 59.7	1.17
2,4,5-trichlorophenol	92.9 ± 47.6	251.7 ± 96.9	2.71 *
Trifluralin	1589.9 ± 510.6	3534.7 ± 813.0	2.22 *
Pyrene	1169.0 ± 514.8	5429.6 ± 1460.9	4.64 *
Pentachlorophenol	297.4 ± 130.6	968.4 ± 440.7	3.26 *
Cyfluthrin	93.8 ± 48.3	514.2 ± 160.2	5.48 *

* values significantly different between them (p>0.05

3.2. Relationships between Accumulation and the Octanol-Water Partition Coefficients (Kow)

Relationships between chemical bioaccumulation from either the aqueous phase or the sample of natural sediment and Kow on a log-log basis are presented in Figure 6 A and B respectively.

A

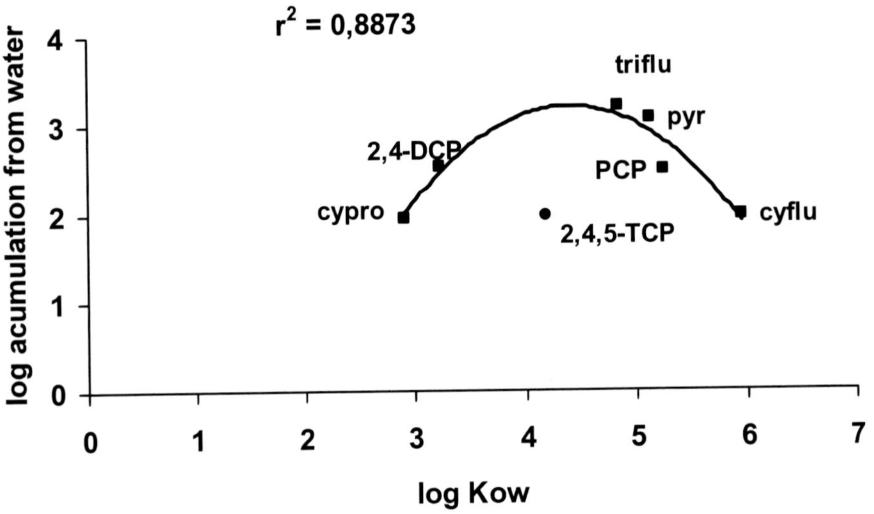

B

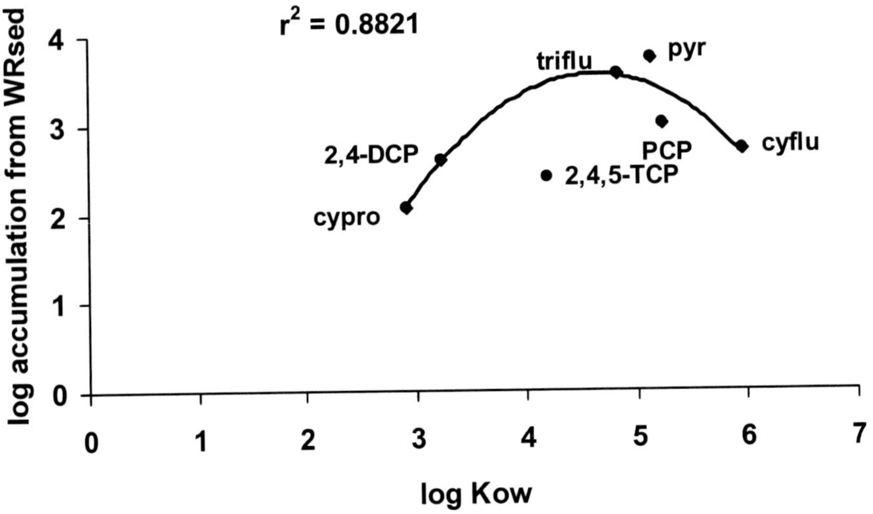

cypro: Ciproconazol, triflu: Trifluralin, pyr: Pyrene, cyflu: Cyfluthrin.

Figure 6. Bioaccumulation of organic contaminants by *S. corneum* (pmol g^{-1} wet tissue) and Kow values on a log-log basis. Clams were exposed for 48 h to (A): systems containing water only, or (B): the sample of WR natural sediment.

When all the substances were included, a very poor correlation was observed between the bioaccumulation from the system containing only water and the Kow, with a correlation coefficient r^2 = 0.2311. However, when removing the data for 2,4,5-TCP the correlation improved significantly, with r^2 = 0.8873. The parabolic type of the curve shows that the uptake increases up to a maximum value after which it starts to decrease as the Kow increases further.

When the uptake occurred from the sediment sample and considering all the compounds, a better correlation was found, with r^2 = 0.5979. Nevertheless, the value improved when removing the data for 2,4,5-TCP and a highly significant parabolic relationship could be obtained (r^2 = 0.8821).

3.3. Relationships between Accumulation and the Particle-Water Partition Coefficients (Kd)

A highly significant correlation (r^2 = 0.9719) was found between the accumulation of all chemicals from the sediment sample and the values of the water-particle partition coefficients (Kd) as shown in Figure 7. This correlation could be described by a log function. At higher Kd values, the higher uptake and accumulation was observed.

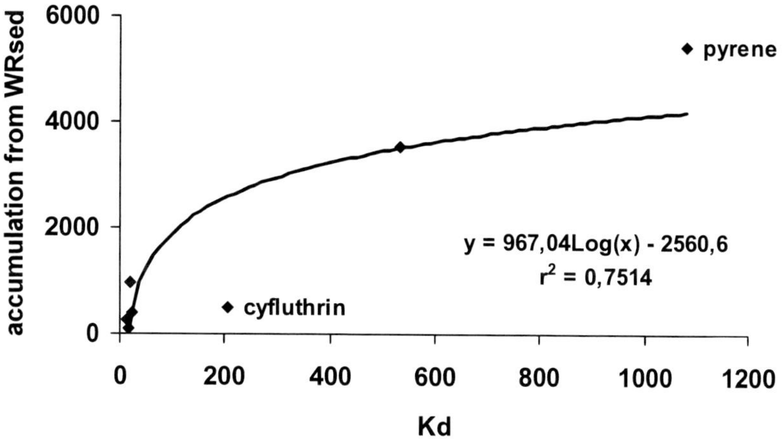

Figure 7. Relationships between bioaccumulation of organic contaminants by *S. corneum* (pmol g^{-1} wet tissue) exposed to systems containing the sample of WR natural sediment and the corresponding Kd values. Clams were exposed for 48 h.

3.4. Influence of the Organic Matter

To investigate the influence of the organic matter on the uptake and accumulation of the contaminants by *S. corneum*, clams were exposed to the substances in the presence of a 20 mg L^{-1} solution of humic acids. Data of chemical bioaccumulated from the WR sediment sample and from the humic acid solution were plotted and shown in Figure 8. A significant correlation was found that could be described by a highly significant regression line for all the compounds.

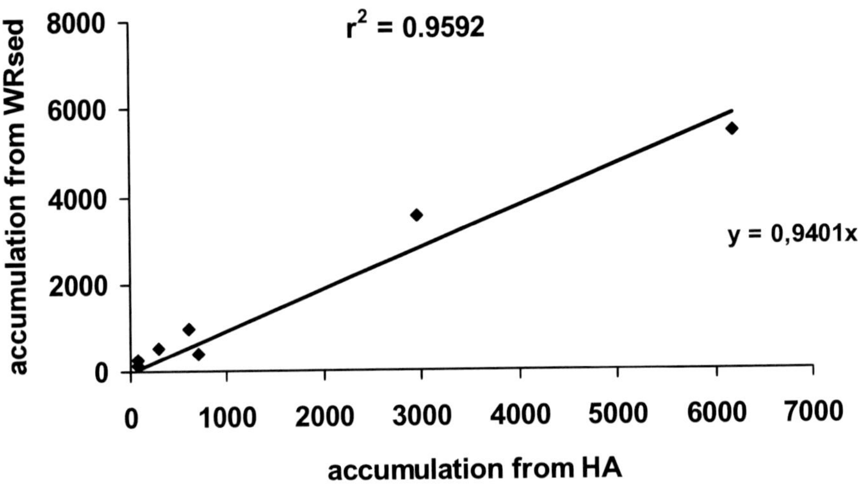

Figure 8. Relationships between bioaccumulation of organic contaminants by *S. corneum* exposed to systems containing the sample of WR natural sediment and a 20 mg L^{-1} humic acid solution. Data are expressed as pmol g^{-1} wet tissue. Clams were exposed for 48 h.

3.5. The experimental model: I. Toyopearl SP particles

Figure 9 shows the data of bioaccumulation from the WR sediment sample versus the values obtained when the clams were exposed to the artificial particles Toyopearl SP. In this case any significant correlation was found.

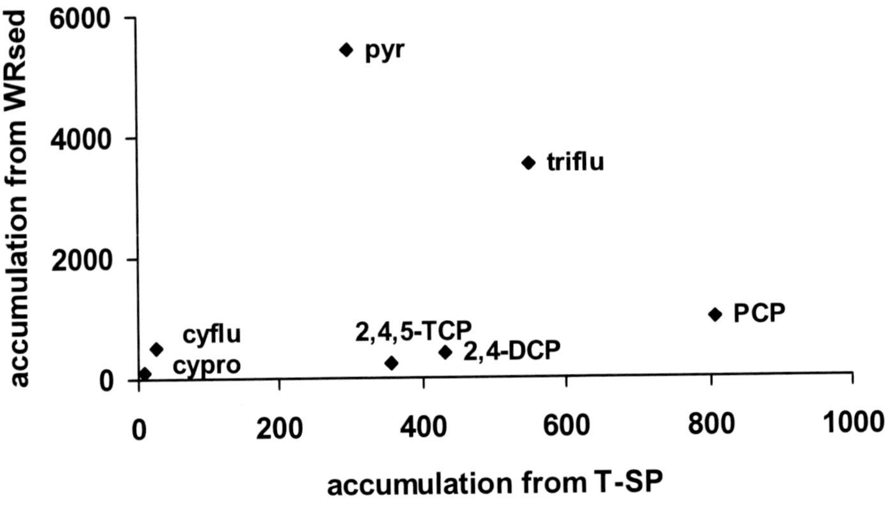

cypro: Ciproconazol, triflu: Trifluralin, pyr: Pyrene, cyflu: Cyfluthrin.

Figure 9. Relationships between bioaccumulation of organic contaminants by *S. corneum* exposed to systems containing the sample of WR natural sediment and Toyopearl SP particles. Data are expressed as pmol g^{-1} wet tissue. Clams were exposed for 48 h.

3.6. The Experimental Model: II. Toyopearl SP Particles Pretreated with Humic Acids

Figure 10 shows the data of bioaccumulation from the natural sediment sample versus the accumulation obtained when the clams were exposed to the system containing Toyopearl SP particles that had been pretreated with a 20 mg L^{-1} solution of humic acids. In this case, the correlation could be described by a highly significant regression line.

In addition, a very good correlationship was observed between bioaccumulation from the Toyopearl SP particles plus humic acids and the Kd values as it is shown in Figure 11.

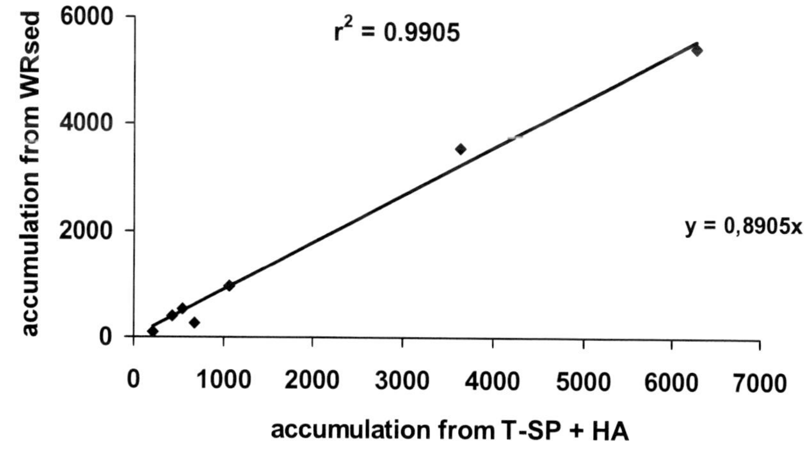

Figure 10. Relationships between bioaccumulation of organic contaminants by *S. corneum* exposed to systems containing the sample of WR natural sediment and Toyopearl SP particles pretreated with a 20 mg L^{-1} humic acid solution. Data are expressed as pmol g^{-1} wet tissue. Clams were exposed for 48 h.

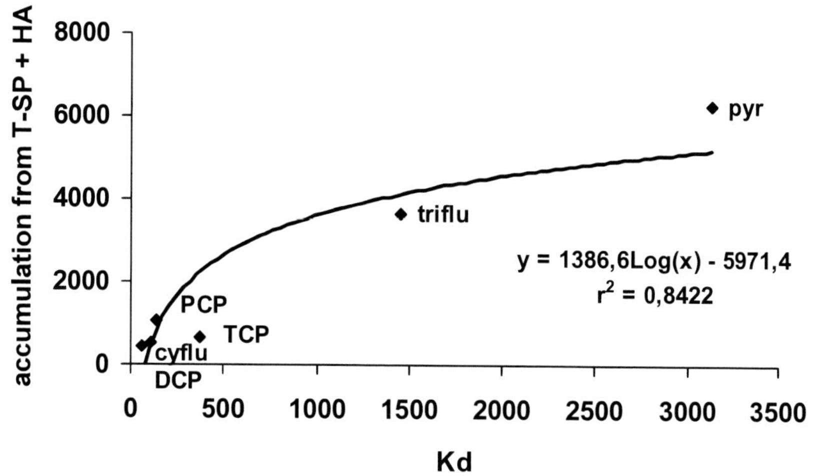

Figure 11. Relationships between bioaccumulation of organic contaminants by *S. corneum* (pmol g^{-1} wet tissue) exposed to systems containing Toyopearl SP particles pretreated with a 20 mg L^{-1} humic acid solution and the corresponding Kd values. Clams were exposed for 48 h.

4. Discussion

4.1. Comparison between Uptake of Contaminants from Water and from Natural Sediments by *S. corneum*

No radioactivity was observed in control clams. This fact was particularly important because the animals had been collected from a natural aquatic system and it had to be ensured that the clams had not been previously exposed to any radioactive source of contamination that could corrupt the results.

All the bioassays were devised using the same nominal chemical concentration (50 nmol L^{-1}) therefore direct comparisons of uptake and accumulation can be done among animals exposed to the different test systems. In addition, it has been previously shown that these clams were not able to metabolize the selected chemicals during the treatments (Verrengia Guerrero et al., 2002). Only in the case of pyrene, less than 10% of the compound was metabolized by *S. corneum*.

The presence of the sample of natural sediments significantly increased the uptake and accumulation by *S. corneum* of the chemicals being studied except for the most polar compounds, such as Cyproconazole and 2,4-DCP. These results constitute clear evidence of the role of sediments, specifically suspended particulate matter, as a potential relevant source of contamination for filter feeding organisms. The influence of sediment particles was greater for the more hydrophobic and non-polar substances, in agreement with previous reports (Mackay et al., 1991; Lamoureux and Browanwell, 1999).

4.2. Influence of the Octanol-water Partition Coefficients (Kow) on Uptake and Accumulation

The use of the octanol-water partition coefficients (Kow) has been long recognized as a useful tool for modeling and predicting the uptake and bioaccumulation of contaminants by aquatic organisms (Spacie et al., 1995). The properties of n-octanol simulate many of those associated with cell membranes and consequently the Kow may reflect the proportion of chemical bioavailable for absorption. These correlations have been extensively studied in fish (Isnard and Lambert, 1988; Schuurman and Klein, 1988; Meylan et al., 1999) and in other aquatic organisms (Legierse et al., 1998), but less information is still available for freshwater bivalves.

Typical parabolic relationships between bioaccumulation factors and Kow on a log-log basis have been established for a great number of substances. According to the literature there is an optimal range of log Kow (usually between 4 and 6) where the influence of hydrophobicity promotes the maximum uptake and accumulation of chemicals (Landrum and Robbins, 1990; Landrum and Faust, 1991). For compounds more hydrophilic, with values of log Kow below than 4, the uptake is lower as a consequence of the difficulties to diffuse across the cell membrane. On the other hand, compounds with higher log Kow than 6 are big molecules that are not able to cross the membranes.

When studying the influence of the Kow, in most of the cases reported, the organisms had been exposed to the chemicals in systems containing water only. Even in these cases there are some controversies in relation to the bioconcentration factors. There are two approaches to calculate them (Lee, 1992). Firstly, as the ratio between the concentration of a given substance in the animals and the concentration of the substance in the aqueous phase once the system has reached the equilibrium. Alternatively, bioconcentration factors may be estimated as the ratio between the rates of uptake and elimination of the substance in the animal. In fact under laboratory conditions, it can be difficult to reach the steady state. To avoid this problem, it was decided to plot directly the values of chemicals present in the animals versus the respective Kow values, on a log-log basis. The data showed that the uptake and accumulation of chemicals was related to the Kow. Parabolic relationships were obtained either in systems containing water only or the sample of WR sediment. Also, in agreement with the literature, the range of higher uptake was 4-5.

However, the data for 2,4,5-TCP had to be excluded from the plots. Therefore, the hydrophobicity of contaminants was not the only important property that influenced the uptake and accumulation. Considering the chlorophenols, if Kow was the only factor, the accumulation should be higher for PCP, followed by 2,4,5-TCP and finally 2,4-DCP. However the experimental data followed an order quite different from that prediction. In those bioassays performed with water only, the levels of 2,4-DCP accumulated by the clams (342.8 ± 98.1 pmol g^{-1}) were similar to those of PCP (297.4 ± 130.6 pmol g^{-1}), and the lowest value was found for 2,4,5-TCP (92.9 ± 47.6 pmol g^{-1}). One of the key factors to explain these experimental results may be the differences in acidity of the chlorophenols used. For the reaction:

$$AH \Leftrightarrow A^- + H^+$$

the equation of Henderson-Hasselbach gives:

$$pH = pKa + \log [A^-]/[AH]$$

and allows the calculations of the non-dissociated [AH] and the dissociated $[A^-]$ chemical species (Timbrell, 2000). The current experiments were performed at pH 8.3 and the values of pKa from literature were 7.85 for 2,4-DCP, 6.94 for 2,4,5-TCP, and 4.75 for PCP (Escher and Schwarzenbach, 1996). Therefore the respective relationships $\log [A^-]/[AH]$ are equal to 0.45; 1.36, and 3.55. These values show that the acidity of these chlorophenols increases with the number of chlorine atoms, indicating that the proportion of the non-dissociated forms, which are able to cross the cellular membranes, decrease in the same order. For instance, it has been estimated that the rate of absorption for the anionic form of PCP is more than ten times lower than that for the non-ionic form (Landner, 1988). This effect makes it possible to understand why the uptake and accumulation of PCP was not higher than that of 2,4-DCP. However, this fact is not sufficient reason to explain the lowest values observed for 2,4,5-TCP. An alternative factor may be due to differences in biotransformation or detoxification processes (De Wolf et al., 1992), but this was not the case in *S. corneum* as we have demonstrated previously (Verrengia Guerrero et al., 2002). Kishino and Kobayshi (1996) also found lower levels of accumulation for 2,4,5-TCP than for 2,4-DCP in fishes. Therefore, it could be hypothesized that the position of the chlorine atoms in 2,4,5-TCP reduces membrane permeation.

It is worth noting that in the presence of natural sediments, the uptake and accumulation of PCP (968.4 ± 440.7 pmol g^{-1}) was higher than for 2,4-DCP (399.9 ± 59.7 pmol g^{-1}) while 2,4,5-TCP presented again the lowest value (251.7 ± 96.9 pmol g^{-1}). In this case, the presence of negative charges on the surfaces of sediment particles may display the acid/base equilibrium to the non-dissociated and more available PCP.

4.3. Influence of the Particle-Water Partition Coefficients (Kd)

A fairly good correlation was observed between the levels of chemical bioaccumulated by the clams and the Kd values. This means that clams were able to desorb the chemicals from the particles accordingly to the amount bound to them. However, two substances failed to fit properly the logarithm function. One of them, cyfluthrin, gave an experimental bioaccumulation level below the value expected by the curve. Possibly, because of its high Kow value. On the contrary, in the case of pyrene, the bioaccumulation found in the clams was above the values predicted.

4.4. Influence of the Organic Matter

The first attempt to investigate the effect of the organic matter on the bioavailability and bioaccumulation of chemicals by *S. corneum* was performed using a solution containing 20 mg L^{-1}. No sediments were present in these experiments. A very good regression line was derived between the levels of substances bioaccumulated by the clams from systems containing the sample of WR sediment and the levels found when the organisms were exposed to the humic acid solution. This result reflects the relevance of the organic matter content in determining the bioavailability and bioaccumulation of organic chemicals.

4.5. Influence of the Artificial Particles

The previous result encouraged the analysis of the problem. However, we are comparing systems very different from each other. On one hand, there are systems containing natural particles that have a very complex and unknown composition. On the other, there are humic acid solutions free of sediments. The next step was to investigate a more representative analogue for natural sediments, consisting in a negatively charged particle, usually used as chromatographical resin for cationic exchange. From the results shown in the Figure 9. any good correlation could be established.

One possible explanation for this result could be due to the absence of organic matter in the artificial particles. In addition, these beads were unsuitable for digestion in the digestive tract of the organisms. Therefore, even when the particles could be able to bind the chemicals they were not able to release them, leading to a decrease in their bioavailability.

As the negatively charged particles were not a good analogue for the natural sediment, the model was improved by pretreating the particles with a 20 mg L^{-1} humic acid solution. After the equilibrium was reached, the organic matter content of the particles was very close to the value measured in the sample of natural sediment. When comparing the results of bioaccumulation in animals exposed to the sample of WR sediment and those obtained with the Toyopearl SP plus humic acids, a very good correlation was found for all the studied compounds. Therefore, this artificial system showed to be a very good experimental model to predict the bioavailability and bioaccumulation in the clams.

Simkiss et al. (2000, 2001) have found that a very different artificial particle, the hydrophobic interaction resin Toyopearl Phenyl, was able to mimic the uptake and accumulation of several organic contaminants in the species *Chironomus riparius* and *Lumbriculus variegatus*. In agreement with this, this resin was also useful to reflect the uptake and accumulation of 2,4-DCP by the clam *S. corneum* (Verrengia Guerrero et al, *in press*). However, when using the Toyopearl Phenyl in bioassays with PCP and pyrene, the resulting uptake in *S. corneum* was much lower that the values obtained for the sample of the natural sediment (Verrengia Guerrero et al., 2001, 2003). Therefore, this resin could not be considered a good analogue of natural sediment in the clams.

There could be many reasons for these discrepancies. The present study differs from those performed by Simkiss et al. (2000, 2001) in several aspects. They used a similar approach but standardized their treatments by using conditions that held the contaminants at a constant concentration in the aqueous phase. In contrast to that approach, the bioassays described here contained the same total amount of contaminants in all the treatments. The experiments of Simkiss et al. were designed to study the uptake from particles in static systems at a solid concentration of 100 mg ml^{-1}, while in the present approach it was ensured homogeneous suspensions that reflected the uptake from suspended particles at a much lower solid concentration (1 mg ml^{-1}). Finally, and perhaps the most important factor, Simkiss et al. (2000, 2001) used two species of dwelling organisms, in contrast with the filter feeding clam, the experimental animal model selected for this work.

5. Conclusion

Studying the bioavailability of sediment-bound contaminants is complicated by many factors. From the inherent complexity and variable composition of sediment particles, which in turn may change over time; the low levels of contaminant concentrations that may be present; their partitioning between diverse aqueous and particulate phases, up to the variety of uptake routes that may be involved with the biota. In fact, aquatic organisms differ on the election, ingestion, digestion and elimination of sediment particles. An integrative approach based on chemical analyses, sediment bioassays and ecological studies might be the most useful tool to reach that goal. However, these studies are costly and time-consuming. Therefore, new simple and low-cost innovative methodologies should also be tested as a promising alternative. The approach adopted in this work has attempted to simplify the complexity in order to test some basic factors that may be of more general application. In order to achieve this a negatively charged particle was used, with well-defined surface properties, pretreated with a humic acid solution, to provide an assimilation route for a filter-feeding bivalve (*S. corneum*). This artificial particle proved to be a good analogue for the sample of natural sediment and could be used to predict the bioavailability of sediment-bound contaminants. Therefore, this experimental model might be a promising tool for the assessment of sediment quality and regulatory purposes.

Acknowledgements

This manuscript was supported by the University of Buenos Aires, Argentina (grant X-147) and the Agencia Nacional de Promoción Científica y Tecnológica, Argentina (PICTR 2002-00203). The experimental work was conducted at the School of Animal and Microbial Sciences, University of Reading, United Kingdom and funded from the Environmental Diagnostic Programme of the Natural Environmental Research Council, United Kingdom. The author is indebted to Prof. Ken Simkiss and Dr. Marina G. Taylor for their valuable encouragement, advice and friendship, and for reviewing parts of the manuscript. Thanks to Prof. Dr. Ana M.P. de D'Angelo and to Lic. M.N. Piol.

References

Accardi-Dey, A; Gschwend, PM. Assessing the combined roles of natural organic matter and Black carbon as sorbents in sediments. *Environ Sci Technol* 2002, 36, 21-29.

Accardi-Dey, A; Gschwend, PM. Reinterpreting literature sorption data considering both absorption into and adsorption onto black carbon. *Environ Sci Technol* 2003, 37, 99-106.

Aiken, GR; McKnight, DM; Wershaw, RL; MacCarthy, P. An introduction to humic substances in soil, sediment and water. In: Aiken GR, MacCarthy P, Malcolm RL, Swift RS, editors. *Humic Substances in Soil, Sediment and Water*. New York: Wiley, 1985, 1-9.

Åkerblom, N; Goedkoop, W. Stable isotopes and fatty acids reveal that *Chironomus riparius* feeds selectively on added food in standardized toxicity tests. *Environ Toxicol Chem* 2003, 22, 1473-1480.

Alexander, M. Aging, bioavailability, and overestimation of risk from environmental pollutants. *Environ Sci Technol* 2000, 34, 4259-4265.

Ankley, GT; Phipps, GL; Leonard, EN; Benoit, DA; Mattson, VR; Kosian, PA; Cotter, AM; Dierkes, JR; Hansen, DJ; Mahony, JD. Acid-volatile sulfide as a factor mediating cadmium and nickel bioavailability in contaminated sediments. *Environ Toxicol Chem* 1991, 10, 1299-1307.

APHA-AWWA-WPCF: American Publish Health Association, American Water Works Association, and Water Pollution Control Federation. *Standard Methods for the Examination of Water and Wastewater*. Washington: 1980.

Arizzi Novelli, A; Losso, C; Libralato, G; Tagliapietra, D; Pantani, C'Volpi Ghirardini, A. Is the 1:4 elutriation ratio reliable? Ecotoxicological comparison of four different sediment:water proportions. *Ecotoxicol Environ Saf, in press*.

Arjonilla, M; Forja, JM; Gómez-Parra, A. Sediment analysis does not provide a good measure of heavy metal bioavailability to *Cerastoderma glaucum* (Mollusca: Bivalvia) in confined coastal ecosystems. *Bull Environ Contam Toxicol* 1994, 52, 810-817.

Ashton, A; Chan, R. Monitoring of microgram per litre concentrations of trace metals in sea water: the choice of methodology for sampling and analysis. *Analyst* 1987, 112, 841-844.

ASTM: American Society for Testing and Materials. Standard guide for conducting 10-d static sediment toxicity tests with estuarine and marine amphipods. E 1364-92. Philadelphia: *American Society for Testing and Materials*, 1992.

ASTM: American Society for Testing and Materials. *Standard guide for conducting sediment toxicity tests with marine and estuarine polychaetous annelids.* **E 1611-94.** Philadelphia: American Society for Testing and Materials, 1994a.

ASTM: American Society for Testing and Materials. *Standard guide for collection, storage, characterization, and manipulation of sediment for toxicological testing.* **E 1391-94.** Philadelphia: American Society for Testing and Materials, 1994b.

ASTM: American Society for Testing and Materials. *Standard methods for measuring the toxicity of sediment-associated contaminants with freshwater invertebrates.* **E 1706-95b.** Philadelphia: American Society for Testing and Materials, 1995.

ASTM: American Society for Testing and Materials. *Standard methods for measuring the toxicity of sediment-associated contaminants with freshwater invertebrates.* In: Annual Book of ASTM Standards, Volume 11.05. Philadelphia: American Society for Testing and Materials, 2002.

Baker, RA. *Organic Substances and Sediments in Water.* Chelsea: Lewis Publishers, 1991, 3 Vol.

Basack, SB; Oneto, ML; Verrengia Guerrero, NR; Kesten, EM. Accumulation and elimination of pentachlorophenol in the freshwater bivalve *Corbicula fluminea.* Bull *Environ Contamin Toxicol* 1997, 58, 497-503.

Baudo, R; Muntau, H. Lesser known in place-pollutants and diffuse source problems. In: Baudo R, Giesy JP, Muntau H, editors. Sediments: *Chemistry and Toxicity of in Place Pollutants.* Michigan: Lewis Publishers; 1990, 1-14.

Belfroid, AC; Sijm, DTHM. Influence of soil organic matter content on elimination rates of hydrophobic compounds in the earthworm: possible causes and consequences. *Chemosphere* 1998, 35, 1221-1234.

Benés, P; Gjessing, ET; Steinnes, E. Interactions between humus and trace elements in fresh water. *Water Res* 1976, 10, 711-716.

Benson, WH; Long, SF. Evaluation of humic-pesticide interactions on the acute toxicity of selected organophosphate and carbamate insecticides. *Ecotoxicol Environ Saf* 1991, 21, 301-307.

Berglund, O; Larsson, P; Okla, EL. Influence of trophic status on PCB distribution in lake sediments and biota. *Environ Poll* 2001, 113, 199-210.

Borchert, J; Karbe, L; Westendorf, J. Uptake and metabolism of benzo(a)pyrene absorbed to sediment by the freshwater invertebrate species *Chironomus riparius* and *Sphaerium corneum. Bull Environ Contamin Toxicol* 1997, 58, 158-165.

Bordalo Costa, MM; Peneda, MC. Heavy metal detection in sediment-water components of the sago estuary by multielemental analysis. *Environ Technol Lett* 1989, 10, 697-705.

Brezonik, PL; King, SO; Mach, CE. The influence of water chemistry on trace metal bioavailability and toxicity to aquatic organisms. In: Newman MC, McIntosh AW, editors. *Metal Ecotoxicology.* Chelsea: Lewis Publishers, 1991, 1-31.

Burton Jr, GA; MacPherson, C. Sediment toxicity testing issues and methods. In: Hoffman DJ, Rattner BA, Allen Burton Jr G, Cairns Jr J, editors. Handbook of Ecotoxicology. Boca Raton: Lewis Publishers, 1995, 71-103.

Caille, N; Tiffreau, C; Leyval, C; Morel, JL. Solubility of metals in an anoxic sediment during prolonged aeration. *Sci Total Environ* 2003, 301, 239-250.

Chapman, PM. Effects of gut sediment contents on measurements of metal levels in benthic invertebrates - a cautionary note. *Bull Environ Toxicol* 1985, 35, 345-347.

Chapman, PM; Ho, KT; Munns, WR Jr.; Solomon, K; Weinstein, M.P. Issues in sediment toxicity and ecological risk assessment. *Mar Poll Bull* 2002, 44, 271-278.

Cornelissen, G; Gustafsson, O. Sorption of phenanthrene to environmental Black carbon in sediment with and without organic matter and native sorbates. *Environ Sci Technol* 2004, 38, 148-155.

Cornelissen, G; Gustafsson, O. Effects of added PAHs and precipitated humic acids coatings on phenanthrene Black carbon. *Environ Poll* 2006, 141, 526-531.

Davies, NA; Taylor, MG; Simkiss, K. The influence of particle surface characteristics on pollutant metal uptake by cells. *Environ Poll* 1997, 96, 179-184.

Davies, NA; Edwards, PA; Lawrence, MAM; Simkiss, K; Taylor, MG. Biocide testing using particles with controlled surface properties (artificial sediments). *Environ Toxicol Chem* 1999a, 18, 2337-2342.

Davies, NA; Edwards, PA; Lawrence, MAM; Taylor, MG; Simkiss, K. Influence of particle surfaces on the bioavailability to different species of 2,4-dichlorophenol and pentachlorophenol. *Environ Sci Technol* 1999b, 33, 2465-2468.

Dawson, TD; Lott, KG; Leonard, EN; Mount, DR. Time course of metal loss in *Lumbriculus variegatus* following sediment exposure. *Environ Toxicol Chem* 2003, 22, 886-889.

De Haas, EM; Paumen, ML; Koelmans, AA; Kraak, MHS. Combined effects of copper and food on the midge *Chironomus riparius* in whole-sediment bioassays. *Environ Poll* 2004, 127, 99-107.

De Lange, HJ; de Haas, EM; Maas, H; Peeters, ETHM. Contaminated sediments and bioassay responses of three macroinvertebrates, the midge larva *Chironomus riparius*, the water louse *Asellus aqauticus* and the mayfly nymph *Ephoron virgo*. *Chemosphere* 2005, 61, 1700-1709.

De Wolf, W; de Bruijn, JHM; Seinen, W; Hermens, JLM. Influence of biotransformation on the relationship between bioconcentration factors and octanol-water partition coefficients. *Environ Sci Technol* 1992, 26, 1197-1201.

Dekker, T; Greve, GD; Ter Laak, TL; Boivin, ME; Veuger, B; Gortzak, G; Dumfries, S; Lücker, SMG; Kraak, MHS; Admiraal, W; van der Geest, HG. Development and application of sediment toxicity test using the benthic cladocerean *Chydorus sphaericus*. Environ Poll 2006, 140, 231-238.

Di Toro, DM; Zabra, CS; Hansen, DJ; Berry, WJ, Swartz, RC; Cowan, CE; Pavlou, SP; Allen, HE; Thomas NA; Paquin, PR. Technical basis for establishing sediment quality criteria for nonionic organic chemicals using equilibrium partitioning. *Environ Contam Toxicol* 1991, 10, 1541-1583.

Dunnivant, FM; Coates, JT; Elzerman, AW. Labile and non-labile desorption rate constants for 33 PCB congeners from lake sediment suspensions. *Chemosphere* 2005, 61, 332-340.

Edwards, VT; McMinn, AL. Biotransformation of pesticides and other xenobiotics in plants and soils - Recent developments. In: Hutson DH, Roberts TR, Editors. Progress in Pesticide Biochemistry and Toxicology, volume 4. New York: John Wiley & Sons Ltd., 1983, 103-164.

Eggleton, J; Thomas, KV. A review of factors affecting the release and bioavailability of contaminants during sediment disturbance events. Environ Intern 2004, 30, 973-980.

Ehlers, GAC; Loibner, AP. Linking organic pollutant (bio)availability with geosorbent properties and biomimetic methodology: A review of geosorbent characterisation and (bio)availability prediction. *Environ Poll* 2006, 141, 494-512.

Elbaz-Poulichet, F; Hollinger, P; Huang, WW; Martin, JM. Lead cycling in estuaries, illustrated by the Gironde estuary, France. *Nature* 1984, 308, 409-414.

Environment Canada. Biological Test Method: Test of reproduction and survival using the cladoceran *Ceriodaphnia dubia. Report EPS 1/RM/21.* Ontario: Environment Canada, 1992.

Environment Canada. Biological Test Method: Test for survival and growth in sediment using the freshwater amphipod *Hyalella azteca. EPS 1/RM/33. Technical Report*, Ottawa: Environment Canada, 1997.

Eriksson Wiklund, AK; Dag Broman, BS. Toxicity evaluation by using intact sediments and sediment extracts. *Mar Poll Bull* 2005, 50, 660-667.

Erstfeld, KM; Park, SS. Dynamic of chlordane under non-constant exposure conditions: A numerical bioaccumulation model. *Bull Environ Contam Toxicol* 1997, 58, 364-371.

Escher, BI; Schwarzenbach, RP. Partition of substituted phenols in liposome – water, biomembrane – water, and octanol – water systems. *Environ Sci Technol* 1996, 30, 260-270.

FAO: Food and Agricultural Organisation of the United Nations, Manual of methods in aquatic environment research, Part 10 – Short term static bioassays. FAO Fisheries *Technical Paper* N° **247**, Rome. 1987.

Farrington, JW. Bioaccumulation of hydrophobic organic pollutant compounds. In: Levin SA, Harwell MA, Kelly JR, Kimball KD, editors. *Ecotoxicology*: Problems and Approaches. New York: Springer-Verlag, 1989, 279-313.

Fisk, AT; Norstrom, RJ; Cymbalisty, CD; Muir, DCG. Dietary accumulation and depuration of hydrophobic organochlorines: Bioaccumulation parameters and their relationship with the octanol/water partition coefficient. *Environ Toxicol Chem* 1998, 17, 951-961.

Fleming, RJ; Holmes, D; Nixon, SJ. Toxicity of permethrin to *Chironomus riparius* in artificial and natural sediments. *Environ Toxicol Chem* 1998, 17, 1332-1337.

Florence, TM; Morrison, GM; Stauber, JL. Determination of trace element speciation and the role of speciation in aquatic toxicity. *Sci Total Environ* 1992, 125, 1-13.

Ford, J. The effects of chemical stress on aquatic species composition and community structure. In: Levin SA, Harwell MA, Kelly JR, Kimball KD, editors. *Ecotoxicology: Problems and Approaches*. New York: Springer-Verlag, 1989, 99-144.

Förstner, U. Chemical forms of metal enrichment in recent sediments. In: Amstutz G, editor. Ore Genesis. New York: Springer-Verlag, 1982, 191-199.

Förstner, U. Inorganic sediment chemistry and elemental speciation. In: Baudo R, Giesy JP, Muntau H, editors. *Sediments: Chemistry and Toxicity of in Place Pollutants*. Michigan: Lewis Publishers, 1990, 61-105.

Geffard, O; His, E; Budzinski, H; Chiffoleau, JF; Coynel, A; Etcheber, H. Effects of storage method and duration on the toxicity of marine sediments to embryos of *Crassostrea gigas* oysters. *Environ Poll* 2004, 129, 457-465.

GESAMP: Joint Group of Experts on the Scientific Aspects of Marine Pollution. *Statement of the Intergovernmental Meeting of Experts on Land Based Sources of Marine Pollution*. Geneva: World Health Organization, 1991.

Gillis, PL; Chow-Fraser, P; Ranville, JF; Ross, PE; Wood, CM. *Daphnia* need to be gut-cleared too: the effect of exposure to and ingestion of metal-contaminated sediment on the gut-clearance patterns of *D. magna*. *Aquat Toxicol* 2005, 71, 143-154.

Griffiths, RA. Sorption and desorption by ideal two-compartment systems: unusual behavior and data interpretation problems. *Chemosphere* 2004, 55, 443-454.

Harkey, GA; Landrum, PF; Klaine, SJ. Preliminary studies on the effect of feeding during whole sediment bioassays using *Chironomus riparius* larvae. *Chemosphere* 1994, 28, 597-606.

Hayes, MHB. Emerging concepts of the compositions and structures of humic substances. In: Hayes MHB, Wilson WS, editors. *Humic Substances, Peats and Sludges*. Health and Environmental Aspects. Cambridge: The Royal Chemistry Society, 1997, 3-30.

Heip, CHR; Goosen, NK; Herman, PMJ; Krompkamp, J; Middelburg, JJ; Soetaert, K. Production and consumption of biological particles in temperate tidal estuaries. *Ocean Mar Biol* 1995, 33, 1-149.

Horowitz, AJ. A review of physical and chemical partitioning of inorganic constituents in sediments. In: Bradford WL, Horowitz AJ, editors. The Role of Sediments in the Chemistry of Aquatic Systems-Proceedings of the Sediment Chemistry Workshop, February 8-12, 1982. Denver: US Geological Survey, 1988, *Circular* 969, 7-22.

Huang, W; Peng, P; Yu, Z; Fu, J. Effects of organic matter heterogeneity on sorption and desorption of organic contaminants by soils and sediments. *Appl Geochem* 2003, 18, 955-972.

Huggett, RJ; Kimerle, RA; Mehrie, PM; Bergman, HL, editors. Biomarkers: Biochemical, Physiological and Histological Markers of Anthropogenic Stress. Boca Raton: Lewis Publishers, 1992.

Ingersoll, CG. Sediment tests. In: Rand GM, editor. Fundamentals of Aquatic Toxicology. Washington: Taylor & Francis; 1995, 231-255.

Isnard, P; Lambert, S. Estimating bioconcentration factors from octanol/water partition coefficient and aqueous solubility. *Chemosphere* 1988, 17, 21-34.

Jager, T; van der Wal, L; Fleuren, RHLJ; Barendregt, A; Hermens, ILM. Bioaccumulation of organic chemicals in contaminated soils: evaluation of bioassays with earthworms. *Environ Sci Technol* 2005, 39, 293-298.

Jepsen, R; Lick, W. Nonlinear and interactive effects in the sorption of hydrophobic organic chemicals by sediments. *Environ Toxicol Chem* 1999, 18, 1627-1636.

Jonker, MTO; Koelmans, AA. Polyoxymethylene solid phase extraction as a partitioning method for hydrophobic organic chemicals in sediment and soot. *Environ Sci Technol* 2001, 35, 3742-3748.

Jonker, MTO; Koelmans, AA. Sorption of polycyclic aromatic hydrocarbons and polychlorinated biphenyls to soot and soot-like materials in the aqueous environment mechanistic considerations. *Environ Sci Technol* 2002, 36, 3725-3734.

Kemble, NE; Dwyer, FJ; Ingersoll, CG; Dawson, TD; Norberg-King, TJ. Tolerance of freshwater test organisms to formulated sediments for use as control materials in whole-sediment toxicity tests. *Environ Toxicol Chem* 1999, 18, 222-230.

Kenworthy, IP; Hayes, MHB. Investigation of some structural porperties of humic substances by fluorescence quenching. In: Hayes MHB, Wilson WS, editors. Humic Substances, Peats and Sludges. *Health and Environmental Aspects*. Cambridge: The Royal Chemistry Society, 1997, 39-45.

Ketchum, BH, editor. Ecosystems of the World 26: Estuaries and Enclosed Seas. Amsterdam: Elsevier Scientific Publishing, 1983.

Kishino, T, Kobayshi, K. Studies on the mechanism of toxicity of chlorophenols found in fish through quantitative structure-activity relationships. Water Res 1996, 30, 393-399.

Koelmans, AA; Jonker, MTO; Cornelissen, G; Bucheli, TD; Van Voort, PCM; Gustafsson, O. Black carbon: The reverse of its dark side. Chemosphere 2006, 63, 365-377.

Kraaij, R; Mayer, P; Busser, FJM; van het Bolscher, M; Seinen, W; Tolls, J; Belfroid, AC. Measured pore-water concentrations make equilibrium partitioning work - A data analysis. *Environ Sci Technol* 2003, 37, 268-274.

Kukkonen, J; Landrum, PF. Toxicokinetics and toxicity of sediment-associated pyrene to *Lumbriculus variegatus* (oligochaeta). *Environ Toxicol Chem* 1994, 13, 1457-1468.

Kukkonen, J; Landrum, PF. Effects of sediment-bound polydimethylsiloxane on the bioavailability and distribution of benzo(a)pyrene in lake sediment to *Lumbriculus variegatus*. *Environ Toxicol Chem* 1995, 14, 523-531.

Lai, TM. Concepts and methods for characterizing the physical properties of sediments. In: Bradford WL, Horowitz AJ, editors. The Role of Sediments in the Chemistry of Aquatic Systems-Proceedings of thc Scdiment Chemistry Workshop, February 8-12, 1982. Denver: US Geological Survey, 1988, *Circular* **969**, 56-64.

Lamoureux, EM; Brownawell, BJ. Chemical and biological bioavailability of sediment-sorbed hydrophobic organic contaminants. *Environ Toxicol Chem* 1999, 18, 1733-1741.

Landner, L. Hazardous chemicals in the environment - Some new approaches to advanced assessment. *Ambio* 1988, 17, 360-366.

Landrum, PF; Robbins, JA. Bioavailability of sediment-associated contaminants to benthic invertebrates. In: Baudo R, Giesy JP, Muntau H, editors. Sediments: *Chemistry and Toxicity of in Place Pollutants*. Michigan: Lewis Publishers; 1990, 237-263.

Landrum, PF; Faust, WR. Effect of variation in sediment composition on the uptake rate coefficient for selected PCB and PAH congeners by the amphipod, *Diporeia sp*. In: MA Mayes, MG Barron, editors. *Aquatic Toxicology and Risk Assessment*: Fourteenth Volume, STP 1124. Philadelphia: American Society for Testing and Materials, 1991, 263-279.

LeBoeuf, EJ; Weber, WJ Jr. Reevaluation of general partitioning model for sorption of hydrophobic organic contaminants by soil and sediment organic matter. *Environ Toxicol Chem* 1999, 18, 1617-1626.

Lee II, H. Models, muddles, and mud: predicting bioaccumulation of sediment-associated pollutants. In: Burton GA, editor. *Sediment Toxicity Assessment*. Chelsea: Lewis Publishers, 1992, 267-293.

Leenheer, JA. Analysis, association, and effects of organic constituents of aquatic sediment. In: Bradford WL, Horowitz AJ editors. The Role of Sediments in the Chemistry of Aquatic Systems-Proceedings of the Sediment Chemistry Workshop, February 8-12, 1982. Denver: US Geological Survey, 1988, *Circular* 969, 22-32.

Leenheer, JA. Organic substances structures that facilitate contaminant transport and transformations in aquatic sediments. In: Baker RA, editor. *Organic Substances and Sediments in Water*. Chelsea: Lewis Publishers, 1991, 3-22.

Legierse, KCHM; Sijm, DTHM; van Leeuwen, CJ; Seinen, W; Hermens, JLM. Bioconcentration kinetics of chlorobenzenes and the organophosphorus pesticide chlorthion in the pond snail *Lymnaea stagnalis* - a comparison with the guppy *Poecilia reticulata*. *Aquat Toxicol* 1998, 41, 301-323.

Leppänen, M.T; Kukkonen, JVK. Relationship between reproduction, sediment type, and feeding activity of *Lumbriculus variegatus* (Müller): implications for sediment toxicity testing. *Environ Toxicol Chem 1998*, 17, 2196-2202.

Leppänen, M.T; Kukkonen, JVK. Evaluating the role of desorption in bioavailability of sediment-associated contaminants using oligochaetes, semipermeable membrane devices and Tenax extraction. *Environ Poll* 2006, 140, 150-163.

Lerma, A; Mackenzie, FT; Geiger, RJ. Environmental chemical stress effects associated with carbon and phosphorus biogeochemical cycles. In: Levin SA, Harwell MA, Kelly JR, Kimball KD, editors. *Ecotoxicology: Problems and Approaches*. New York: Springer-Verlag, 1989, 315-350.

Luoma, SN. Bioavailability of sediment-bound metals. In: Bradford WL, Horowitz AJ editors. The Role of Sediments in the Chemistry of Aquatic Systems-Proceedings of the Sediment Chemistry Workshop, February 8-12, 1982. Denver: US Geological Survey, 1988, *Circular* **969**, pp. 32-56.

Luoma, SN; Ho, KT. Appropriate uses of marine and estuarine sediment bioassays. In: Calow P, editor. Handbook of Ecotoxicology. London: Blackwell Science Ltd., 1995, 193-223.

Luthy, RG; Aiken, GR; Brusseau, ML; Cunninham, SD; Gschwend, PM; Pignatello, JJ; Reinhard, M; Traina, SJ; Weber, WJ Jr; Westall, JC. Sequestration of hydrophobic organic contaminants by geosorbents. *Environ Sci Technol* 1997, 31, 3341-3347.

Lyman, WJ. Transport and transformation processes. In: Rand GM, editor. Fundamentals of Aquatic Toxicology. Washington: Taylor & Francis, 1995, 449- 492.

Mackay, D; Diamond, M: Stiver, W. The case of modeling sediment-water interactions in aquatic and marine systems. In: Baker RA, editor. Organic Substances and Sediments in Water. Chelsea: Lewis Publishers, 1991, 63-97.

Marcus, WA. Regulating contaminated sediments in aquatic environments: a hydrologic perspective. Environ Manage 1989, 13, 703-713.

Martín-Díaz, ML; Blasco, J; Sales, D; DelValls, TA. Biomarkers as tools to assess sediment quality. Laboratory and field surveys. *Trends Anal Chem* 2004, 23, 807-818.

Masala, O; O'Brien, P; Rainbow, PS. Analysis of metal-containing granules in the barnacle *Tetraclita squamosa*. J Inorg Chem 2004, 98, 1095-1102.

Mayer, P; Vaes, WHJ; Wijnker, F; Legierse, KCHM; Kraaij, R; Tolls, J; Hermens, JLM. Sensing dissolved sediment porewater concentrations of persistence and bioaccumulative pollutants using disposable solid-phase microextractions fibers. *Environ Sci Technol* 2000, 34, 5177-5183.

Means, JC; Wijayaratne, R. Role of natural colloids in the transport of hydrophobic pollutants. *Science* 1982, 215, 968-970.

Menzer, RE. Water and soil pollutants. In: Amdur MO, Doull J, Klaassen CD, editors. Casarett and Doull's Toxicology: *The Basic Science of Poisons*. New York: Pergamon Press, 1991, 872-902.

Meylan, WM; Howard, PH; Boethling, RS; Aronson, D; Printup, H; Gouchie, S. Improved method for estimating bioconcentration/bioaccumulation factor from octanol/water partition coefficient. *Environ Toxicol Chem* 1999, 18, 664-672.

Millward, GE; Liu, YP. Modelling metal desorption kinetics in estuaries. *Sci Total Environ* 2003, 314/316, 613-623.

Miño, LA; Folco, S; Pechén de D'Angelo, AM; Verrengia Guerrero, NR. Modeling lead bioavailability and bioaccumulation by *Lumbriculus variegatus* using artificial particles. Potential use in chemical remediation processes. *Chemosphere* 2006, 63, 261-268.

Moffett, JW; Ho, J. Oxidation of cobalt and manganese in seawater via a common microbially catalized pathway. *Geochim Cosmochim Acta* 1996, 60, 3415-3424.

Mount, DR; Dawson, TD; Burkhar, LP. Implications of gut purging for tissue residues determined in bioaccumulation testing of sediment with *Lumbriculus variegatus*. *Environ Toxicol Chem* 1999, 18, 1244-1249.

Naylor, C; Rodrigues, C. Development of a test method for *Chironomus riparius* using a formulated sediment. *Chemosphere* 1995, 31, 3291-3303.

Neal, C; House, WA; Whitton, BA; Leeks, GJL. Water quality and biology of United Kingdom rivers entering the North Sea: The land ocean interaction study (LOIS) and associated work. *Sci Total Environ* 1998, 210-211, 585-594.

NRWI: National Water Research Institute. *Direct Sediment Toxicity Testing Procedures. Contribution* N° 91-90, Burlington, Ontario, 1991.

OECD Guidelines for the testing of chemicals. *Proposal for a new guide guideline 218. Sediment-water chironomid toxicity test using spiked sediments*. 2001.

Pardo, R; Barrado, E; Pérez, L; Vega, M. Determination and speciation of heavy metals in sediments of the Pisuerga river. *Water Res* 1990, 24, 373-379.

Park, SS; Erstfeld, KM. A numerical kinetic model for bioaccumulation of organic chemicals in sediment-water systems. *Chemosphere* 1997, 34, 419-427.

Park, SS; Erstfeld, KM. The effect of sediment organic carbon content on the bioavailability of hydrophobic compounds in aquatic ecosystems. *Environ Poll* 1999, 105, 9-15.

Parkerton, TF; Stone, MA; Letinsky, DJ. Assessing the aquatic toxicity of complex hydrocarbons using solid-phase microextraction. *Toxicol Lett* 2000, 112/113, 273-282.

Peakall, D. Animal Biomarkers as Pollution Indicators. London: Chapman & Hall, 1992.

Peijnemburg, WJGM; Jager, T. Monitoring approaches to assess bioaccessibility and bioavailability of metals: Matrix issues. *Ecotoxicol Environ Saf* 2003, 56, 63-77.

Perminova, IV; Grechishceva, NY; Petrosyan, Relationships between structure and binding affinity of humic substances for polycyclic aromatic hydrocarbons: Relevance of molecular descriptors. *Environ Sci Technol* 1999, 33, 3781-3787.

Phipps, GL; Ankley, GT; Benoit, DA; Mattson, VR. Use of aquatic oligochaete *Lumbriculus variegatus* for assessing the toxicity and bioaccumulation of sediment-associated contaminants. *Environ Toxicol Chem* 1993, 12, 269-279.

Pignatello, JJ. Soil organic matter as a nanoporous sorbent of organic pollutants. *Adv Colloi Interf Sci* 1998, 76-77, 445-467.

Pignatello, JJ; Xing, B. Mechanisms of slow sorption of organic chemicals to natural particles. *Environ Sci Technol* 1996, 30, 1-11.

Podoll, RT; Mabey, WR. Factors to consider in conducting laboratory sorption/desorption tests. In: Dickson KL, Maki AW, Brungs WA, editors. *Fate and Effects of Sediment-Bound Chemicals in Aquatic Systems*. New York: Pergamon, 1987, 99-108.

Power, EA; Munkittrick, KR; Chapman, PM. An ecological impact assessment framework for decision-making related to sediment quality. In: MA Mayes, MG Barron, editors. *Aquatic Toxicology and Risk Assessment:* Fourteenth Volume, STP 1124. Philadelphia: American Society for Testing and Materials; 1991, 48-64.

Quiroga, JM; Sales, D; Gómez-Parra, A. Experimental evaluation of pollution potential of anionic surfactants in the marine environment. *Water Res* 1989, 23, 801-807.

Rainbow, PS. Trace metal concentrations in aquatic invertebrates: why and so what? *Environ Poll* 2002, 120, 497-507.

Rand, GM; Wells, PG; McCarty, LS. Introduction to aquatic toxicology. In: Rand GM, editor. *Fundamentals of Aquatic Toxicology*. Washington: Taylor & Francis; 1995, 3-67.

Reid, BJ; Jones, KC; Semple, KT. Bioavailability of persistent organic pollutants in soils and sediments - a perspective on mechanisms, consequences and assessment. *Environ Poll* 2000, 108, 103-112.

Reinfelder, JR; Wang, WX; Luoma, SN; Fisher, NS. Assimilation efficiencies and turnover rates of trace elements in marine bivalves: a comparison of oysters, clams and mussels. *Mar Biol* 1997, 129, 443-452.

Reynoldson, TB; Day, KE. Freshwater sediments. In: Calow P, editor. *Handbook of Ecotoxicology*. London: Blackwell Science Ltd., 1995, 83-103.

Riget, F; Johansen, P; Asmund, G. Uptake and release of lead and zinc by blue mussels. Experience from transplantation experiments in Greenland. *Mar Poll Bull* 1997, 10, 805-815.

Ruus, A; Schaanning, M; Øxnevad, S; Hylland, K. Experimental results on bioaccumulation of metals and organic contaminants from marine sediments. *Aquat Toxicol* 2005, 72, 273-292.

Samoiloff, MR. Toxicity testing of sediments: problems, trends and solutions. In: Nriagu JO, Lakshminarayana JSS, editors. *Aquatic Toxicology and Water Quality Management*. New York: John Wiley & Sons, 1989, 143-152.

Schuurman, G; Klein, W. Advances in bioconcentration prediction. *Chemosphere* 1988, 17, 1551-1574.

Sheedy, BR; Mattson, VR; Cox, JS; Kosian, PA; Phipps, GL; Ankley, GT. Bioconcentration of polycyclic aromatic hydrocarbons by the freshwater oligochaete *Lumbriculus variegatus*. *Chemosphere* 1998, 36, 3061-3070.

Simkiss, K. Ecotoxicants at the cell-membrane barrier. In: Newman MC, Jagoe CH, editors. *Ecotoxicology*. A Hierarchical Treatment. Boca Raton: CRC Press, 1996, 59-83.

Simkiss, K; Taylor, M. Cellular mechanisms of metal detoxification and some new indices of pollution. *Aquatic Toxicol* 1981, 1, 279-290.

Simkiss, K; Taylor, MG. Calcium magnesium phosphate granules: atomistic simulations explaining cell death. J Exp Biol 1994, 190, 131-139.

Simkiss, K; Edwards, PA; Lawrence, MAM; Davies, NA; Taylor, MG. The use of hydrophobic resins as analogues for sediment testing. *Environ Sci Technol* 2000, 34, 2388-2392.

Simkiss, K; Davies, NA; Edwards, PA; Lawrence, MAM; Taylor, MG. The use of sediment analogues to study the uptake of pollutants by chironomid larvae. *Environ Poll* 2001, 115, 89-96.

Sokal, RR; Rohlf, FJ. Biometry. The Principles and Practice of Statistic in Biological Research. New York: WH Freeman & Company, 1997.

Song, J; Peng, P; Huang, W. Black carbon and kerogen in soils and sediments: 1. Quantification and characterisation. *Environ Sci Technol* 2002, 29, 1000-1007.

Spacie, A; Hamelink, JL. Bioaccumulation. In: Rand GM, editor. *Fundamentals of Aquatic Toxicology*. Washington: Taylor & Francis, 1995, 1052-1082.

Spacie, A; McCarty, LS; Rand, GM. Bioaccumulation and bioavailability in multiphase systems. In: Rand GM, editor. *Fundamentals of Aquatic Toxicology*. Washington: Taylor & Francis, 1995, 493-521.

Stapleton, JM; Mihelcic, JR; Lueking, DR. Adsorption and desorption kinetics of pyrene onto a Great Lake sediment. *J Great Lakes Res* 1994, 20, 561-568.

Steinberg, CEW; Sturm, A; Kelbel, J; Kyu Lee, S; Hertkorn, N; Freitag, D; Kettrup, AA. Changes of acute toxicity of organic chemicals to *Daphnia magna* in the presence of dissolved humic material (DHM). *Acta Hydrochim Hydrobiol* 1992, 20, 326-332.

Suedel, BC; Rogers, JH Jr. 1994. Development of formulated reference sediments for use in freshwater and estuarine sediment tests. *Environ Toxicol Chem* 1994, 13, 1163-1175.

Talley, JW; Ghosh, U; Tucker, SG; Furey, JS; Luthy, RG. Particle-scale understanding of the bioavailability of PAHs in sediment. *Environ Sci Technol* 2002, 36, 477-483.

Taylor, G; Baird, DJ; Soares, AMVM. Surface binding of contaminants by algae: consequences for lethal toxicity and feeding to *Daphnia magna* Straus. *Environ Toxicol Chem* 1998, 17, 412-419.

Timbrell, JA. *Principles of Biochemical Toxicology*. London: Taylor and Francis, 2000.

Tomlin, CDS (editor). The Pesticide Manual. 11th Edition. Farnham, Surrey,: British Crop Protection Council, 1997.

Tracey, GA; Hansen, DJ. Use of biota-sediment accumulation factors to assess similarity of nonionic organic chemical exposure to benthically-coupled organisms of different trophic mode. *Arch Environ Contam Toxicol* 1996, 30, 467-475.

Turner, A; Millward, GE. Suspended particles: Their role in estuarine biogeochemical cycles. *Estuar Coast Shelf Sci* 2002, 55, 857-883.

US EPA: United States Environmental Protection Agency. *Methods for measuring the toxicity and bioaccumulation of sediment-associated contaminants with freshwater invertebrates*, 600/R-94/024. Washington: United States Environmental Protection Agency, 1994.

US EPA: United States Environmental Protection Agency. *Methods for measuring the toxicity and bioaccumulation of sediment-associated contaminants with freshwater invertebrates,* second edition. EPA/600/R-99/064. Washington: United States Environmental Protection Agency. 2000.

Valsaraj, KT; Thibodeáux, LJ. Relationships between micelle-water and octanol-water partition constants for hydrophobic organics of environmental interest. *Water Res* 1989, 23, 183-189.

van Noort, PCM; Cornelissen, G; ten Hulscher, TEM; Vrind, BA; Rigterink, H; Belfroid, A. Slow and very slow desorption of organic compounds from sediment: influence of sorbate planarity. *Water Res* 2003, 37, 2317-2322.

Verrengia Guerrero, NR. Contaminantes metálicos en el Rio de La Plata: Monitoreo del sistema acuático y estudio de algunos efectos tóxicos en moluscos bivalvos por medio de bioensayos. PhD thesis. Facultad de Ciencias Exactas y Naturales, Universidad de Buenos Aires, Buenos Aires, Argentina. 1995.

Verrengia Guerrero, NR; Nahabedian, DE; Wider, EA. Analysis of some factors that may modify the bioavailability of cadmium and lead by *Biomphalaria glabrata*. *Environ Toxicol Chem* 2000, 19, 2779-2787.

Verrengia Guerrero, NR; Taylor, MG; Wider, EA; Simkiss, K. Modeling pentachlorophenol (PCP) bioavailability and bioaccumulation by the freshwater fingernail clam *Sphaerium corneum* using artificial particles and humic acids. *Environ Toxicol Chem* 2001, 20, 2910-2915.

Verrengia Guerrero, NR; Taylor, MG; Davies, NA; Lawrence, MAM; Edwards, PE; Simkiss, K; Wider, EA. Evidence of differences in the biotransformation of organic contaminants in three species of freshwater invertebrates. *Environ Poll* 2002, 117, 523-530.

Verrengia Guerrero, NR; Taylor, MG; Wider, EA; Simkiss, K. Influence of particle characteristics and organic matter content on the bioavailability and bioaccumulation of pyrene by clams. *Environ Poll* 2003, 121, 115-122.

Verrengia Guerrero, NR; Taylor, MG; Simkiss, K. Modelling 2,4-dichlorophenol bioavailability and bioaccumulation by the freshwater fingernail clam *Sphaerium corneum* using artificial particles and humic acids. *Environ Poll, in press.*

Volparil, IM; Burgess, RM; Mayer, LM; Tien, R; Cantweel, MG; Ryba, S.A. Digestive bioavailability to a deposit feeder (*Arenicola marina*) of polycyclic aromatic hydrocarbons associated with anthropogenic particles. *Environ Toxicol Chem* 2004, 23, 2618-2626.a

Walker, CH; Hopkin, SP; Sibly, RM; Peakall, DB. Principles of Ecotoxicology, 2nd Edition. London: Taylor & Francis, 2001.

Wang, FY; Chen, JS; Chen, HL; Forsling, W. Surface properties of natural aquatic sediments. *Water Res* 1997, 31, 1796-1800.

Wang, F; Goulet, RR; Chapman, PM. Testing sediment biological effects with the freshwater amphipod *Hyalella azteca*: the gap between laboratory and nature. *Chemosphere* 2004, 57, 1713-1724.

Weber, WJ Jr; LeBoeuf, EJ; Young, TM; Huang, W. Contaminant interactions with geosorbent organic matter: insights drawn from polymer sciences. *Water Res* 2001, 35, 853-868.

Weston, DP; Mayer, LM. *In vitro* digestive fluid extraction as a measure of the bioavailability of sediment-associated polycyclic aromatic hydrocarbons: sources of variations and implications for partitioning models. *Environ Toxicol Chem* 1998a, 17, 820-829.

Weston, DP; Mayer, LM. Comparison of *in vitro* digestive fluid extraction and traditional *in vivo* approaches as measures of polycyclic organic hydrocarbon bioavailability from sediments. *Environ Toxicol Chem* 1998b, 17, 830-840.

Witkoswki, PJ; Smith, JA; Fusillo, TV; Chiou, CT. A Review of Surface-Water Sediment Fractions and their Interactions with Persistent Manmade Organic Compounds. Denver: US Geological Survey, 1987, *Circular* 993, 39 p.

Wong, CS; Capel, PD; Nowell, IH. National-scale, field-based evaluation of biota-sediment accumulation factor model. *Environ Sci Technol* 2001, 35, 1709-1715.

Xia, GS; Ball, WP. Adsorption-partitioning uptake of nine low-polarity organic chemicals on a natural sorbent. *Environ Sci Technol* 1999, 33, 262-269.

Yang, Y; Ratté, D; Smets, BF; Pignatello, JJ; Grasso, D. Mobilization of soil organic matter by complexing agents and implications for polycyclic aromatic hydrocarbon desorption. *Chemosphere* 2001, 43, 1013-1021.

Young, TM; Weber, WJ. A distributed reactivity model for sorption by soils and sediments. 3. Effects of diagenetic processes on sorption energetics. *Environ Sci Technol* 1995, 29, 92-97.

In: Environmental Pollution: New Research
Editor: Rachel H. Plattenberg, pp. 185-224

ISBN: 1-60021-285-9
© 2007 Nova Science Publishers, Inc.

Chapter 6

STEPS IN THE POLLUTION OF SOILS AFTER A PYRITE MINE ACCIDENT IN A MEDITERRANEAN ENVIROMENT

M. Simón[1], F. Martín[2], I. García[1], C. Dorronsoro[2] and J. Aguilar[2]
[1]Dpto. Edafología y Química Agrícola. Univ. Almería, CITE IIB, 04120 Almería, Spain
[2]Dpto. Edafología y Química Agrícola. Univ. Granada,
Fac. Ciencias, 18002 Granada, Spain

Abstract

Aznalcollar pyrite mine is located on the bank of the middle to upper stretch of the Guadiamar River (SW Spain). On 25 April 1998, the walls of the two contiguous ponds containing the ore-processing residues from of this pyrite mine broke open, spilling some 45 x 10^5 m^3 of polluted water and toxic tailings, affecting roughly 46 km^2. Most of the Cu, Zn, and Cd penetrated the soils in the solution phase of the spill (polluted water); meanwhile, the As, Pb and Tl penetrated mostly as part of the solid phase (tailings). Subsequently, drying and aeration of the tailings deposited over the soils rapidly oxidized sulphides to sulphates, lowered the pH, and increased the solubility of the pollutants. These soluble elements infiltrated the soils with the rainwater, swiftly augmenting the soil pollution. Consequently, a rapid removal of the tailings was recommended. In an extensive clean-up operation (May to December 1998) the tailings and 10 cm of the upper soil were removed using heavy machinery, which resulted in a deficient clean-up, as part of the tailings remained mixed with the soil. The oxidation of the leftover tailings lowered the pH and solubilized the contaminants of the cleaned soils. In non-carbonate soils, the pH fell to 3.0, the elements solubilized, rising to the soil surface and forming a crust that lowered the infiltration rate and thus encouraged the spread of the pollutants dissolved in the liquid phase of the runoff water. In carbonate soils, the acidity was largely neutralized ($pH \geq 6.5$), causing the pollutants to precipitate without forming a surface crust and reducing the spread of pollutants by the liquid phase of the runoff water. Nevertheless, the soil-erosion rate increased, intensifying the spread of the pollutants in the solid phase of the runoff. In both soil types, though for different reasons, 15 months after spill, the pollutants had not penetrated deeper than 10 cm. This implies that the speed of the clean-up was not as critical as a thorough removal of the spill without leaving behind remains of tailings mixed with soils. In addition, a random systematic sampling and analysis of the soils at 3 depths (0-10 cm, 10-30 cm, and 30-50 cm) at the end of

1998 (just after of the clean-up), enabled an estimate of the contamination level of the soils before and after the spill. The soils were contaminated prior to the accident, which raised the concentration in the first 10 cm of 60% of the soils. In 24% of the soils, the contaminant infiltrated to 10-30 cm in depth and only in the 12% reached 30-50 cm in depth. Because of the heavy pollution of the soils after tailings were removed, different remediation measures were taken (new cleaning of the most polluted soils, liming with sugar-refinery scum, addition of organic matter plus iron-rich clayey materials, and tilling the upper 20-25 cm of the soils). The analysis of the soils after each cleaning and remediation operation revealed that, although the measures were effective in reducing the concentration of contaminating elements in the soils, 6 years after the spill, 65% of the soils analysed continue to exceed the level of 50 mg $kg^{-1}$of As and 30% double this concentration.

1. Introduction

The Iberian pyrite belt covers nearly 8000 km^2 in the south-western corner of the Iberian Peninsula. This area, which includes massive sulphide deposits of the Rio Tinto, Aljustrel, Neves Corvo, Tharsis, Sotiel, and Aznalcóllar mines, is considered to be one of the largest metal reserves in western Europe. The origin of this belt can be found in the volcanic sediments produced by submarine volcanic intrusions, with alternating acidic and basic episodes. These intrusions emerged during the Hercinian period through of pre-existing pre-Cambrian and Palaeozoic sedimentary materials, precipitating finally as metallic sulphides on the marine bed. The mineral phase, located mainly in the upper part of the layer formed during the acidic episodes, consists of different sulphides (Schermerhorn, 1982; Almodovar et al., 1998): pyrite (83.1%), sphalerite (5.4%), galena (2.1%), chalcopyrite (1.4%), arsenopyrite (0.9%) and non-productive materials (7.1%). Mining activities (crushing, grinding, washing, flotation) produce the largest quantities of wastes, consisting primarily of pyrite ore refuse and acidic waters, which are deposited in holding ponds.

On 25 April 1998 the retention walls broke open in a holding pond belonging to the Aznalcóllar mine (Seville province, southern Spain), operated by the Boliden-Apirsa Company. Approximately 36 10^5 m^3 of polluted waters and 9 10^5 m^3 of toxic tailings were spilled into the Agrio and Guadiamar River basin (Figure 1). The tailings spread some 40 km in a down-river direction, stopping at Puente Don Simón; meanwhile, the polluted waters reached the Guadalquivir River, affecting the National Park of Doñana (proclaimed by UNESCO in 1994 as part of World Heritage). Nevertheless, the low dry-season flow of the Agrio and Guadiamar Rivers made possible the rapid construction of a retention dam at Lucio del Cangrejo, some 50 km downstream of the pond, minimizing the damage of the toxic wastes in the wildlife reserve. The total surface area affected by the toxic spill was approximately 45 km^2, forming a patch about 50 km in length and 0.9 km in wide (Figure 2).

2. Steps in the Soil Pollution

2.1. First Step: Infiltration of the Water and Tailings

On 4 May 1998 (9 days after spill) seven sectors were studied along the basins of the Agrio and Guadiamar rivers: near the mine (M), Soberbina (SO), Puente de las Doblas (D), Aznalcazar (A), Quema (Q), Pescante (P) and Los Pobres (LP). In each sector, analyses were made of contaminated and uncontaminated soils at two depths (0-10 and 10-30 cm), of tailings covering the soils, as well as of isolated pools of water from the toxic spill (Simón et al., 1999). All soils (Table 1) were neutral or slightly alkaline (pH between 7.2 and 8.1) and with relatively narrow ranges of organic-carbon content (0.5-1.7 %) and iron oxides (Fed: 0.8-1.5 %). By contrast, there was a broader range of $CaCO_3$ equivalent (0-20%), gravel (0-43%) and particle size of the fine-earth fraction (< 2 mm), ranging from clay loam (P and LP) to sandy loam (SO).

Table 1. Analytical data and structure of the contaminated soils (0-10 and 10-30 cm in depth) by sectors.

Soils (depth)	pH	$CaCO_3$	Organic Carbon	Gravel	Sand	Silt	Clay	Fed	Structure type /size /grade	SDI
		%								
M(0-10)	7.6	2.9	0.50	12.3	39.6	22.0	26.1	1.29	sbk / c / 2	10
M(10-30)	7.9	2.5	0.43	9.6	51.3	16.6	22.5	1.15	sbk / vc / 1	3
SO(0-10)	7.3	0.3	0.80	0.8	68.0	18.5	12.7	0.89	sg / - / 0	9
SO(10-30)	7.4	0.0	0.65	1.2	71.2	16.6	11.0	1.00	sg / - / 0	0
D(0-10)	7.6	18.6	0.93	29.7	22.5	21.4	26.4	0.96	sbk / m / 2	14
D(10-30)	7.4	19.6	0.61	42.8	16.1	22.2	18.9	1.12	sbk / m / 2	14
A(0-10)	7.7	14.8	1.65	0.0	1.9	52.5	45.6	0.81	abk / vc / 1	3
A(10-30)	7.8	14.6	1.34	0.0	1.7	54.8	43.5	0.92	m / - / 0	0
Q(0-10)	7.9	9.4	1.36	0.0	21.9	37.7	40.4	1.15	sbk / m / 1	7
Q(10-30)	8.1	7.0	0.92	0.0	19.4	45.1	35.5	1.44	sbk / c / 1	5
P(0-10)	7.2	14.3	1.02	2.1	7.7	27.1	63.1	1.10	abk / f / 3	30
P(10-30)	7.8	14.7	0.61	1.9	6.3	28.2	63.6	1.18	abk / vc / 1	3
LP(0-10)	7.8	15.3	0.98	6.9	1.1	29.7	62.3	0.84	abk / vc / 1	3
LP(10-30)	7.8	15.5	0.46	5.8	0.9	31.0	62.3	0.88	m / - / 0	0

Abbreviations of structure are from Soil Survey Staff (1951). M= Mine, SO= Soberbina, D= Puente de las Doblas, PA= Pilas-Aznalcazar, Q= Quema, P= Pescante and LP= Los Pobres.

SDI (Structural-development index) = grade x size, where the size of the structure take the following values: fine = 10, medium = 7, coarse = 5, very coarse = 3.

Figure 1. Break in the retention wall of the pond that spilled the tailings into the Agrio River (Junta de Andalucía).

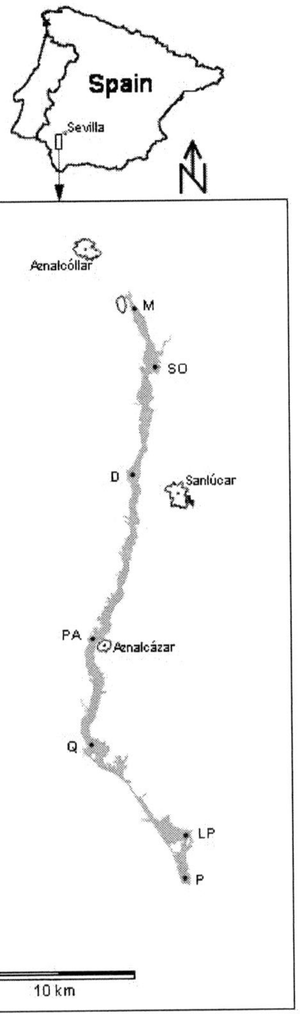

Figure 2. Map of the zone affected by the spill and situation of the seven study sectors.

Figure 3. White salty crust formed on the tailings.

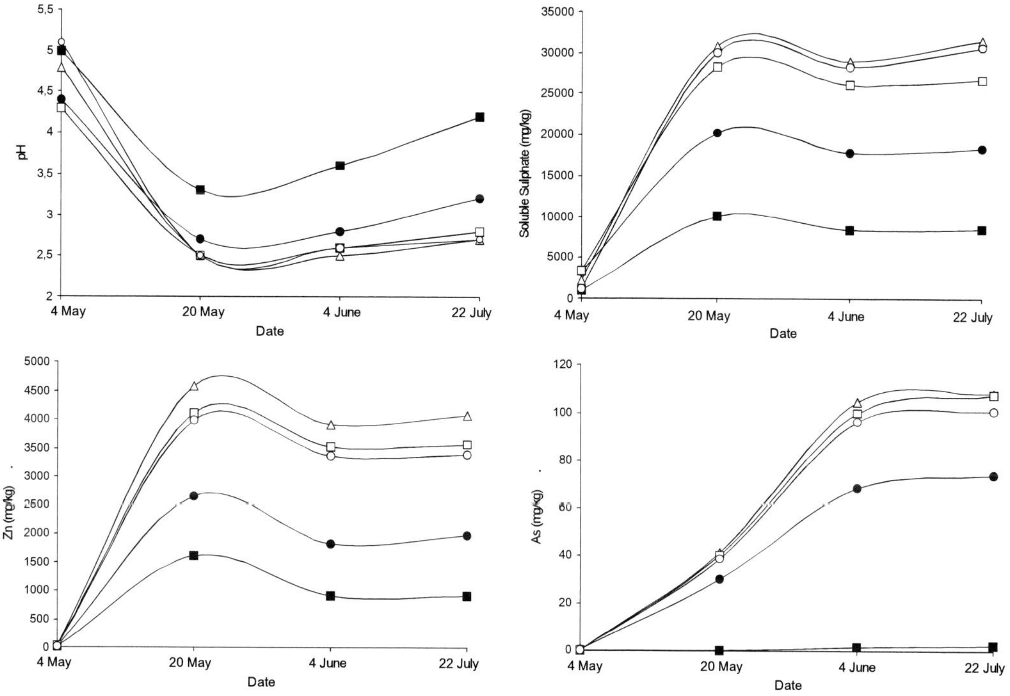

Figure 4. The pH and water-soluble sulphates, Zn and As in the tailings on different dates of 1998 by sectors: ■ M, ● SO, △ A, □ Q, ○ P.

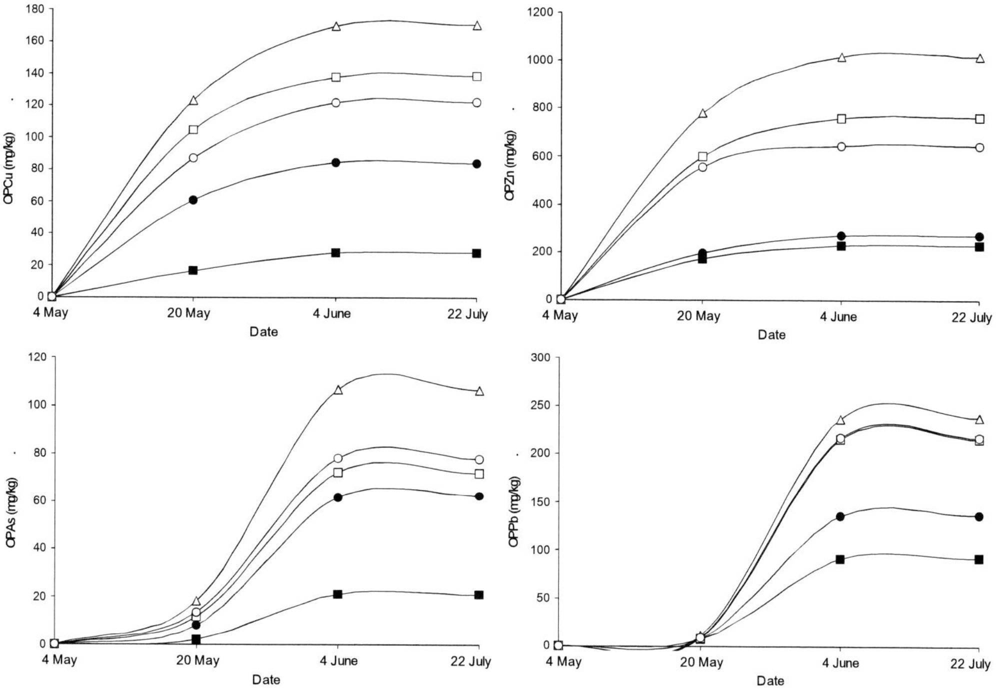

Figure 5. Oxidative pollution for Cu, Zn, As and Pb in the soils between 0 and 10 cm in depth on different dates of 1998 by sectors: ■ M, ● SO, △ A, □ Q, ○ P.

The infiltration of the water and tailings from the spill considerably increased the concentration in Zn, Pb, Cu, As, Sb, Bi, Tl and Cd in the first 10 cm of the soils affected in relationship to uncontaminated soils (Table 2). The elemental contents of the polluted waters were relatively homogeneous (low standard-deviation values), exceeding the maximum concentration in As, Cd, and Cu allowed in water to be used for irrigation (Crook and Bastian, 1992, in Brieva Romero, 1995). On the contrary, the standard deviation of these pollutants increased considerably in the tailings and especially in the contaminated soils (in some cases surpassing the mean value), indicating that the contamination had highly irregular effects on the soils of the different sectors.

Table 2. Mean values (mg kg^{-1}) and standard deviation (SD) of pollutants in tailings (T), polluted waters (W) and contaminated (CS) and uncontaminated (UCS) soils between 0 and 10 cm in depth.

Element	T (mg kg^{-1})		W (mg dm^{-3})		CS (mg kg^{-1})		UCS (mg kg^{-1})	
	Mean	SD	Mean	SD	Mean	SD	Mean	SD
Zn	8063.4	613.5	72.1	1.8	647.2	444.9	230.8	123.1
Pb	9635.9	2364.1	2.1	0.3	361.1	630.5	41.8	16.4
Cu	2175.8	381.1	0.3	0.08	152.7	129.8	40.1	17.1
As	4122.6	1066.4	0.002	0.001	121.7	211.0	18.1	6.5
Sb	957.9	214.6	0.003	0.001	28.4	52.5	2.2	0.7
Bi	86.6	20.7	nd	nd	2.5	4.4	0.4	0.2
Cd	36.8	5.3	0.65	0.06	2.3	1.8	0.5	0.3
Tl	66.8	12.6	0.07	0.00	2.3	3.2	0.6	0.2

nd = not detected

Because the water from the toxic spill contained no Bi (Table 2), the total Bi contamination of the soils must have come from the tailings. Thus, the quantity of tailings that penetrated the soil in each sector (Z) could be calculated by the equation:

$$Z\ (\text{g kg}^{-1}) = (CS_{Bi} - UCS_{Bi})\ 10^{3} / T_{Bi}$$

where T_{Bi} is the Bi concentration in the tailings and CS_{Bi} and UCS_{Bi} are the Bi concentration in the contaminated and uncontaminated soils of each sector, respectively, all expressed in mg kg^{-1}. The results (Table 3) indicated that the penetration of the tailings into the soil was highly irregular, varying considerably from one sector to another according to the characteristics of each soil, especially particle size and structure. In general, the penetration of the tailings diminished as structure size increased and structure grade decreased (Table 1).

From the quantity of tailings that penetrated the soil (Z) and from the concentration of each element i in the tailings (Ti, in mg kg^{-1}) of each sector, estimations were made of the quantity of each element i which entered the soil from the solid phase (Si) of the spill, using the equation:

$$Si\ (\text{mg kg}^{-1}) = Z\ Ti\ 10^{-3}$$

Table 3. Quantity of tailings (Z) that penetrated the soil (0-10 cm in depth) in each sector.

Sector	Z (mg kg^{-1})
M	7.4
SO	1.7
D	9.8
A	2.4
Q	8.3
P	150.8
LP	24.0

In addition, since the total pollution of the soils was caused both by water and the tailings from the toxic spill, the quantity of each element i that entered as part of the solution phase (Wi) was possible to estimate by the equation:

$$Wi\ (mg\ kg^{-1}) = (CSi - UCSi) - Si$$

where CSi-UCSi (in mg kg^{-1}) is the total contamination of each soil for each element i.

Table 4. Median and range of the total contamination (CSi-UCSi) for all soils pooled by each element, and percentage of this total belonging to the solid phase (Si) of the spill.

	0-10 cm in depth			
	CSi – UCSi (mg kg-1)		Si (%)	
Element	Median	(Range)	Median	(Range)
Zn	416.4	(59.9 -1266.4)	34	(5-87)
Pb	319.3	(20.9 -1710.9)	69	(40-85)
Cu	112.1	(16.8 - 325.9)	36	(8-87)
As	103.2	(4.3 - 566.7)	96	(88-100)
Sb	25.2	(1.3 -144.9)	96	(90-100)
Bi	2.1	(0.1 - 11.5)	100	(100)
Cd	1.7	(0.3 - 5.2)	34	(6-71)
Tl	1.7	(0.1- 8.7)	85	(73-96)

The results (Table 4) indicate that the range of total contamination in the first 10 cm on the soils for each element was very broad, as a consequence of the markedly different quantities of tailings that penetrated each soil (Table 3). The highest values corresponded to the sector where the penetration was the greatest (P), and the lowest where the tailings penetrated the least (SO and A). The percentage of this total corresponding to the solid phase of the spill clearly varied from one element to another. In the case of Cu, Zn and Cd, only about 35% of the mean total contamination in the first 10 cm of the soil penetrated as part of the solid phase of the spill, and thus the remaining 65% must have penetrated as part of the solution phase of the spill. On the contrary, As and Sb entered the soil mainly through the solid phase (mean of 95%) and to a far lesser degree (5%) through the solution phase. Meanwhile, Pb and Tl registered intermediate values, penetration through the solid phase reaching 70 and 85%, respectively.

Table 5. Principal components analysis (Load matrix)

Variable	F1	F2	F3
S_{Zn}	0.98		
S_{Cd}	0.99		
S_{As}	0.99		
S_{Cu}	0.98		
S_{Sb}	0.98		
S_{Bi}	0.98		
S_{Tl}	0.99		
S_{Pb}	0.98		
W_{Zn}		0.91	
W_{Cd}		0.82	
W_{As}			0.82
W_{Cu}			0.58
W_{Sb}	0.79		
W_{Tl}	0.98		
W_{Pb}	0.93		
pH	-0.51	0.54	
CaCO3 (%)		0.72	
O.C. (%)		0.64	
Sand (%)		-0.82	
Silt + Clay		0.75	
Fed			0.87
SDI	0.87		
% cumulative variance	52	74	83

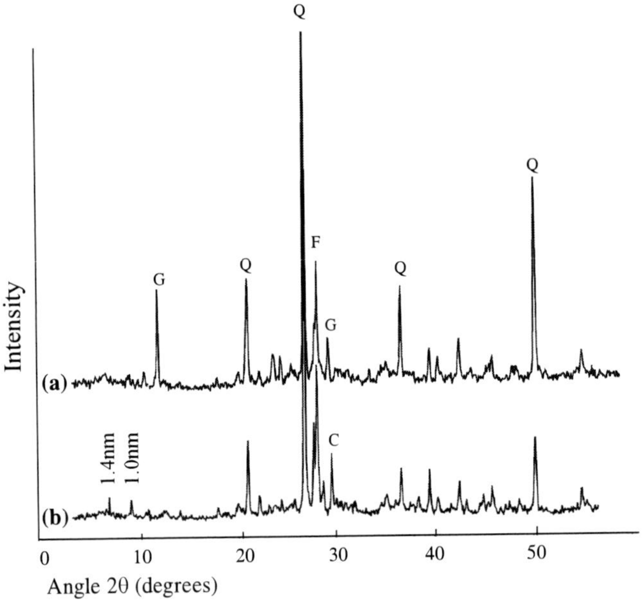

Figure 6. X-ray diffraction diagrams of the soil-powder samples within (a) and beneath (b) the first 12 mm in the soil sampled on 25 July 1999: G = gypsum, F = feldspars, C = calcite, Q = quartz.

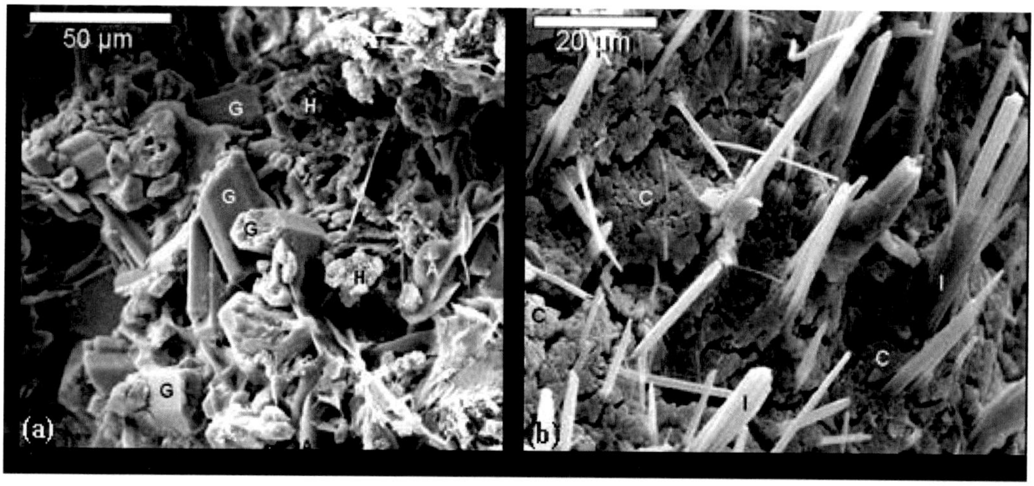

Figure 7. SEM-EDX analysis of certain minerals present in the first 10 cm of the soil sampled on 25 July 1999: (a) gypsum (G), raised scaly mineral composed of S and Al (A), aggregate microcrystals made up mainly of Fe (H); (b) needle-like minerals composed of S and Fe (I) projecting from a mass composed of S, Zn, Fe, Mn and Al (C).

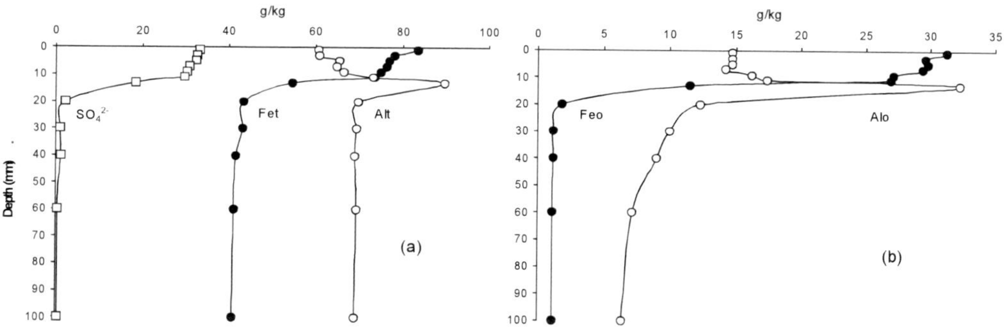

Figure 8. (a) Concentrations of SO_4^{2-}, Fe_t and Al_t, and (b) concentrations of iron (Fe_o) and aluminium (Al_o) extracted with ammonium oxalate of the soil sampled on 25 July 1999.

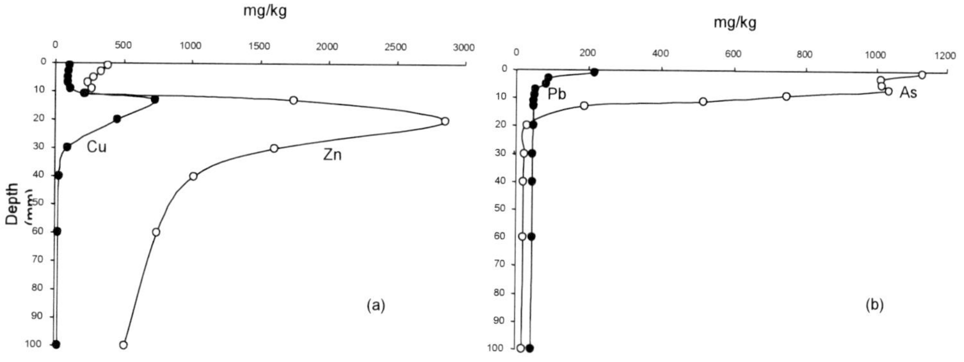

Figure 9. (a) Concentrations of Zn and Cu, and (b) concentrations of As and Pb of the soil sampled on 25 July 1999.

Figure 10. Morphological characteristics of the soil sampled in July 2001.

A principal-component analysis was performed with Si and Wi values, structural-development index (SDI), and the analytical characteristics of the soils (Table 1). Three factors explain more than 80% of the variance (Table 5). Factor 1 confirmed the structure (SDI) to be the soil characteristic which best determined the penetration of the tailings; meanwhile, the negative sign of the pH indicated that the pollution tended to acidify the soil, although this trend was not strongly evident (low load) due to the buffering effect of the $CaCO_3$ in most of the soils. Factor 2 indicates that the sorption of Zn and Cd was caused by the formation of insoluble precipitates as well as organic complexation and cation exchange (Alloway, 1995). The rise in the pH tended to favour these sorption processes, although the effect was not pronounced (low load) because of the above-noted narrow range of the pH values of the soils. Factor 3 reveals that the specific adsorption of As and to a lesser extent Cu was due primarily to the iron oxides (Alloway, 1995).

In any case, in the first 10 cm of the soils, only As and Zn in the same sector exceeded the maximum permitted concentrations, for agricultural soils, set by Canada (Sheppard et al., 1992), Belgium (Stringer, 1990) and Holland (NMHPPE, 1991, in Alloway (ed), 1995), as well as the ecotoxicological level set by Van Den Berg et al. (1993). Nevertheless, in the not-too-distant future, all the elements could exceed the maximum permitted levels, given that, as the tailings dry and aerate, a complex process oxidizes the sulphides to sulphates (Nordstrom, 1982) from reactions such as:

$$FeS_{2(s)}\ (\text{pyrite}) + 3O_{2(g)} + 2H_2O \longleftrightarrow Fe^{2+}_{(aq)} + 2SO_4^{2-}{}_{(aq)} + 4H^+_{(aq)}$$

$$AsFeS_{2(s)} \text{ (arsenopyrite)} + 3O_{2(g)} + 2H_2O \Leftrightarrow As^{3+}_{(aq)} + Fe^{2+}_{(aq)} + 2SO_4^{2-}_{(aq)} + 4H^+_{(aq)}$$

$$ZnS_{(s)} \text{ (sphalerite)} + O_{2(g)} + 2H_2O \longleftrightarrow Zn^{2+}_{(aq)} + SO_4^{2-}_{(aq)} + 4H^+_{(aq)}$$

These processes would lower the pH (Stumn and Morgan, 1981) and solubilize part of the pollutants that had formerly remained insoluble (Förstner and Wittmann, 1983). Thus, on 4 May 1998, pH fell and the concentrations of soluble SO_4^{2-} increased logarithmically in the tailings as moisture declined:

$$pH = 3.36 + 0.99 \log (\text{moisture } \%) \qquad r^2 = 0.923$$

$$SO_4^{2-} (\text{mg kg}^{-1}) = 7031.6 - 3916.1 \log (\text{moisture } \%) \qquad r^2 = 0.984$$

2.2. Second Step: Oxidation of thc Tailings

Given that between 25 April (date of the spill) and 4 May (date of sampling) no rain fell, these solubilized elements remained in the solution phase of the tailings and, with evaporation, rose by capillary action to the surface, forming a white salty crust (Figure 3) constituted by hydroxysulphate complexes such as: hexahydrite $MgSO_4.6H_2O$, bianchite $(Zn,Cu)SO_4.6H_2O$, and beaverite $(Pb,Cu)(Fe^{III},Al)_2(SO_4)_2(OH)_6$. Subsequent rains would be expected to dissolve these soluble salts, which would then infiltrate the soil and thereby raise the pollution level. This process, and its speed (in scarcely 10 days multiplying 10-fold the content in soluble SO_4^{2-} in the driest tailings in relation to the wettest), underscores the urgency of removing the tailings from the soil surfaces in these types of spills. In an extensive clean-up operation (May to December 1998) the tailings and approximately 10 cm of the upper soils were removed using heavy machinery.

To ascertain the oxidation rate of the tailings as well as the increase in soil contamination over time, on three different sampling data: 20 May, 4 June and 22 July 1998, we continued monitoring the concentrations of the pollutants in the tailings and soils in five sectors (of the seven mentioned above) where the tailings had not been removed (M, SO, A, Q, and P).

In all the sectors the oxidation process was very rapid (Simón et al., 2001), reaching the highest concentration of sulphates and lowest pH values on 20 May (25 days after spill). After 20 May, the sulphate content stopped rising, indicating a slowing of the oxidation rate, and in fact this content fell between 20 May and 4 June, implying a greater loss of sulphate ions by leaching than those formed by oxidation during this period (Figure 4). This pattern of concentration in sulphate ions paralleled that of water-soluble Zn, Cu, and Cd, with the solubilization on 20 May reaching between roughly 20% (sector M) and 65% (sectors A, Q and P) of the total Zn and Cd present in the tailings, and between about 8% (sector M) and 45% (sectors A, Q and P) of the total Cu. This reflects a rapid and intense oxidation of sulphides bonded to these elements (such as sphalerite or chalcopyrite). On the contrary, the oxidation of the sulphides containing As and Pb, such as arsenopyrite or galena, was quantitatively far lower, the soluble contents of these elements in the tailings representing only a minor fraction with respect to the total (with peak values of around 2.5% for As and 0.1% for Pb). In addition, the oxidation was slower, increasing more in the second interval (20 May - 4 June).

Table. 6. Particle size, bulk density and sulphur content of the tailings sampled on 4 May by sectors

Sector	% Sand (2-0.005 mm)	Coarse silt (0.005-0.002 mm)	Fine silt (0.002-0,0002 mm)	Clay (<0.0002 mm)	Bulk density (g/cm3)	Sulphur (g/kg)
M	9.6	46.3	26.6	17.5	2.51	275.2
SO	9.6	17.2	58.7	14.5	2.33	353.1
A	0.6	9.5	73.5	16.4	2.19	382.1
Q	1.2	10.6	71.3	16.9	2.08	380.0
P	1.8	8.9	72.6	16.7	1.96	382.4

On the other hand, the oxidation rate of the tailings, estimated from the ratio between soluble sulphur (Ss) and total sulphur (St), clearly accelerated with the distance from spill due to the different physical and chemical properties of the tailings in each sector (Table 6). Thus, there was a significant positive relationship with the fine-silt and total-sulphur contents, and a negative one with the sand, coarse-silt, and bulk density (Table 7).

The elements solubilized during the oxidation and acidification of the tailings leached with the rainwater, accumulating in the underlying soil and thereby exacerbating the soil pollution (hereafter termed "oxidative pollution"). The climate of the affected area is typically Mediterranean (hot, dry summers; cold, wet winters; temperate autumns and springs with variable rainfall), with a mean annual rainfall of 613 mm, a mean temperature of 17.7 °C, and a potential evapotranspiration of 900 mm (taken from 7 weather stations near the spill, over 27 years). Given that between 25 April (date of the spill) and 4 May (date of the first sampling) no rain fell (Table 8), the increase of these elements in the soil sampling on 20 May or later, can be ascribed to oxidative pollution. Thus, the rate of oxidative pollution of each element (OPi) can be calculated by the difference in the concentration between the first soil sampling (on 4th May) and any subsequent one.

Table 7. Correlation matrix including oxidation rate (S_s:S_t), particle size (%), bulk density (g/cm^3) and sulphur concentration (g/kg) of the tailings sampled on 4 May 1998.

	S_s:S_t	Sand	CS	FS	Clay	BD	Sulphur
S_s:S_t	1.00						
Sand	-0.91**	1.00					
CS	-0.95**	0.76	1.00				
FS	0.98***	-0.82	-0.99***	1.00			
Clay	-0.09	-0.31	0.36	-0.28	1.00		
BD	-0.90**	0.84*	0.87*	-0.88**	-0.04	1.00	
Sulphur	0.97***	-0.99***	-0.99***	0.99***	-0.29	-0.89**	1.00

CS=coarse silt, FS= fine silt, BD= bulk density

Significance levels: *P < 0.1; **P < 0.05; ***P < 0.01

The majority of the oxidative pollution in Zn, Cd, Cu (65-90%) occurred by 20 May (Figure 5), as a consequence of the strong oxidation and solubilization of the sulphides bonded to these elements in the tailings during the first 25 days after the spill (as discussed above). Meanwhile, most of this contamination in As and Pb (70-90%) occurred between 20

May and 4 June, when the oxidation and solubilization of the sulphides bonded to these elements in the tailings were highest. Between 4 June and 22 July the oxidative pollution did not increase because of the absence of rainfall (Table 8).

Table 8. Rainfall on different dates of 1998 by sectors. The rainfall data for 4 May refer to the period between the spill and the first sampling, and the rest of the dates refer to the period between samplings

Sectors	Rainfall (mm)			
	4 May	20 May	4 June	22 July
M and SO	0.0	45.6	15.2	0.0
A, Q, and P	0.0	63.1	25.6	0.0

Table 9. Oxidative pollution in the soils from 0 to 10 cm and 10 to 30 cm in depth on 22 July 1998 by sectors

Sector (cm)	(mg/kg)				
	Cu	Zn	Cd	As	Pb
M (0-10)	27.6	229.6	2.32	21.2	91.6
M (10-30)	5.7	28.8	0.27	5.2	18.1
SO (0-10)	84.0	270.5	1.73	62.4	136.1
SO (10-30)	11.2	37.1	0.20	17.9	21.3
A (0-10)	171.1	1019.1	5.00	106.5	237.3
A (10-30)	20.2	125.8	0.74	24.5	43.1
Q (0-10)	139.1	760.3	4.62	71.8	217.5
Q (10-30)	18.9	119.7	0.63	20.8	39.9
P (0-10)	122.2	644.3	4.83	77.8	217.0
P (10-30)	14.6	109.8	0.70	15.4	32.8

As a result of this intense oxidative pollution, on 22 July 1998 the first 10 cm of the soils of M, SO, A and Q reached or exceeded the maximum concentration permitted for agricultural soils in As (50 mg kg^{-1}; NMHPPE, 1991, in Alloway, 1995) and Tl (1 mg kg^{-1}; Kabatas Pendias and Pendias, 2000). The soils of A and Q exceeded the maximum permitted concentrations in Zn and Cu (720 mg kg^{-1} and 190 mg kg^{-1}, respectively; NMHPPE, 1991, in Alloway, 1995; Van Den Berg et al., 1993). In the soils nearest the spill (M and SO), where the initial pollution and subsequent oxidation and solubilization rates of the tailings were lower, the increases in the Zn and Cu content were weaker and the above-mentioned toxicity limits were not reached. Meanwhile, the soil concentration in Cd and Pb was considerably raised by the oxidative pollution, but in no sector, except P (highly contaminated in the first step), did Cd and Pb exceed the maximum level allowed for agricultural soils (12 and 530 mg kg^{-1}, respectively; NMHPPE, 1991, in Alloway, 1995; Van Den Berg et al., 1993). In any case, the oxidative pollution from 10 to 30 cm in depth was considerably less than that of the first 10 cm of the soils (Table 9), decreasing an average of 80-85% in the case of Cd, Zn, Cu and Pb, and 70-80% in As. Consequently, the pollutants tended to remain in the first 10 cm of the soils without seriously contaminating either the subsoil or the groundwater. Nevertheless, the continued infiltration of the acidic solution with time would increase the concentration of pollutants and could also affect other soil properties.

2.2.1. Oxidative Pollution 15 Months after the Spill

Although the clean-up operation of the tailings ended on December 1998, a plot (50 x 50 m length) in the Quema sector (Q) was preserved without being cleaned, with the aim of ongoing research. In this plot, the effect of the continued infiltration of the solution in the soils was studied with a new sampling performed 15 months after spill.

On July 1999, the first 12 mm of the soil had a reddish-yellow colour (7.5YR 6/8), clearly different of the uniform pale brown colour (10YR 6/3) showed in the prior sampling (22 July 1998). In this soil, a detailed study was made of the soil properties and their pollutant concentrations (Simón et al., 2002), sampling the reddish-yellow layer every 2 mm (0-2, 2-4, 4-6, 6-8, 8-10, 10-12 mm in depth) and sampling the underlying soil at different depths to 101 mm (12-14, 19-21, 29-31, 39-41, 59-61 and 99-101 mm in depth).

The infiltration of the acidic solution (of which the chemical characteristics can be estimated from extracts of tailings:water; Table 10) into the soil resulted in the total decarbonation and strong acidification of the reddish-yellow layer (Table 11). The pH fell to a low of 3.2 compared with a high of 8.1 in the underlying calcareous soil. The X-ray of the soil-powder samples (Figure 6) revealed that, within the first 12 mm, the reflections of the calcite disappeared, the relative intensities of feldspars diminished, and the intensity and sharpness of the basal reflections at 1.0 (illite) and 1.4 nm (smectite) also decreased; indicating that the acidity encouraged the total weathering of carbonates and the partial weathering of silicates. Consequently, the $CaCO_3$ present in the soils partially neutralized the acidity of the solution, the oxidation of Fe^{2+} to Fe^{3+} proceeded rapidly (Singer and Stumm, 1968), iron precipitated to form the layer of reddish-yellow colour, and the calcium and sulphate ions formed gypsum by the reaction (Ritsema and Groenenberg, 1993; Kashir and Yanful, 2000):

$$Fe^{3+}_{(aq)} + 2SO_4^{2-}{}_{(aq)} + H^+_{(aq)} + 2CaCO_{3(s)} + 5H_2O \longrightarrow Fe(OH)_{3(s)} + 2CaSO_4.2H_2O_{(s)} + 2CO_{2(g)}$$

Table 10. The pH and soluble-pollutants concentrations on 25 July 1999 in the extract of tailings:water (1:10).

	mg/dm^3
pH	2.51
SO_4^{2-}	2305.9
Fe	312.6
Al	1.0
Cu	58.6
Zn	332.1
Cd	1.0
As	28.6
Pb	3.8

The SEM-EDX analysis in the first 12 mm, in addition to crystals composed of S and Ca and tabular habit typical of gypsum, also showed raised scaly crystals composed of S and Al and aggregate microcrystals made up of Fe (Figure 7a), as well as needle-like crystals composed of S and Fe projecting from a mass made up of S, Zn, Fe, Mn and Al (Figure 7b). These results indicate that, in addition to the precipitation of gypsum, sulphate may also be adsorbed by various Fe and Al oxides and hydrous oxides, this being assumed to be the primary mechanism for SO_4^{2-} retention in a variety of soils (Parfitt & Smart, 1978; Johnson & Todd, 1983; Singh, 1984).

The chemical analysis of the soil samples confirmed the accumulation of sulphate ions in the reddish-yellow layer (Figure 8a), where their concentration increased by approximately 30 g kg^{-1} with respect to the underlying soil at 100 mm in depth. The iron in the acid solution also precipitated in this layer, increasing the total iron (Fe_t) by approximately 36 g kg^{-1} with respect to the soil at 100 mm in depth. Meanwhile, the Al diminished in the first 12 mm of the soil (by approximately 2.5 g kg^{-1}) and increased fundamentally between 12-14 mm in depth (pH = 6), reflecting the hydrolysis mineral and leaching processes that affected this reddish-yellow layer. From the 36 g kg^{-1} of the iron accumulated on the first 12 mm, 28 g kg^{-1} were extracted with ammonium oxalate (Fe_o), indicating that the iron precipitated basically in the amorphous or poorly crystallized form (Figure 8b). Similarly, almost all aluminium accumulated between 12-14 mm was extracted with ammonium oxalate (Al_o), and therefore also precipitated in the amorphous or poorly crystallized form. The relatively high sulphate content in these layers and the predominance of the poorly crystallized form, appear to indicate that the precipitation could have occurred fundamentally as schwertmannite in the case of iron (Nordstrom, 1982; Bigham et al., 1996; Bigham and Nordstrom, 2000), and basaluminite in the case of aluminium (Nordstrom and May, 1996). The colour 7.5YR 6/8 of the uppermost 12 mm coincides with that characteristic of schwertmannite (Scheinost and Schwertmann, 1999).

All the pollutant elements dissolved in the acidic solution also tended to accumulate in the first 100 mm of the soil, although with clear differences between them. Thus, the mean concentrations of Zn and Cd in the first 12 mm of the soil were equal to or even less than between 99-101 mm depth (Figure 9a), and therefore these elements did not accumulate in this highly acidic layer of soil ($pH<4.5$); it even appears that the acidic solution from the tailings solubilized and leached part of the Zn already present in the soil. The accumulation began in the soil layer with pH=6.0 (12-14 mm in depth) and reached its maximum in the soil layer with pH>7.0 (19-21 mm in depth). Deeper than 21 mm, concentrations progressively decreased. The Cu partially accumulated in the strongly acidic layer ($pH < 4.5$), where the mean concentration was higher than between 99-101 mm, reaching its highest accumulation at pH=6.0 (12-14 mm in depth) and accumulating almost totally in the first 40 mm of the soil. The As tended to accumulate more or less uniformly in the reddish-yellow and strongly acidic layer (Figure 9b); meanwhile, the Pb accumulated in the first 2 mm of the soil, values falling sharply below this depth.

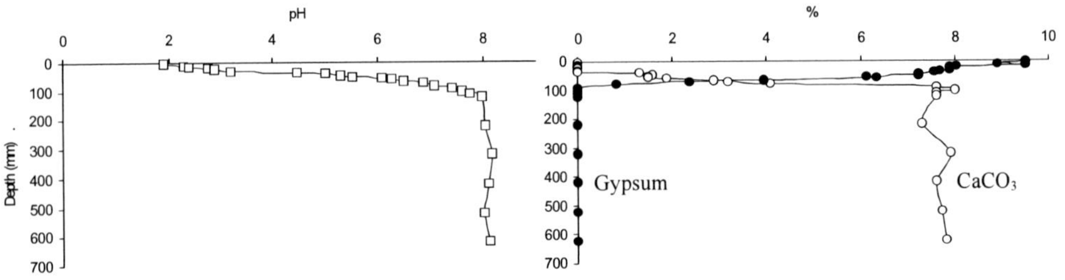

Figure 11. The pH and content of gypsum and $CaCO_3$ of the soil sampled in July 2001.

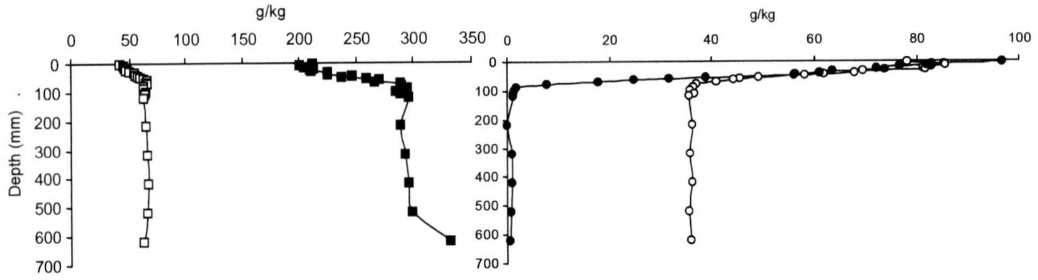

Figure12. Total concentrations of Al and Si (a), and Fe and S (b) with depth in the soil sampled in July 2001.

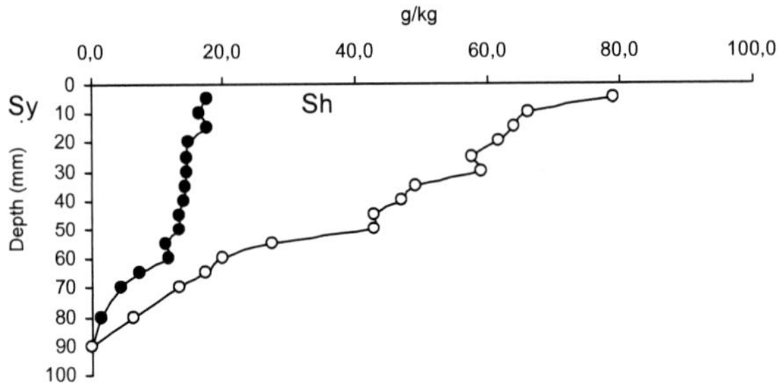

Figure 13. Sulphur precipitated as gypsum (Sy) and as iron and aluminium hydroxysulphates (Sh) with depth of the soil sampled in July 2001.

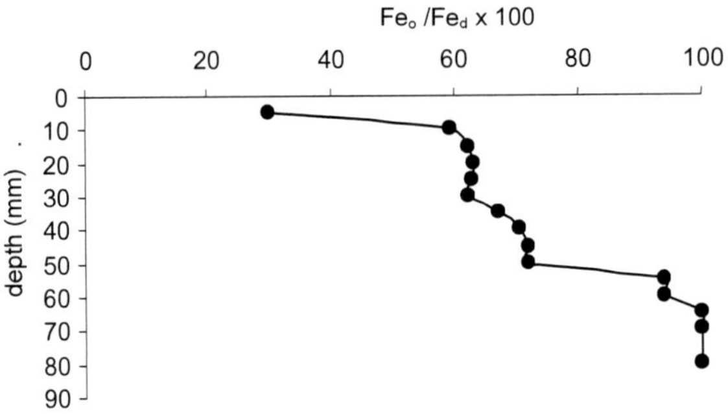

Figure 14. Percentage of total iron oxides (Fe_d) which were extracted with ammonium oxalate (Fe_o) with depth of the soil sampled in July 2001.

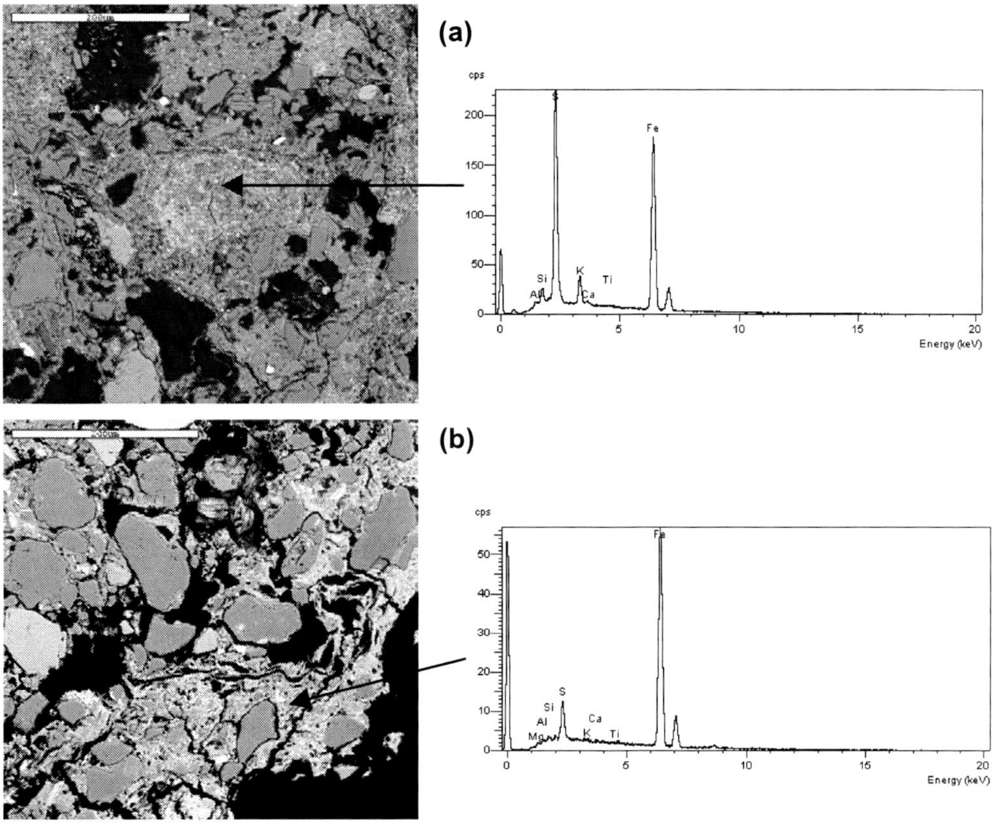

Figure 15. SEM-EDX analysis of the ground mass between 0-5 mm (a) and between 55-80 mm (b) of the soil sampled in July 2001.

The concentrations of the pollutant elements between 99 and 101 mm in depth of the soils sampled on 25 July 1999 was very similar to the concentration between 0 and 10 cm in depth in the soil sampled on 4 May, so the difference in concentration at each depth and between 99-101 mm can be considered oxidative pollution (OP). The oxidative pollution of Zn (OP_{Zn}) and Cd (OP_{Cd}) proved to be significantly related to the clay content and pH ($P < 0.05$), indicating that they were adsorbed onto or diffused through the clay structure, this process increasing with the pH (Harter, 1983; Shuman, 1985; Brummer, 1986). The oxidative pollution of Cu (OP_{Cu}) proved to be significantly related to Al_o ($P < 0.001$), indicating that Cu was adsorbed fundamentally by amorphous or poorly crystallized hydrous Al oxides (Baker and Senft, 1995). The oxidative pollution of As (OP_{As}) was highly correlated to Fe_o and SO_4^{2-} ($P < 0.001$), indicating that the As, probably as H_2AsO^- (the major species in this pH range), was highly adsorbed by iron hydroxysulphates (O'Neill, 1995; Waychunas et al., 1995). Meanwhile, the oxidative pollution of Pb (OP_{Pb}) proved not to be related to the hydrous oxides (Al_o or Fe_o), despite that the amorphous Fe oxides have a high affinity for Pb (Kinniburgh et al., 1976; McKenzie, 1980). Nevertheless, the presence of SO_4^{2-} might have caused the Pb to accumulate in the first 2 mm of the soil as $PbSO_4$ (Lindsay, 1979; Fuge et al., 1994), while the Fe_o content was distributed more or less uniformly throughout the first 12 mm, masking any statistical relationship between the two.

To estimate the mobility of the different pollutants in the soil, a mobility index (M) was calculated by the equation:

$$M = \sum_{n=1}^{12} (OP_{in} / TOP_i) d_n$$

where n represents the layers sampled, OP_{in} is the oxidative pollution of each element i in each layer sampled n, TOP_i is the sum of the oxidative pollution of each element i of all layers sampled, and d_n the depth (mm) of the lower limit of each layer sampled n. The value of this index would range from 2 (if the element accumulated entirely in the first 2 mm) to 101 (if the element accumulated entirely between 99 and 101 mm). The mobility in descending order was (M values in parenthesis):

Zn (29.7) > Cd (28.1) >> Cu (14.7) >> As (6.6) > Pb (4.8)

In any case, after 15 months, the solution from the oxidation of the tailings acidified only the first 20 mm of the soils, and most of the pollutants penetrated no deeper than 100 mm, without contaminating the subsoil or groundwater. This implies that under a Mediterranean climate in calcareous soils, the speed of the cleanup is not as important as a thorough removal of the spill without leaving behind remains of tailings mixed with soils.

Nevertheless, the more mobile elements as Zn as well as the less mobile as As, were progressively increasing the concentration in the first 10 cm of the soils with time. In all cases, the concentration was significantly related to the square root of the elapsed time from the spill in days:

$$\text{Zn (mg/kg)} = 400.6 + 96.7 \sqrt{t(days)} \quad r^2 = 0.925 \qquad P < 0.001$$

$$\text{As (mg/kg)} = -1.56 + 15.73 \sqrt{t(days)} \quad r^2 = 0.978 \qquad P < 0.001$$

Consequently, it could be expected that, with time, the pollutant elements infiltrate below the uppermost 10 cm, affecting a larger soil volume. To continue studying the effect of this infiltration in the soils, a new sampling was undertaken in the same plot on July 2001 (39 months after spill).

2.2.2. Oxidative Pollution 39 Months after the Spill

Three years after the accident, the soil morphology was strongly changed (Figure 10). Just below the tailings, the upper 5 cm of the soil was discoloured to a pale grey (2.5Y 7/2), while the reddish-yellow layer (7.5YR 6/8) considerably augmented its thickness, including at depths of between 5 and 70 mm. From 70 mm in depth, the reddish-yellow colour tended to become progressively pale brown (10YR 6/3), as in the original soil; the unaffected soil began at 110 mm in depth. For a detailed analysis of these changes, we carried out a systematic study of the soil properties at different depths, taking one sample every 5 mm to the depth of 70 mm, one sample every 10 mm between 70 and 120 mm and, finally, one sample every 100 mm between 120 and 620 mm.

The results indicated that the soil surface in contact with the tailings continued decarbonating and acidifying (Figure 11), with the simultaneous precipitation of gypsum. The weathering of the carbonates was total to 35 mm in depth, and partial to 80 mm (higher in comparison to the depths of 10 and 21 mm, respectively of the previous sampling). In the first 30 mm the pH was below 4.0 (similar value to the first 10 mm of the previous sampling), and the pH in the uppermost part of the soil (pale-grey layer) decreased to 1.9 (much more acidic than the previous sampling, pH=3.2). The percentage of gypsum (Figure 11b), estimated by precipitation with acetone (Bower and Huss, 1948), was between 6-10% to a depth of 60 mm, decreasing progressively to the 90 mm in depth.

The total concentration of silicon (Si_t) and aluminium (Al_t) clearly declined in the first 70 mm of the soil (Figure 12a), confirming that the acidification partially weathered the primary minerals such as feldspars and phyllosilicates (Preda and Cox, 2004). The mobility of these elements seems to be closely related to soil pH, although with some differences. Thus, Al accumulated mainly from 50 to 90 mm in depth, where the pH was higher than 5.5; whereas Si presented higher relative mobility, accumulating at a depth slightly greater than in the case of Al (70 - 120 mm), coinciding with the zone of pH > 7.5. On the contrary, the iron dissolved in the acidic solution continued precipitating in the upper part of the soil, especially in the first 30 mm where the total iron content (Fe_t) increased by around 40 g kg^{-1} in relation to the original soil (> 320 mm in depth). Deeper than 30 mm, the iron accumulation progressively diminished to a depth of 90 mm, where the concentration was similar to the one in the original soil.

The total sulphur concentration (S_t) presented a distribution in depth very similar to that of Fe (Figure 12b) and gypsum (Figure 11b), accumulating in the uppermost 80 mm, indicating that the precipitation was be due mainly to the formation of Ca hydrate sulphates and Fe hydroxysulphates (Figure 13). Therefore, the uppermost 55 mm of the soil contained 18–24 % of the S_t precipitated as gypsum (S_y); meanwhile between 55 and 80 mm in depth, this percentage clearly increased (24-40%). The remaining S_t that had not precipitated as gypsum (S_t-S_y), from now on S_h, was significantly related ($P < 0.001$) to the total iron oxides accumulated (Fe_d, extracted with citrate-dithionite) by the linear regression:

$$S_h\,(\text{g kg}^{-1}) = 1.52\ Fe_d\,(\text{g kg}^{-1}) + 13.36 \qquad r^2 = 0.990$$

This indicates that these elements must be precipitated as iron hydroxysulphates. Given that 60% or more of the iron oxides accumulated in the reddish-yellow layer were also extracted with ammonium oxalate (Figure 14), the iron hydroxysulphates must precipitate mainly as amorphous forms such as schwertmannite (the same as occurred in the previously sampling). However, the percentage of amorphous iron tended to decrease clearly towards the soil surface, especially in the pale-grey layer (< 30%), apparently indicating that the amorphous iron hydroxysulphates tend to transform to more crystalline forms (e.g. jarosite) with time. The SEM-EDX analysis of the uppermost 80 mm of the soil confirmed the previous results. In part of the ground mass of the soil sampled between 55 and 80 mm (Figure 15b), the ratio between Fe and S precipitated was very high (≈ 5), indicating the formation of schwertmannite (Bigham et al., 1996); meanwhile, in the pale-grey layer of the surface (Figure 15a) this ratio was lower (≈ 1) and presented K in its composition, reflecting that the mineral formed was jarosite (Filipeck et al., 1987).

On the other hand, the iron oxides had a powerful agglomerating effect on the smaller soil particles (Colombo and Torrent, 1991). This effect was clearly visible under the light microscope, revealing numerous aggregates the size of sand grains in which the clay had been bound by iron compounds (Figure 16). For an estimate of the intensity of the effect, a textural analysis was made with the samples previously treated with dithionite-citrate-bicarbonate (DCB) to eliminate iron oxides (Figure 17b), and the results were compared with those of the same samples left untreated (Figure 17a). Thus, the samples from the reddish-yellow layer treated with DCB clearly increased in the quantity of clay and decreased in the quantity of sand with respect to the untreated ones, reflecting the agglomerating power of the iron oxides present. This effect diminished towards the soil surface, mainly in the pale-grey layer, indicating that the crystallization of the iron oxides tended to reduce its capacity to aggregate the finest particles.

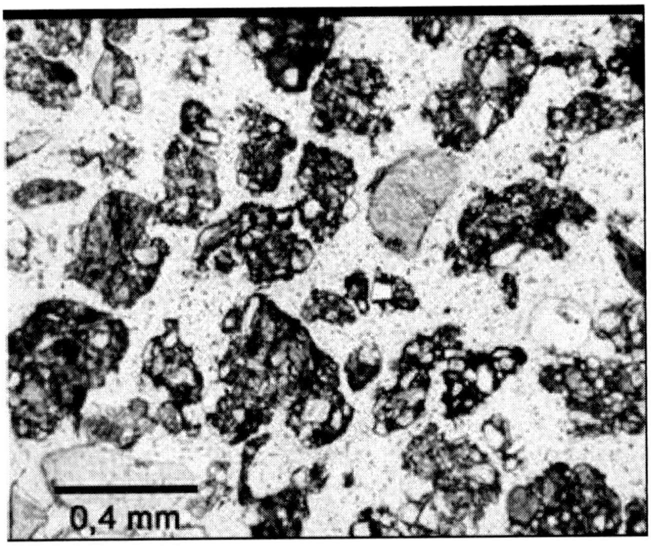

Figure 16. Sand-sized aggregates in the reddish-yellow layer of the soil sampled in July 2001.

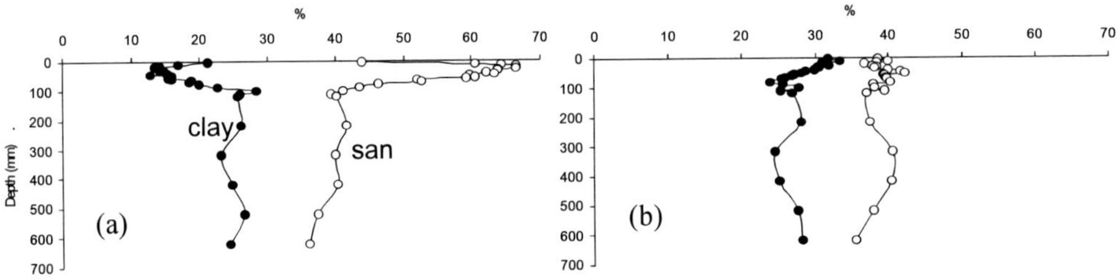

Figure 17. Textural analysis without (a) and with (b) treatment with DCB at the different depths of the soil sampled in July 2001.

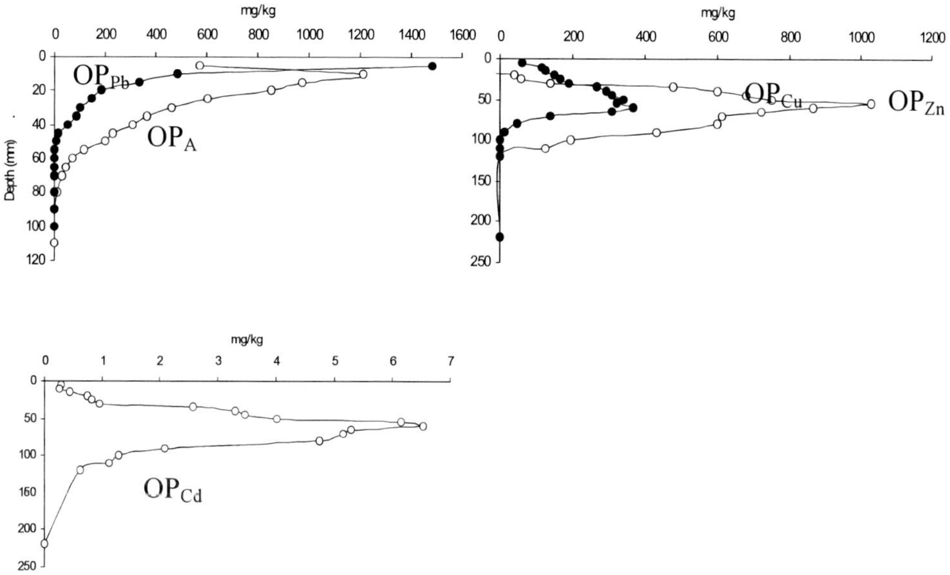

Figure 18. Oxidative pollution of Pb (OP_{Pb}), As(OP_{As}), Cu(OP_{Cu}), Zn (OP_{Zn}), and Cd (OP_{Cd}) with depth of the soil sampled in July 2001.

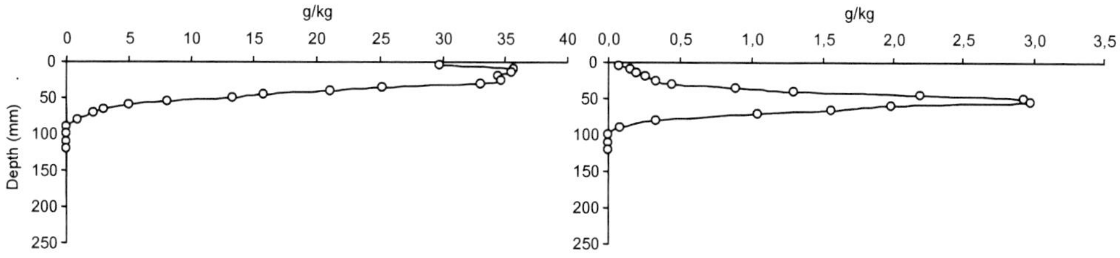

Figure 19. Total oxides of Fe (OP_{Fed}) and Al (OP_{Ald}) accumulated with depth of the soil sampled in July 1999.

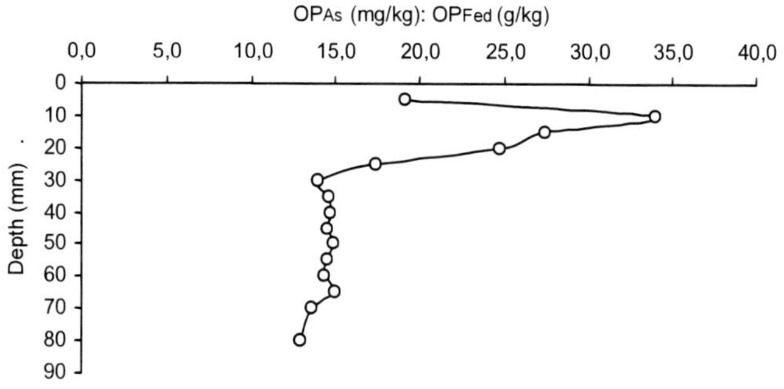

Figure 20. Ratio between the oxidative pollution of As (OP_{As}) and the accumulation of iron oxides (OP_{Fe}) with depth of the soil sampled in 2001.

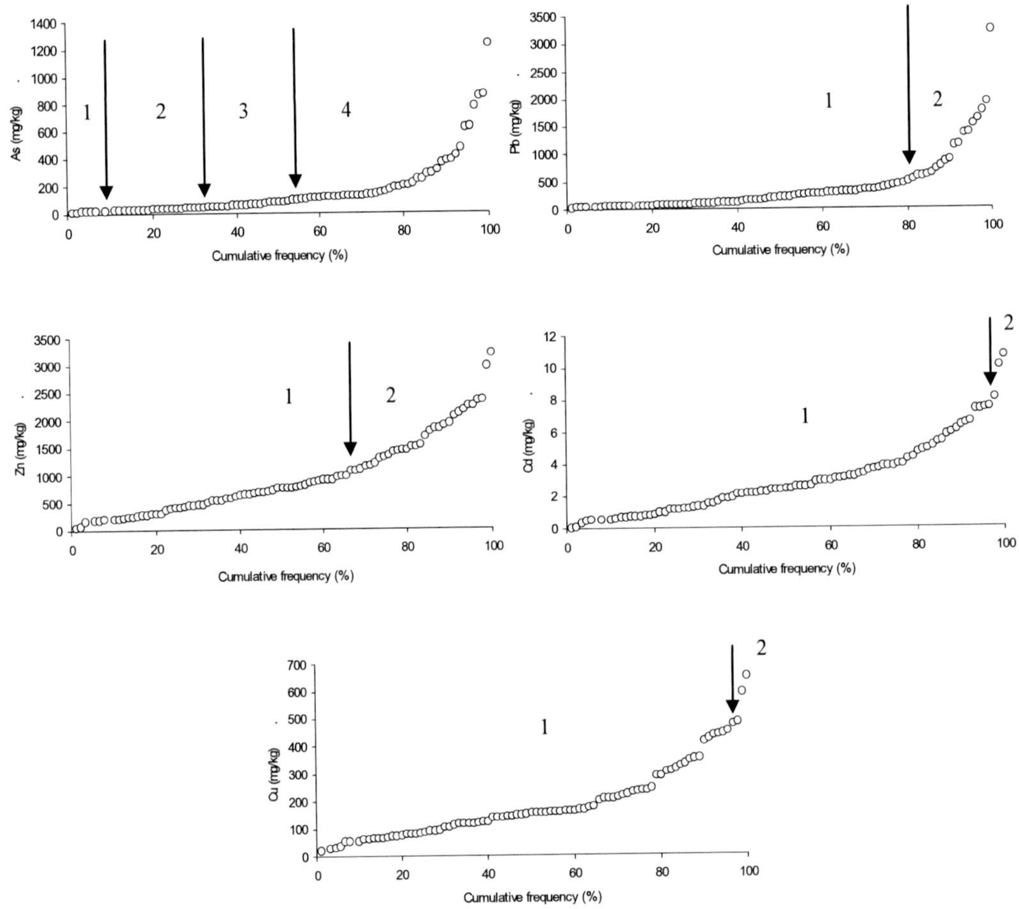

Figure 21. Relative cumulative frequency curves and limits of each population of As, Pb, Zn, Cd and Cu concentrations of the first 10 cm of the soils sampled immediately after clean-up of the tailings.

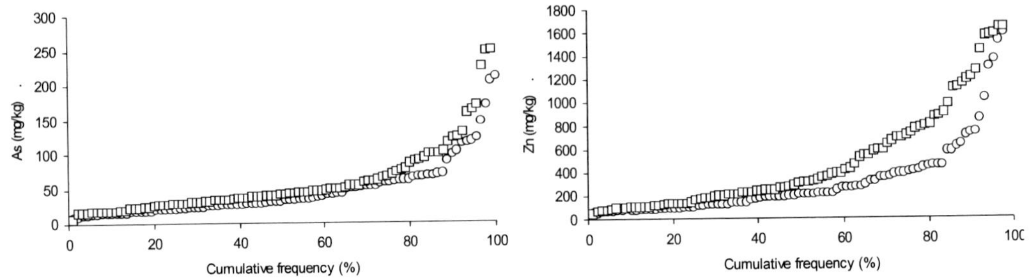

Figure 22. Relative cumulative frequency curves of As and Zn between 10-30 cm (□) and 30-50 cm (○) of the soils sampled immediately after clean-up of the tailings.

Like the major elements, all the pollutants that had dissolved in the acidic solution continued to accumulate in the uppermost 100 mm of the soil, although with clear differences between them (Figure 18). Thus, the Pb continued accumulating in the first 5 mm of the soil where the concentration reached the 1500 mg kg^{-1}, clearly higher than the previous sampling (< 250 mg kg^{-1}; Figure 9) and, as in the former case, values sharply declined below this depth. Like the first one, in this second sampling the Pb accumulation (determined by the difference between the Pb concentration at each depth and the mean concentration between 320 and 620 mm; previously termed "oxidative pollution", OP_{Pb}), proved not to be related to the hydrous oxides (Al_o or Fe_o), probably for the same reasons as in the previous sampling (accumulation in the first 5 mm of the soil as $PbSO_4$).

Figure 23. Accumulation of the pollutants in the small dips in the landscape, forming white, yellowish and reddish efflorescences.

The As (OP_{As}) tended to accumulate in the reddish-yellow layer, reaching its highest concentration of 1200 mg kg^{-1} between 5-10 mm depth (Figure 18), very similar to the 1100 mg/kg reached in the previous sampling between 0-2 mm depth (Figure 9); from this depth, the amount of accumulated As progressively declined to become negligible at 90 mm in depth. This decrease in depth of the accumulated As was very similar to the distribution of the total accumulated iron oxides (Figure19, determined by the difference between the concentration of iron extracted with dithionite-citrate at each depth and the mean concentration between 320 and 620 mm, termed "oxidative pollution", OP_{Fed}, as in the case of the pollutants). Both constituents were significantly ($P<0.001$) and linearly related by the equation:

$$OP_{As}\ (\text{mg kg}^{-1}) = 22.27\ OP_{Fed}\ (\text{g kg}^{-1}) - 30.43 \qquad r^2 = 0.838$$

This indicates that, each gram of Fe precipitated tend to adsorb a mean of around 22.3 mg of As. In any case, the ratio OP_{As}:OP_{Fed} at different depths (Figure 20), indicates that the amount of As adsorbed by the iron oxides amply varied. This was highest between 5-10 mm ($\approx$ 34 mg As/g Fe_d) and similar to values found in the previous sampling between 0-2 mm ($\approx$ 31 mg As/g Fe_d), progressively decreasing to 14 mg As/g Fe_d at 35 mm in depth and, from here, remaining relatively constant to 80 mm (maximum depth at which both elements accumulated). Therefore, the comparison of the results of the two samplings indicates that the maximum amount of iron oxides that can precipitate in this soil will approach 35 g Fe_d/kg dry soil. The maximum amount of As that can adsorb precipitated iron oxides will be some 35 mg As/g Fe_d; and the maximum amount of total As that can accumulate in the soil will be roughly 1200 mg As/kg dry soil.

On the other hand, in relation to the depth of 5-10 mm, both As and iron-oxide accumulation clearly decreased (Figure 18) in the uppermost 5 mm of the soil (pale-grey layer), indicating that the continuous infiltration into the soil of the acidic solution came from the tailings, while the decrease in the soil pH to below 2.2 signified the partial solubilization of the precipitated iron oxides and, consequently, also of the adsorbed As. The amorphous forms of iron, because of their higher solubility (Linsay, 1979), leached more intensely than the crystalline ones, and thus their percentage clearly decreased on the surface (Figure 14). Also, in the first 5 mm of the soil, the sharp decline of the ratio OP_{As} (mg kg^{-1}) : OP_{Fed} (g kg^{-1}) indicates (Figure 20) that the crystalline forms of iron tended to adsorb less As than did the amorphous ones.

As in the previous sampling, Zn, Cd and Cu had higher mobility and tended to accumulate mainly between 45 and 80 mm (pH between 5.3 and 7.1), peaking at around 60 mm (pH=6.2). In this sampling, the maximum concentrations of Zn, Cu and Cd accumulated were 1050, 370 and 6.7 mg kg^{-1} dry soil, respectively (Figure 18), clearly lower than in the first sampling (2700, 700 and 11 mg kg^{-1}, respectively; Figure 9); indicating that, over time, these elements may decrease their concentration at a certain position in the soil, but can redistribute and affect a shallower soil depth. The distribution of these elements in depth was very similar to the total distribution of the aluminium oxides accumulated (Figure 19, difference between the concentration of aluminium extracted with dithionite-citrate at each depth and the mean concentration between 320 and 620 mm; also called "oxidative pollution", OP_{Ald}), and therefore should be related. Thus, the levels of oxidative pollution of Zn (OP_{Zn}) and Cd (OP_{Cd}) were significantly related ($P < 0.001$) with OP_{Ald} and pH by the equations:

$$OP_{Zn}(\text{mg kg}^{-1}) = 253.9\ OP_{Ald}(\text{g kg}^{-1}) + 33.6\ \text{pH} \qquad r^2 = 0.900$$

$$OP_{Cd}(\text{mg kg}^{-1}) = 1.17\ OP_{Ald}(\text{g kg}^{-1}) + 0.33\ \text{pH} \qquad r^2 = 0.873$$

Meanwhile, the oxidative pollution of Cu (OP_{Cu}) was significantly ($P < 0.001$) related with OP_{Ald} and OP_{Fed} by the equation:

$$OP_{Cu}(\text{mg kg}^{-1}) = 126.1\ OP_{Ald}(\text{g kg}^{-1}) + 3.35\ OP_{Fed}(\text{g kg}^{-1}) \qquad r^2 = 0.947$$

This signifies that these elements either were adsorbed fundamentally by the aluminium oxides, or else precipitated in the same pH interval (between 5.0 and 7.0).

2.3. Third Step: Residual Tailings

The social alarm that this toxic spill provoked throughout Europe prompted the Spanish authorities to launch a very quick clean-up of the tailings deposited on the soil surface. In an extensive clean-up operation (May to December 1998) the tailings and approximately 10 cm of the upper soil were removed using heavy machinery. This rapid work (in eight months, 45 km^2 were cleaned) resulted in a deficient clean-up, as part of the tailings remained mixed with the soil, appearing residual tailings below 30 cm in depth. Just after the clean-up of the soils, we undertook a systematic random sampling, dividing the affected area into a grid of squares 400 m x 400 m and selecting 100 georeferenced sampling points (GPS) uniformly distributed through-out the area. At each sampling point, three soil samples were taken at the depths of 0-10 cm, 10-30 cm, and 30-50 cm. In addition, 30 unaffected soils near the border of the spill and uniformly distributed through-out the area, were also sampled at the same depths.

In the unaffected soils, the concentration of the pollutant elements did not significantly differ in depth, so that all the samples (90) were used to establish the background level of the area. The background ranges were determined using the method of the calculated distribution function (Matschullat et al., 2000), the results being: 13.3 – 25.3 mg kg^{-1} for As, 52.6 – 101.4 mg kg^{-1} for Zn, 17.4 – 34.4 mg kg^{-1} for Cu, 15.7 – 37.3 mg kg^{-1} for Pb, and 0.11 – 0.28 mg kg^{-1} for Cd.

With the background ranges and the relative-cumulative-frequency curves (RCFC) of the pollutants in the soils affected by the spill, we estimated the degree of pollution after the clean-up of the tailings. The RCFC of As in the first 10 cm of the contaminated soils (Figure 21) revealed that only 10% of the sampled soils had an As concentration lower than the upper limit of the background (population 1), meanwhile the other 90% presented different degrees of pollution: 24% with low contamination (population 2) and below the maximum permitted level for agricultural soils of 50 mg kg^{-1} (Sheppard et al., 1992; Godin, 1983); 21% (population 3) with moderate contamination (50-100 mg kg^{-1}); and 45% (population 4) with high contamination (> 100 mg kg^{-1}) and with an extreme value of 1227 mg As/kg dry soil.

Meanwhile, the RCFC of Pb, Zn, Cd and Cu indicated that most of the sampled soils (> 98%) surpassed the upper limits of the background, although the degree of pollution appeared to be less serious than in the case of As, because only 33% for Zn, 20% for Pb, and 2% for Cu and Cd exceeded the maximum permitted levels (1000 mg kg^{-1}, 500 mg kg^{-1}, 500 mg kg^{-1}, and 10 mg kg^{-1}, respectively) for agricultural soils (Godin, 1983; Sheppard et al., 1992; Adriano et al., 1997; Baize, 1997).

Therefore, the spill seriously polluted the soils and, after the enormous efforts of cleaning up the tailings, almost all the soils were still contaminated between 0 and 10 cm in depth, mainly in As, Zn and Pb. However, the studies of Cabrera et al. (1984, 1987) and Ramos et al. (1994) indicated that, prior to the spill, the soils of the area had abnormally high contents in Cu, Cd, Zn and Pb, with a tendency to increase over the years. This contamination appears to have originated from the periodic discharges of acidic and metal-rich waters into the Guadiamar River from the mining companies of Aznalcóllar working for decades, and by the use of these waters by the farmers.

2.3.1. Were the Soils Contaminated Prior to the Spill?

Given that in the two first steps of the pollution the soils were not significantly affected under the first 10 cm, with the exception of the a few sectors in which the tailings had penetrated in depth through structural cracks in the first step of pollution, the RCFC of concentration of pollutant elements between 10-30 cm and 30-50 cm can be used to estimate the degree of pollution before the spill and, consequently, to calculate the effect of the spill in the total pollution level of the soils.

Thus, the RCFC of As and Zn between 30 and 50 cm show the two populations of data clearly differ (Figure 22). The first part of the curves (≈ 88% of the soils in the case of As, and 85% of the soils in the case of Zn), has a relatively uniform slope, and represents the range of As and Zn concentrations in soils before the spill. This first part can also be divided into two sectors: the first one, with As and Zn concentrations below the upper limit of the background, the non-polluted soil group (≈ 40% in the case of As, and 25% in the case of Zn); and the second one, with As and Zn concentrations above the upper limit of the background, the contaminated soil group prior to the spill (≈ 48% in the case of As, and 60% in the case of Zn). The second part of the curve, characterized by an abrupt change in the slope, according to Bauer and Bor (in Matschullat et al., 2000) represents the soils contaminated in the accident, both by the initial infiltration of the acidic waters and tailings as well as by the mixing of the tailings below the first 30 cm during the clean-up (12% in the case of As, and 15% in the case of Zn). The higher percentage of polluted soils in Zn before the spill in relation to the As, may be due to the higher oxidation rate of the sulphides containing Zn and to a higher concentration of this element in the acidic waters coming from the mine, as stated above.

The comparison of the RCFC of As between 10-30 cm and 30-50 cm (Figure 22) shows that both curves are parallel and very close each other up to 75 percentile, where they clearly separate, indicating that, between 10 and 30 cm in depth, the spill on 25 April increased the As concentration in 25% of the soils studied, more than double the amount that affected depths between 30-50 cm. In the case of Zn, the RCFC are also very close and parallel up to the 50 percentile, indicating that, between 10-30 cm in depth, the spill increased the Zn content in 50% of the soils, more than three-fold the amount that affected depths between 30-50 cm. On the other hand, aside from the fact that the first part of the RCFC of As and Zn are parallel and very close each other, the curve between 10-30 cm was consistently above the curve for the depth between 30-50 cm, indicating that before the spill the As and Zn concentrations trended to increase towards the soil surface. Therefore, although the soils were contaminated prior to the accident, the spill clearly increased the concentration of the different pollutant elements, mainly in the first 30 cm of the soils.

2.3.2. Element Mobility and Hydrological Properties of Contaminated Soils

After the 1998-99 winter rains, the soils of the northern sector (between the Mine and Sanlúcar, Figure 2), without carbonates and with a loamy texture, formed a thin surface crust; meanwhile, in the small dips and hollows in the landscape, pollutants began to accumulate, forming white, yellowish and reddish efflorescences (Figure 23). On the contrary, in the soils of the middle and southern sectors (from Sanlúcar to the end of the spill), with similar texture but with a mean carbonate content of roughly 10%, no crusts or accumulations of pollutants appeared. In carbonated and non-carbonated contaminated soils, the mobility of pollutant elements in the soil profile were studied together with the hydrological properties of the soils and the distribution of pollutants in the landscape.

In May 2000 (17 months after the cleanup), three plots were sampled in the northern sector, Vicario (VIC-1, VIC-2 and VIC-3), and three in the middle sector, Aznalcázar (AZN-

1, AZN-2 and AZN-3). In each plot, a detailed study was made of the soil properties and their pollutant concentrations at different depths (0-1, 1-5, 5-10, 10-20, 20-40 and 40-60 cm). The infiltration rate into the soils of the different plots was estimated by a rain simulator consisting basically of a square water chamber (50 x 50 cm) perforated by a total of 1225 droplet holes and regulated by a peristaltic pump to provide the desired rain intensity; runoff, directed through a triangular pan sealed to the ground, was collected in bottles of 150 cm^3 (Figure 24). For more details concerning the rain simulator, see Simón et al. (1998). All experiments were conducted with deionised water. The rain intensity was 50 mm h^{-1}, and the runoff was estimated by the filling time per bottle. Each experiment lasted 72 min, during which 12 bottles were filled (at roughly 5-min intervals after an initial 12 min of rain). The filling time in all plots reached constant values in less than one hour. The infiltration rate for each interval was calculated as the difference between the rain intensity and runoff rate. Equivalent experiments were made in three uncontaminated soils of the northern sector and three in the middle sector, with textures similar to that of the contaminated soils.

Table 11. The pH and $CaCO_3$ content at different depths of the soil sampled on 25 July 1999.

Depth (mm)	pH	$CaCO_3$ (%)
0-2	3.2	0
2-4	3.3	0
4-6	3.4	0
6-8	3.4	0
8-10	3.7	0
10-12	4.1	0.2
12-14	6.0	1.1
19-21	7.4	5.8
29-31	7.8	6.4
39-41	8.0	6.8
59-61	8.1	6.7
99-101	8.1	6.8

During the 17 months since the remediation to the present sampling, the oxidation of the residual tailings (Figure 27B) considerably acidified the uppermost few cm of the soils (Table 12), and soil pH was related to the sulphur content. In the soils of the northern sector (Figure 25), the exchangeable bases and the weathering of the silicate minerals neutralized most of the acidity caused by the oxidation of the S content of up to 8 g kg^{-1}, the pH remaining between 6.2 and 6.0. When the S content exceeded 8 g kg^{-1}, the pH fell sharply, tending to stabilize at around pH 3.0. In the soils of the middle sector, the $CaCO_3$ content neutralized most of the acidity from the oxidation of the S content of up to 16 g kg^{-1}, the pH remaining between 8.3 and 7.9. When the S content surpassed 16 g kg^{-1}, the pH declined more rapidly, although, as the S content in no case exceeded 22 g kg^{-1}, the pH did not drop below 6.5. The strong acidification of the uppermost 5 cm of the soils in the northern sector hydrolysed the finest particles, substantially diminishing the clay and, to a lesser extent, silt content in relation to the underlying 5 cm (Table 12). In the soils of the middle sector, the acidity was largely neutralized by $CaCO_3$, and the hydrolysis of the fine particles was far less pronounced.

Figure 24. Rainfall simulator (A) and details of the perforated square water chamber (B) and triangular pan sealed to the ground (C)

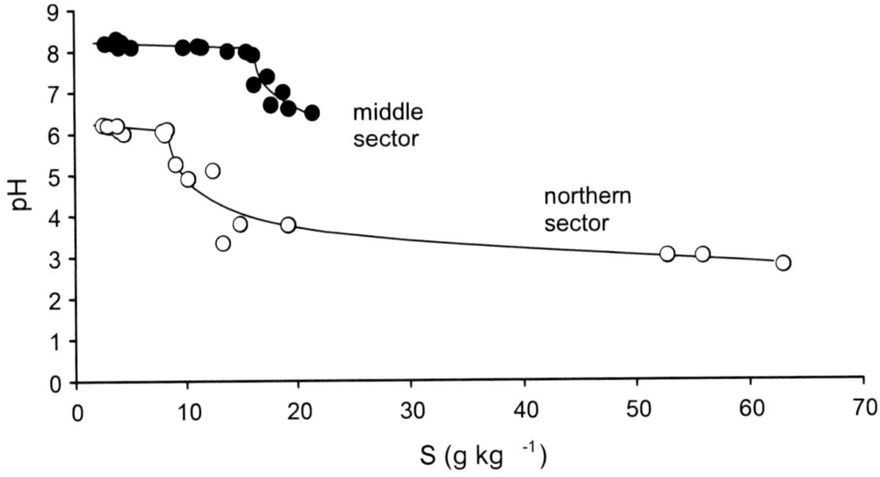

Figure 25. Soil pH plotted against sulphur (S) content in the soils of the northern sector and middle sector.

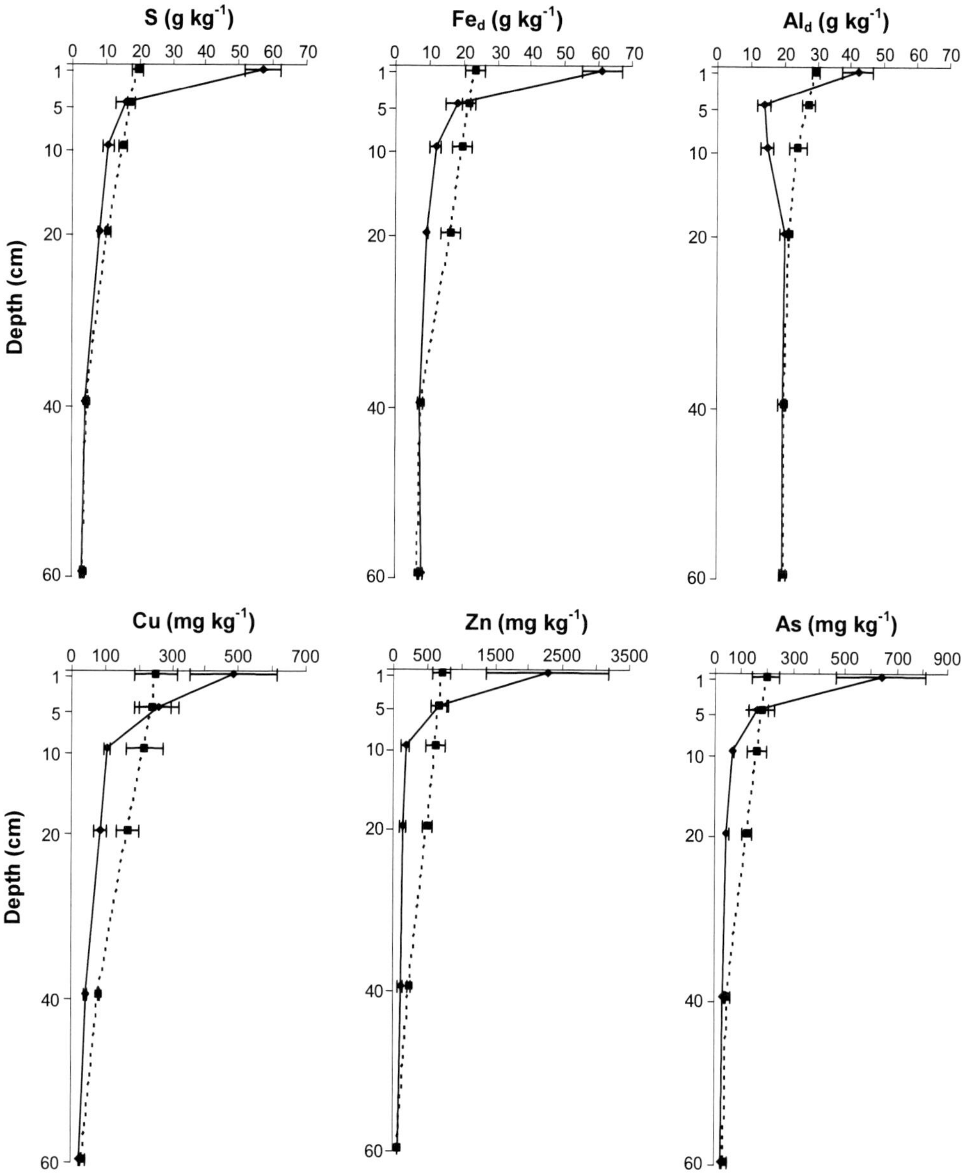

Figure 26. Content of major and trace elements in the soils of the northern sector (solid line) and middle sector (broken line). Horizontal bars represent standard deviation.

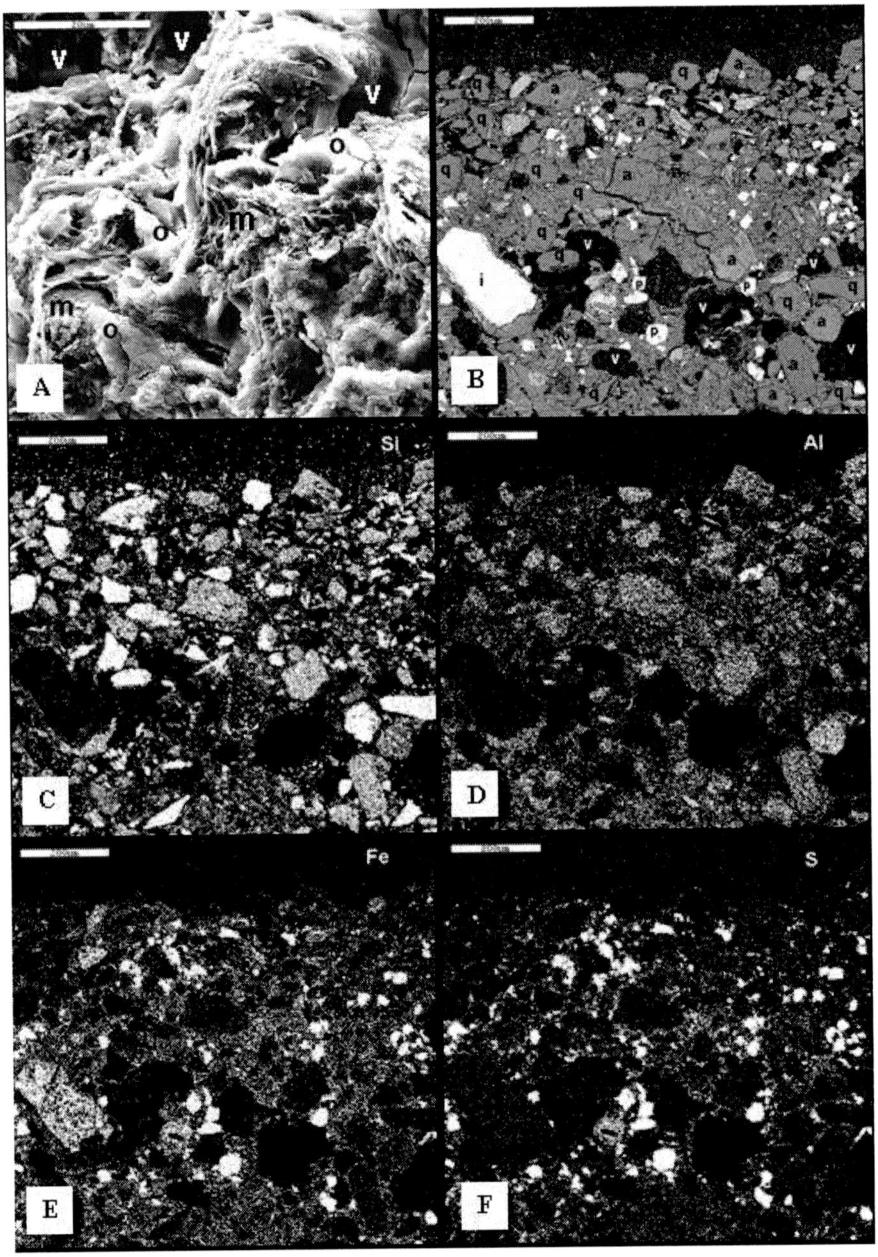

Figure 27. (A) SEM-EDS analysis of the crust, showing vesicles (v) and mineral grains (m) enveloped by a amorphous mass (o) composed principally of Fe with S and Al. (B) Backscattered electron scanning (BES) micrograph from a polished section of the crust, showing vesicles (v) and mineral grains (a = aluminosilicates, i = ilmenite, p = pyrite, q = quartz). (C) Distribution of Si by EDS analysis in the polished section of the (B) micrograph; the light-grey areas are quartz grains and the heterogeneous dark-grey areas correspond to aluminosilicates (feldspars and clay particles). (D) Distribution of Al in the (B) micrograph; the heterogeneous dark-grey areas correspond to aluminosilicates with interspersed amorphous aluminium. (E) and (F) Distribution of Fe and S, respectively, the light-grey areas are pyrite grains and the heterogeneous dark-grey areas correspond to interspersed amorphous Fe and S.

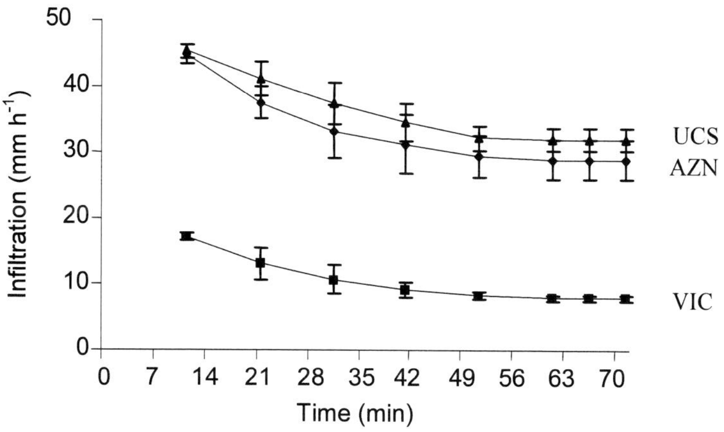

Figure 28. Infiltration rate versus time in rain-simulation experiments in uncontaminated soils (UCS) as well as in contaminated soils of the northern (VIC) and middle (AZN) sectors.

Table 12. Mean and standard deviation (SD) of pH, $CaCO_3$ content, and particle size of the soils in the northern (VIC) and middle (AZN) sectors.

Sector	Prof. (cm)	pH		$CaCO_3$ (%)		Gravel (>2mm) (%)		Sand (2-0.5 mm) (%)		Silt (0.5-0.002 mm) (%)		Clay (< 0.002 mm) (%)	
		Mean	SD	Mean	SD	Mean	SD	Mean	SD	Mean	SD	Mean	SD
VIC	0-1	2.9	0.1	nd	-	15.8	1.7	77.5	3.3	17.9	3.1	4.6	1.6
	1-5	3.6	0.3	nd	-	15.2	2.8	74.7	0.9	18.3	3.0	7.0	2.1
	5-10	5.1	0.2	nd	-	18.4	3.0	65.1	1.1	24.7	2.9	10.3	2.8
	10-20	6.0	0.1	nd	-	18.7	1.8	61.2	3.4	25.5	2.5	13.3	4.2
	20-40	6.1	0.1	nd	-	20.3	2.3	61.6	2.2	25.4	1.6	12.9	1.3
	40-60	6.2	0.1	nd	-	19.7	2.5	62.4	3.4	25.6	2.5	12.0	2.7
AZN	0-1	6.6	0.1	4.8	1.0	3.8	2.2	56.4	9.1	31.1	7.7	12.5	1.6
	1-5	7.2	0.2	8.3	0.4	4.7	4.3	49.2	8.1	35.1	7.4	15.7	0.9
	5-10	8.0	0.1	9.5	0.8	4.6	3.4	45.0	7.2	39.0	6.0	16.0	2.9
	10-20	8.1	0.1	9.3	0.6	4.7	5.0	45.3	9.1	38.9	7.8	15.9	3.1
	20-40	8.1	0.1	10.0	0.6	6.0	5.8	47.6	8.5	37.9	6.1	14.5	4.2
	40-60	8.3	0.1	9.7	0.4	5.8	4.8	46.2	8.4	38.3	6.5	15.5	5.7

nd = not detected

In the soils of the northern sector, the S, Fe and Al released by the oxidation of the tailings, as well as the hydrolysis of the finest soil particles, partially soluble under the acidic conditions, rose to the soil surface by the high evapotranspiration of the Mediterranean climate. Thus, the top 1 cm contained far greater concentrations (between 3- and 4-fold) of sulphur and iron (Fe_d) and aluminium (Al_d) oxides than in the underlying 4 cm (Figure 26). On the contrary, in the carbonate soils of the middle sector, the higher pH of the soils caused most of the elements to precipitate wherever they were released, bringing about a more homogeneous distribution throughout the upper 20 cm of the soils.

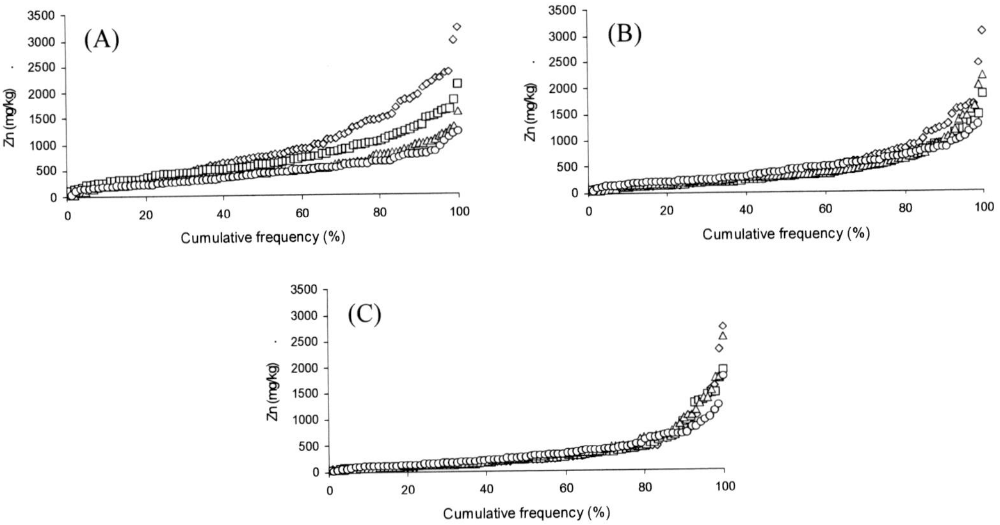

Figure 29. Relative cumulative frequency curves of Zn between 0-10 cm (A), 10-30 cm (B), and 30-50 cm (C) on 1998 (◇), 1999 (△), 2001 (□), and 2004 (○).

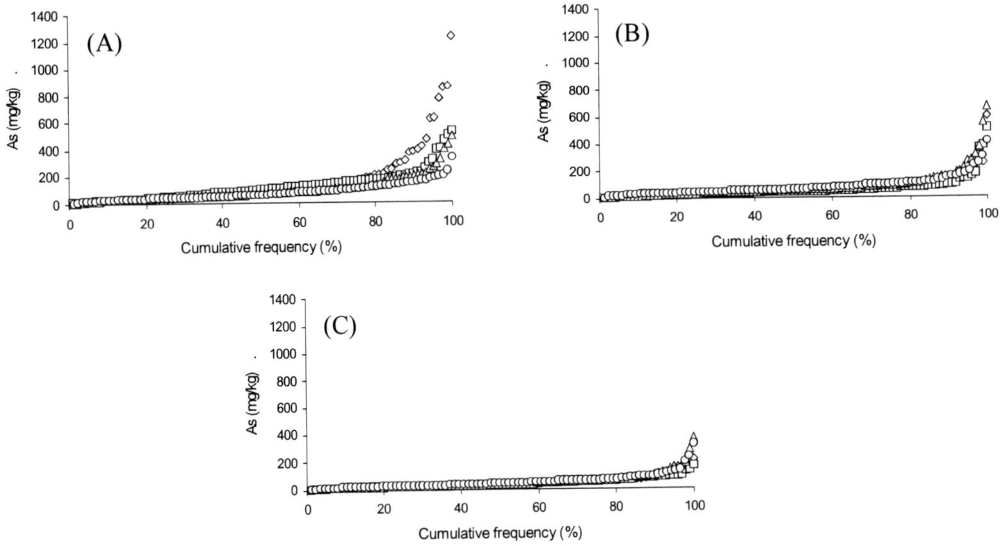

Figure 30. Relative cumulative frequency curves of As between 0-10 cm (A), 10-30 cm (B), and 30-50 cm (C) on (◇), 1999 (△), 2001 (□), and 2004 (○)

In the soils of the northern sector, the S content was significantly related only to the Fe_d content ($p<0.001$) by the linear-regression equation:

$$S\ (g\ kg^{-1}) = -2.344 + 0.976\ Fe_d\ (g\ kg^{-1}) \qquad r^2 = 0.993$$

Meanwhile, in the soils of the middle sector, the S content was significantly related to the Fe_d ($p<0.001$) and Al_d ($p<0.01$) content by the multiple-regression equation:

$$S\ (g\ kg^{-1}) = -8.894 + 0.713\ Fe_d\ (g\ kg^{-1}) + 0.392\ Al_d\ (g\ kg^{-1}) \qquad r^2 = 0.970$$

These results imply that in the acidic soils, where aluminium was relatively soluble, S and Fe precipitated, forming fundamentally iron hydroxysulphates; meanwhile, in the neutral and basic soils, in which aluminium tended also to precipitate, iron and aluminium hydroxysulphates were formed.

The trace elements also considerably increased in the first cm of the soils in the northern sector and had a more homogeneous distribution in depth in the soils of the middle sector (Figure 26), supporting the idea that the elements were adsorbed by the iron and aluminium hydroxysulphates. Nevertheless, the multiple regression between each trace element, on the one hand, and the major elements, on the other hand, split the trace elements into two groups.

The first group was comprised of Zn, Cd and As, which, considering the two sectors together, were significantly related only to S ($p<0.001$) by the linear regressions:

$$Zn\ (mg\ kg^{-1}) = -56.43 + 41.20\ S\ (g\ kg^{-1}) \qquad r^2 = 0.885$$

$$Cd\ (mg\ kg^{-1}) = -0.250 + 0.153\ S\ (g\ kg^{-1}) \qquad r^2 = 0.841$$

$$As\ (mg\ kg^{-1}) = -19.6 + 11.55\ S\ (g\ kg^{-1}) \qquad r^2 = 0.947$$

indicating that these elements were adsorbed fundamentally by the iron hydroxysulphates formed (Kinniburgh et al., 1976; O'Neill, 1995). When the soils of the two sectors were considered separately, the Zn content in the soils of the middle sector proved to be significantly related not only to S ($p<0.01$) but also to Fe_d ($p< 0.01$) and Al_d ($p< 0.001$) by the multiple regression:

$$Zn\ (mg\ kg^{-1}) = 439.7 + 33.0\ S\ (g\ kg^{-1}) + 18.2\ Fe_d\ (g\ kg^{-1}) - 27.3\ Al_d\ (g\ kg^{-1}) \qquad r^2 = 0.973$$

This indicates that, in carbonated soils, the adsorption of Zn by iron and aluminium hydroxysulphates was partially blocked by Al sorption (Anderson and Benjamin, 1990).

In the second group, Cu and Pb, in the soils of the northern sector were significantly related to Fe_d ($p<0.001$) and pH ($p<0.001$) by the multiple regressions:

$$Cu\ (mg\ kg^{-1}) = 345.38 + 5.07\ Fe_d\ (g\ kg^{-1}) - 56.83\ pH \qquad r^2 = 0.945$$

$$Pb\ (mg\ kg^{-1}) = 722.00 + 10.11\ Fe_d\ (g\ kg^{-1}) - 122.33\ pH \qquad r^2 = 0.968$$

In the soils of the middle sector, Cu and Pb exhibited different partitioning behaviour. That is, the Cu content was significantly related to Fe_d ($p<0.001$), Al_d ($p> 0.05$) and $CaCO_3$ ($p<0.05$); meanwhile, the Pb content was significantly related to S ($p<0.001$), Fe_d ($p<0.05$) and Al_d ($p<0.001$), by the multiple regressions:

$$Cu\ (mg\ kg^{-1}) = -274.50 + 12.09Fe_d\ (g\ kg^{-1}) + 5.63\ Al_d\ (g\ kg^{-1}) + 13.28\ CaCO_3\ (g\ kg^{-1}) \quad r^2 = 0.937$$

$$Pb\ (mg\ kg^{-1}) = 467.89 + 43.74\ S\ (g\ kg^{-1}) - 12.46\ Fe_d\ (g\ kg^{-1}) - 23.14\ Al_d\ (g\ kg^{-1}) \quad r^2 = 0.933$$

These relationships imply that in the acidic soils, the adsorption of Cu and Pb by iron hydroxysulphates decreased as the pH increased. In the carbonated soils, Cu was adsorbed by iron and aluminium hydroxysulphates and probably also precipitated as carbonates (McBride and Bouldin, 1984). Meanwhile, Pb was adsorbed fundamentally by S, the adsorption declining as the iron and aluminium oxides augmented. Consequently, the adsorption of S by Fe and Al to form hydroxysulphates appears to block the precipitation of Pb as PbSO4 (Fuge et al., 1994).

SEM-EDX analysis of the top 1 cm of the soil in the northern sector revealed that the Fe, S and Al precipitated as amorphous masses enveloping and cementing the mineral grains, forming a crust with vesicular voids (Figure 26A). Backscattered electron scanning (BES) micrographs of a polished section (Figure 26B) and an EDX analysis of Si, Al, Fe and S (Figure 26C, D, E and F, respectively) confirm the vesicle formation (between 5 and 100 μm of diameter) and the precipitation of Fe, S and Al between the mineral grains (mainly quartz, aluminosilicate and pyrite particles) forming the crust. In the soils of the middle sector, crust formation was not detected.

On dry soil, the simulated rain (50 mm h^{-1}) showed an infiltration rate for contaminated and uncontaminated soils which was higher at first and slower as the soil moistened (Figure 27), tending asymptotically to a constant value (Dunin, 1976) called final infiltration rate (FIR). In all cases, the FIR was reached in less than one hour after the beginning of the experiment. The mean of the FIR in the uncontaminated soils (32.0 mm h^{-1}) was similar to that of the contaminated soils of the middle sector (28.9 mm h^{-1}), but clearly higher than that of the contaminated soils of the northern sector (7.9 mm h^{-1}). This clearly lower FIR was presumably related to the surface-crust formation which, though highly porous (Figure 3B), had vesicular voids (i.e. not interconnected) and therefore limited water infiltration into the soil.

Table 13. Mean and standard deviation (SD) of the element concentration (mg dm^{-3}) in the liquid phase of the runoff-water in both sectors.

	Northern sector		Middle sector	
Element	Mean	SD	Mean	SD
Cu	74.8	21.0	12.2	1.7
Zn	483.1	78.3	61.3	14.9
As	133.1	41.1	3.97	0.34
Cd	2.32	0.55	0.37	0.03
Pb	1.93	0.25	18.0	0.8
SO_4^{2-}	3680	1358	389.9	46.9
Fe	548.0	165.3	0.27	0.04

In addition, the lower infiltration results in higher runoff. Thus, of the 60 mm of water from 72 min of simulated rain, a mean of 48.0 mm (SD = 1.2) was lost as runoff in the soils of the northern sector, while in the middle sector the mean runoff was only 20.2 mm (SD = 3.7). Nevertheless, the greater runoff resulted in less erosion; that is, the northern sector lost 129.3 g (SD = 9.1) of solids in suspension, compared to 201.8 g (SD = 5.3) in the middle sector. Consequently, though the crust encouraged runoff, it discouraged erosion.

The mean concentrations of elements in the liquid phase of the runoff water, except Pb, were clearly higher in the soils of the northern sector than in those of the middle sector (Table 13), especially in the cases of Fe (2000-fold), and to a lesser extent As (33-fold) as well as SO_4^{2-}, Zn, Cd and Cu (between 6 and 9-fold). Only the Pb concentration was higher in the liquid phase from the soils of the middle sector (9-fold). The solid phase in all sectors also showed a high concentration in pollutants (Table 14), higher than in the uppermost cm of the soil (Figure 25), implying that the pollutant accumulation was greater at the soil surface; the only exceptions were Zn, Cd and Cu in the northern sector, probably because the high solubility of these three elements under acidic conditions caused them to concentrate in the liquid phase. Of the total concentration of each pollutant in the runoff, a far greater proportion belonged to the liquid phase than to the solid. Thus, the strong runoff water from the northern sector, being especially charged with the above elements, spread to less contaminated areas, where pollutants accumulated in hollows or affected surface water, thereby posing a threat to nearby ecosystems.

Table 14. Mean and standard deviation (SD) of the element concentration (mg kg^{-1}) in the solid phase of the runoff-water in both sectors.

	VIC		AZN	
Element	**Mean**	**SD**	**Mean**	**SD**
Cu	226.4	9.6	270.4	29.4
Zn	455.1	24.4	1069.7	89.4
As	1171.5	270.5	755.7	101.7
Cd	1.35	0.18	3.71	0.23
Pb	1121.2	228.8	1107.4	108.5

Table 15. Mean of the Zn concentration (mg kg^{-1}) in all soils (TS) and in the most polluted (between percentiles 60 and 100, MPS) at the three depths and in the different samplings.

	0-10 cm		10-30 cm		30-50 cm	
Sample	**TS**	**MPS**	**TS**	**MPS**	**TS**	**MPS**
1998	955	1600	534	1015	366	693
1999	719	1128	412	736	359	686
2001	507	802	455	830	401	761
2004	460	703	453	832	392	752

Because of the high pollution of the soils after tailings were removed, and the possibility of spreading the pollutants to less contaminated areas, the most polluted sectors were cleaned again on spring-summer 1999, followed by a liming with sugar-refinery scum and superficial tilling of the soils (5-10 cm in depth). In spring-summer 2000, organic matter plus iron-rich clayey materials were added to the soils, followed by tilling the upper 20-25 cm of the soils. After each clean-up and remediation action, the soils were sampled and analysed in autumn 1999 and 2001 at the same points sampled before. The tilling of the upper 20-25 cm of the soils was repeated, although not in a generalized way, in 2002 and 2003. A final sampling was made in autumn 2004.

2.4. Pollutant Concentrations in Soils over Time

The results indicate that the different clean-up and remediation actions progressively decreased the concentration in pollutant elements in the soils. The repetition of the clean-up in spring-summer 1999 clearly decreased the Zn concentration in the uppermost 10 cm of the soil (Figure 29A) in relation to 1998 (mean concentration decreasing ≈ 25%, Table 15), and mainly in the most polluted soils, where the mean concentration diminished around 30%. In 2001, the Zn concentration fell again, mainly by the tilling and homogenization of the uppermost 25 cm of the soil in spring-summer 2000 (mean concentration in 2001 declined again by 30% in relation to 1999). Tilling after 2000 slightly affected the Zn concentration in the soils and the RCFC in the samplings in 2001 and 2004 were very close each other and registered similar mean concentrations; only in the most polluted soils (Table 15) was there a clear decrease in Zn concentration between 2004 (mean = 703 mg kg^{-1}) and 2001 (mean = 802 mg kg^{-1}), probably due to a better homogenization of the uppermost 20-25 cm of the soils.

The effect of these remediation actions was clearly weaker below the first 10 cm. Therefore, between 10 and 30 cm in depth (Figure 29B), the repetition of the clean-up in the most polluted sectors in spring-summer 1999, although decreasing the Zn concentration, this decrease was less and affected mainly the most polluted soils (mean concentration decreasing 27% on 1999 respect to 1998, Table 15). In the samplings after 1999, both the RCFC of Zn as the mean concentrations were very close each other. Between 30 and 50 cm in depth (Figure 29C) these actions were almost negligible, and the RCFC of Zn in the different samplings were practically superimposed. In any case, the mean Zn concentration below the first 10 cm increased in the samplings after 1999, possibly due to the tilling and homogenization of the uppermost 25 cm of the soil.

Consequently, both the clean-up of the most polluted sectors as well as the tilling and homogenization of the uppermost 25 cm of the soils were effective actions to lower the Zn concentration, mainly in the uppermost 10 cm. The percentage of total soils surpassing 1000 mg kg^{-1} decreased from a 34% in 1998 to only 5% in 2004 (6 years after the spill).

In the case of As (Figure 30), the repetition of clean-up in the most polluted sectors also decreased the As concentration in the uppermost 10 cm of the soils (mean concentration in 1999 decreased the 40% in relation to 1998). The tilling and homogenization of the uppermost 25 cm of the soil decreased the mean As concentration between 15-20 % in 2001 in relation to 1999 (Table 16), in all soils as well as in the most polluted ones. Tilling after this year proved less effective, so that the As concentration remained practically constant in the samplings in 2001 and 2004.

Table 16. Mean of the As concentration (mg kg^{-1}) in all soils (TS) and in the most polluted (between percentiles 80 and 100, MPS) at the three depths and in the different samplings.

	0-10 cm		10-30 cm		30-50 cm	
	TS	MPS	TS	MPS	TS	MPS
Sample	Mean	Mean	Mean	Mean	Mean	Mean
1998	160	457	63	161	47	109
1999	124	273	57	144	42	88
2001	97	235	81	176	51	127
2004	92	226	79	169	55	121

The remediation actions were clearly less effective below the first 10 cm (as in the case of Zn), and thus the RCFC of As were superimposed in the different samplings (Figure 29B and 29C). The decline in the mean As concentration in 1999 in relation to 1998, in all soils as well as in the most polluted ones, was only 10% (Table 16). Tilling and homogenization of the first 25 cm, as in the case of Zn, increased the mean As concentration below the first 10 cm in the samplings after 1999.

Because of the narrow difference between the upper limit of the background of As (25.3 mg kg^{-1}) and the maximum permitted level for agricultural soils (50 mg kg^{-1}), notably less than in the case of Zn (101.3 mg kg^{-1} and 1000 mg kg^{-1}, respectively), the main difference in the soil pollution of the two elements was their risk. Therefore, although the remediation measurements decreased the As concentrations, the percentage of soils exceeding the maximum permitted level for agricultural soils was almost the same in the different samplings (around 65% of the total analysed soils). In 2004, even around 30% of the soils had double the maximum permitted (100 mg kg^{-1}). Consequently, 6 years after the spills and despite of the many remediation actions, most of the soils still had excessively high, and hence potentially toxic, As concentrations.

3. Conclusion

The waters and tailings from the Aznalcollar mining spill, as well as the subsequent oxidation of the tailings, considerably raised the concentration of toxic elements on the surface of the soils affected. These contaminants, three years after the accident, continued to be concentrated fundamentally in the uppermost 10 cm of the soils. Therefore, the removal of the tailings with the first 10 cm of the contaminated soils should have reduced the contamination to lower than the toxic limits. However, a very hasty clean-up using heavy machinery, caused by the social alarm throughout Europe, resulted in a deficient clean-up, as part of the tailings remained mixed with the soil, and residual tailings appeared below 30 cm in depth. Consequently, the soils remained highly polluted after the clean-up operation. The conclusion to be drawn is that, for this type of pollution in a Mediterranean environment, the speed of the clean-up is not as critical as a thorough removal of the spill without leaving behind remains of tailings mixed with soils.

References

Adriano, D.C., Chlopecka, A., Kaplan, D.I., Clijsters, H., & Vangronsveld, J. (1997). Soil contamination and remediation: philosophy, science and technology. In Prost (Ed.). Contaminated soils. *INRA*. París.

Alloway B.J. (Ed.). (1995). *Heavy Metals in soils.* Blackie Academic and Profesional, London, 235-274.

Alloway B.J. (1995). Soil processes and the behaviour of metals. In B.J. Alloway (Ed.). *Heavy Metals in soils.* Blackie Academic and Profesional, London, 11-37.

Almodovar, G.R., Saez R., Pons, J.M., & Maestre, A. (1998). Geology and genesis of the Aznalcollar massive sulphide deposits, Iberian Pyrite Belt, Spain. *Miner. Deposita*; **33**, 111-136.

Anderson, P. R., & Benjamin, M. M. (1990). Surface and bulk characteristics of binary oxide suspensions. *Environ. Sci. Technol.*, **24**, 692-698.

Baize, D. (1997). Teneur totales en éléments traces métalliques dans le sols (France). In *INRA* (Ed.). Paris.

Baker, D.E., & Senft, J.P. (1995). Copper. In: B.J. Alloway (Ed.). *Heavy Metals in soils.* Blackie Academic and Professional, London, 179-205.

Bigham, J.M., Schwertmann, U., Traina, S.J., Winland, R.L., & Wolf, M. (1996). Schwertmannite and the chemical modeling of iron in acid sulfate waters. *Geochimica et Cosmochimica Acta*, **60**, 2111-2121.

Bigham, J.M., & Nordstrom, D.K. (2000). Iron and aluminum hydroxysulfates from acid sulfate waters. In: C.N. Alpers, J.L. Jambor, & D.K. Nordstrom (Eds.). Sulfate Minerals: Crystallography, Geochemistry, and Environmental Significance,. *Mineral. Soc. Am.*, Washington, DC, 351-403.

Bower, C.A., & Huss, R.B. (1948). Rapid conductometric method for estimating gypsum in soils. *Soil Sci.*, **141**, 99-104.

Brieva Romero C. (1995). Reutilización de aguas. In: M. Alvarez Cobelas, & F. Cabrera Capitan (Eds.). La calidad de las aguas continentales españolas. Geoforma, Logroño, España, 251-262.

Brummer, G.W. (1986). Heavy metal species, mobility and availability in soils. In: M. Bernhard, F.E. Brinckman, & P.J. Sadler (Eds.). The Importance of Chemical Speciation in Environmental Processes, Springer-Verlag, Berlin, 169-192.

Cabrera, F., Toca, C., Díaz, E., & Arambarri, P. (1984). Acid mine-water and agricultural pollution in a river skirting the Doñana National Park (Guadiamar river, south west Spain). *Water Res.*, **18**, 1469-1482.

Cabrera, F., Soldevilla, M., Cordón, R., & Arambarri, P. (1987). Heavy metal pollution in the Guadiamar river and the Guadalquivir estuary (south west Spain). *Chemosphere*, **16**, 463-468.

Colombo, C., & Torrent, J. (1991). Relationship between aggregation and iron oxides in Terra Rossa soils from Southern Italy. *Catena*, **18**, 51-59.

Dunin, F. X. (1976). Infiltration: its simulation for field conditions. In J. C. Rodda (Ed.). Facets of hydrology. John Willey and Sons, Bristol, UK, 199-227.

Filipeck, L.H., Nordstrom, D.K., & Flicklin, W.H. (1987). Interaction of acid mine drainage with waters and sediments of West Squaw Creek in the West Shasta mining distrit, California. *Environ. Sci. Tech.*, **21**, 388-396.

Förstner, U., & Wittmann, G.T.W. (1983). Metal pollution in the aquatic environment. Springer-Verlag, Berlin, 486pp.

Fuge, R., Pearce, F. M., Pearce, N. J., & Perkins, W. T. (1994). Acid mine drainage in Wales and influence of ochre precipitation on water chemistry. In: C. N. Alpers, & D. W. Blowes (Eds.). The Environmental Geochemistry of Sulfide Oxidation. *Am. Chem. Soc. Symp. Series* **550**, American Chemestry Society, Whashington DC, 261-274.

Godin, P. (1983). Les sources de pollution des sols: essai de quantification des risques dus aux elements traces. *Sci. Sol*, **2**, 73-87.

Harter, R.D. (1983). Effect of soil pH on adsorption of lead, copper, zinc and nickel. Soil Sci. Soc. Am. J., 47, 47-51.

Johnson, D.W., & Todd, D.E. (1983). Relationships among iron, aluminum, carbon, and sulfate in a variety of forest soils. *Soil Sci. Soc. Am. J.*, **50**, 776-783.

Kabata Pendias, A., & Pendias, H. (2000). Trace elements in soils and plants. CRC Press, third edition. Boca Raton, Florida, 432pp.

Kashir, M., & Yanful, E.K. (2000). Compatibility of slurry wall backfill soils with acid mine drainage. *Adv. Environ. Res.*, **4**, 252-268.

Kinniburgh, D.G., Jackson, M.L., & Syers, J.K. (1976). Adsorption of alkaline earth, transition, and heavy metal cations by hydrous oxide gels of iron and aluminum. *Soil Sci. Soc. Am. J.*, **40**, 796-799.

Lindsay, W.L. (1979). Chemical Eqilibria in Soils. TheBlackburn Press, Caldwell, New Jersy, 449pp.

Matschullat, J., Ottenstein, R., & Reimann, C. (2000). Geochemical background - can we calculated it? *Environ. Geol.*, **39**, 990-1000.

McBride, M. B., & Bouldin, D. R. (1984). Long-term reactions of copper (II) in a contaminated calcareous soil. *Soil Sci. Soc. Am. J.*, **41**, 1073-1077.

McKenzie, R.M. (1980). The adsorption of lead and other heavy metals on oxides of manganese and iron. *Aust. J. Soil Res.*, **18**, 61-73.

Nordstrom D.K. (1982). Aqueous pyrite oxidation and consequent formation of secundary iron minerals. In: J.A. Kitrick., D.S. Fanning., & L.R. Hossner. (Eds.). Acid sulfate weathering. *Soil Sci. Soc. Am.*, Madison, WI, 37-56.

Nordstrom, D.K., & May, H.M. (1996). Aqueous equilibrium data for mononuclear aluminum species. In: G. Sposito (Ed.). *The Environmental Chemistry of Aluminum*. CRC Press/Lewis Publishers, Boca Raton, Florida, 39-80.

O'Neill. P. (1995). Arsenic. In: B.J. Alloway (Ed.). *Heavy Metals in Soils*. Blackie Academic and Profesional, London, 105-121.

Parfitt, R.L., & Smart, R.St.C. (1978). The mechanism of sulphate adsorption on iron oxides. *Soil Sci. Soc. Am. J.*, **42**, 48-50.

Preda, M., & Cox, M.E. (2004). Temporal variations of mineral character of acid-producing pyritic coastal sediments, Southeast Queensland, Australia. *Sci. Total Environ.* **326**, 257-269.

Ramos, L., Hernández, M., & González, M.J. (1994). Sequential fractionation of koper, lead, cadmium and zinc in soils from or near Doñana National Park. *J. Environ Qual.*, **23**, 50-57

Ritsema, C.J., & Groenenberg, J.E. (1993). Pyrite oxidation, carbonate weathering, and gypsum formation in a drained potential acid sulfate soil. *Soil Sci. Soc. Am. J.,* **57**, 968-976.

Scheinost, A.C., & Schwertmann, U. (1999). Color identification of iron oxides and hydroxysulfates: use and limitations. *Soil Sci. Soc. Am. J.,* **63**, 1463-1471.

Schermerhorn, L.J.G. (1982). Framework and evolution of hercynian mineralization in the Iberian Meseta. *Comun. Serv. Geol. Portugal*; **68**, 91-140.

Sheppar S.C., Gaudet C., Sheppar P.I., Cureton P.M., & Wong M.P. (1992). The development of assessment and remediation guidelines for contaminated soils: a review of the science. *Can. J. Soil Sci.*, **72**, 259-394.

Shuman, L.M. (1985). Zinc, manganese and copper in soil fractions. *Soil Sci.,* **140**, 11-22.

Simón, M., Iriarte, A., Guardiola, J. L., Barahona, E., & García, I. (1998). Soil-infiltration rate under different vegetation cover using a rain simulator of variable intensity. In: A. Rodríguez, C.C. Jiménez, & M.L. Tejedor (Eds.). The soil as a strategic resource: degradation processes and sonservation measures. Geoforma, Logroño, Spain, 177-188.

Simón, M., Ortiz, I., García, I., Fernández, J., Fernández, E., Dorronsoro, C., & Aguilar, J. (1999). Pollution of soils by the toxic spill of a pyrite mine (Aznalcollar, Spain). *Sci. Total Environ.*, **242**, 105-105.

Simón, M., Martín, F., Ortiz, I., García, I., Fernández, J., Fernández, E., Dorronsoro, C., & Aguilar, J. (2001). Soil pollution by oxidation of tailings from toxic spill of a pyrite mine. *Sci. Total Environ.*, **279**, 63-64.

Simón, M., Dorronsoro, C., Ortiz, I., Martín, F., & Aguilar, J. (2002). Pollution of carbonate soils in a Mediterranean climate due to a tailings spill. *Eur. J. Soil Sci.*, **53**, 321-330.

Simón, M., Iriarte, A., García, I., Martín, F., Aguilar, J., & Dorronsoro, C. (2005). Mobility of heavy metals in pyritic mine spills from an accident in Aznalcóllar, SW Spain. In: A. Faz Cano, R. Ortíz Silla, & A.R. Mermut (Eds.). *Advances in Geoecology* **36**, CATENA VERLAG, 467-476.

Singer, P.C., & Stumn, W. (1968). *Kinetics of the oxidation of ferrous iron.* Second Symposium on Coal Mine Research, Mellon Institute, Pittsgurgh, PA, 12-34.

Singh, B.R. (1984). Sulfate sorption by acid forest soils: 2. Sulfate adsorption isotherms with and without organic matter and oxides of aluminum and iron. *Soil Sci.*, **138**, 294-297.

Soil Survey Staff. (1951). Soil survey manual, Handbook 18. US Department of Agriculture, Washington DC, 227 pp.

Stringer, D.A. (1990). Hazard assessment of chemical contaminants in soil. *ECETOC Technical. Rep.* No. **40**. Aneme. Louise 250. Bruselas, Bélgica, 121 pp.

Stumm W.Y., & Morgan, J.J. (1981). Aquatic Chemestry: An introduction emphasizing chemical equilibria in natural waters. Jonh Wiley and Sons, New York, 218 pp.

Van Den Berg R., Dennenman C.A., & Roels J.M. (1993). Risk assessment of contaminated soil: Proposal for adjusted, toxicologically based Dutch soil clean-up criteria. In: F. Arendt, G.J. Annokkée, R. Bosman, & W.J. van der Brink (Eds.). Contaminated Soils'93. Kluwer Academic Publisher, London, 349-364.

Waychunas G.A., Xu, N., Fuller, C.C., Davis, J.A., & Bigham, J.M. (1995). XAS study of arsenate and selenate substituted schwertmannite. *Physica B,* **208**, 481-483.

In: Environmental Pollution: New Research
Editor: Rachel H. Plattenberg, pp. 225-261
ISBN: 1-60021-285-9
© 2007 Nova Science Publishers, Inc.

Chapter 7

EVALUATING THE BURDEN OF HYDROPHOBIC ORGANIC CONTAMINANTS IN SOILS AND SEDIMENTS: AN APPRAISAL OF METHODOLOGIES

Annika L. Swindell and Brian J. Reid*
School of Environmental Sciences, University of East Anglia,
Norwich, UK, NR4 7TJ

Abstract

Over the course of many years different methods for the extraction of solid phase environmental samples have been developed. Traditionally these methods have been concerned with maximising analyte recovery with a view to determining the total contaminant concentration. However, in light of the body of knowledge relating to compound availability and ageing these extraction methods are unsuitable in determining the fraction that may be bioavailable. It is evaluation of available contaminant concentrations that are most significant with respect to meaningful contaminated land assessment under present day risk based frameworks of legislation. However, to date no review has been made of extraction methodologies of solid phase environmental samples despite this shift in paradigm. This review appraises both traditional exhaustive techniques and more contemporary non-exhaustive techniques against this back drop of evolving perspectives in contaminated land assessment and legislation. The review highlights the variability in techniques considered to be exhaustive thereby questioning their value in meaningful assessment of contaminated land. In complement the review draws comparison between non-exhaustive extraction techniques and biological endpoints, specifically compound biodegradability, that supports the use of certain non-exhaustive extraction methods for inference of risk and estimation of bioremediation potential.

* E-mail address: b.reid@uea.ac.uk; Tel +44 1603 592357; Fax +44 1603 591327. Corresponding author.

1. Introduction

This manuscript reviews methodologies applicable to the extraction of polycyclic aromatic hydrocarbons (PAHs) associated with soil and sediment matrices and in particular how these extraction methodologies reflect compound partitioning and (bio)availability/ (bio)accessibility (terms defined in section 3). Occasionally and where appropriate, other compound classes are mentioned to introduce a methodology,. PAHs were selected because of their widespread occurrence in the environment and, in particular, because of their common occurrence at contaminated sites (Peters et al., 1999). PAHs are predominantly formed during the incomplete combustion of organic materials, both from natural events and anthropogenic activities (Wild and Jones, 1995). The majority of the PAHs released to the environment originate from anthropogenic sources through the combustion of fossil fuels and waste incineration (Wild and Jones, 1995). PAHs enter the soil by atmospheric deposition or by accidental or deliberate application (i.e. spillage or leakage). PAHs form a large group of hydrophobic organic compounds, consisting of two or more fused benzene rings arranged in a variety of ways (i.e. linear, angular, cluster) (WHO, 1998). They encompass a wide range of physicochemical properties and as such may be considered as representative hydrophobic organic contaminants. In addition to their ubiquity, PAHs are a cause for concern due to their genotoxic properties. Consequently, the United States Environmental Protection Agency (USEPA) listed 16 PAHs as 'priority pollutants'. The structures and properties of these PAHs are presented in Table 1.

Table 1. Structures and properties of the 16 USEPA PAHs (Environment Agency, 2003)

PAH name	Chemical structure	Mol. weight, g mol^{-1}	Aqueous solubility at 25 °C, mg L^{-1}	Octanol-water partition coefficient (log K_{ow})
naphthalene	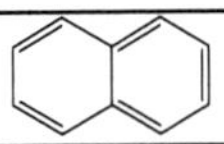	128.17	31.0	3.37
acenaphthylene	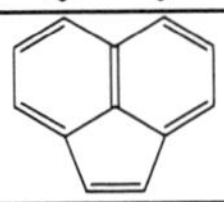	152.20	16.1	4.00
acenaphthene	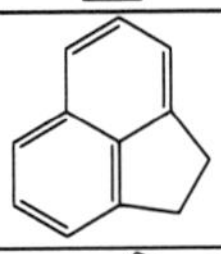	154.20	3.8	3.92
fluorene	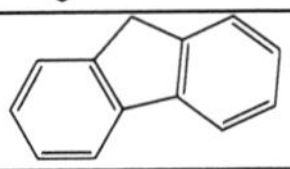	166.22	1.9	4.18
phenanthrene	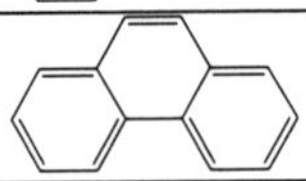	178.23	1.1	4.57
anthracene	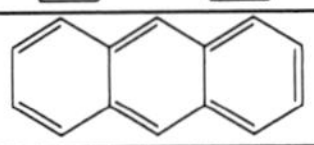	178.23	0.045	4.54
fluoranthene	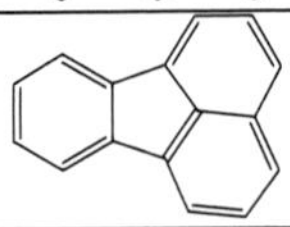	202.26	0.26	5.22

Table 1. continued

PAH name	Chemical structure	Mol. weight, g mol^{-1}	Aqueous solubility at 25 °C, mg L^{-1}	Octanol-water partition coefficient (log K_{ow})
pyrene	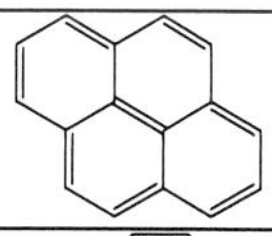	202.26	0.13	5.18
benzo(a)anthracene	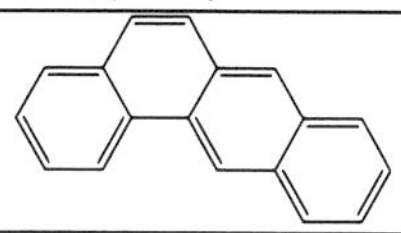	228.30	0.011	5.91
chrysene	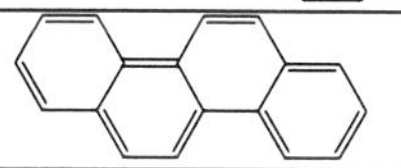	228.30	0.0016	5.70
benzo(b)fluoranthene	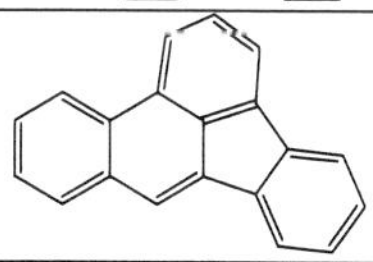	252.32	0.0015	5.80
benzo(k)fluoranthene	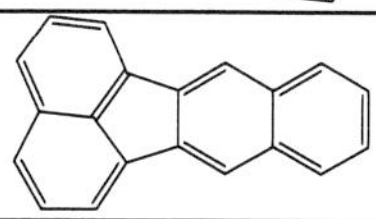	252.32	8.00E-04	6.00
benzo(a)pyrene	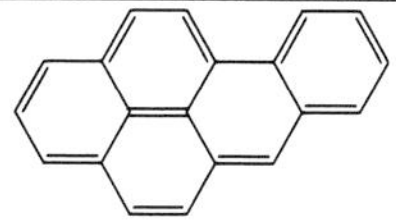	252.32	0.0038	6.04
dibenzo(a,h)anthracene	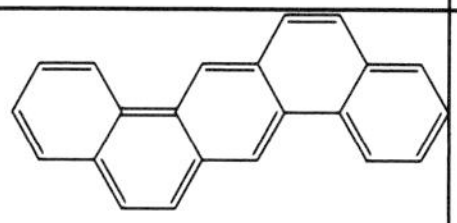	278.36	6.0E-04	6.75
benzo(g,h,i)perylene	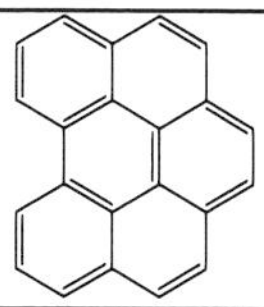	276.34	2.6E-04	6.50
indeno(1,2,3-c,d)pyrene	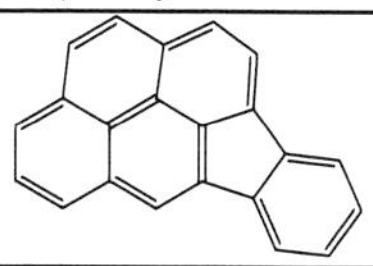	276.34	2.2E-05	6.65

2. Interactions between Soil/Sediment and PAHs

Before considering how PAHs might be extracted from soils and sediments, it is appropriate to consider how they interact with these media and how these interactions influence compound partitioning and (bio)availability/(bio)accessibility; the process of 'ageing'. The mechanisms inherent to ageing have been extensively reported (Hatzinger and Alexander, 1995, 1997; Mader et al., 1997; Nam and Alexander, 1998; Piatt and Brusseau, 1998; Xing and Pignatello, 1997). In summary, ageing has been hypothesised to result from the slow diffusion of organic contaminants through soil organic matter (Nam et al., 1998) and / or as a result of entrapment within soil nanopores (Nam and Alexander, 1998), where they become sequestered (Figure 1). Soil organic matter (SOM) has been described as a polymeric material consisting of flexible 'rubbery' phases and condensed 'glassy' phases (Xing and Pignatello, 1997; Xing et al., 1996) (or as 'soft' and 'hard' domains (Huang and Weber, 1997; McGinley et al., 1993, 1996; Young and Weber, 1995)).

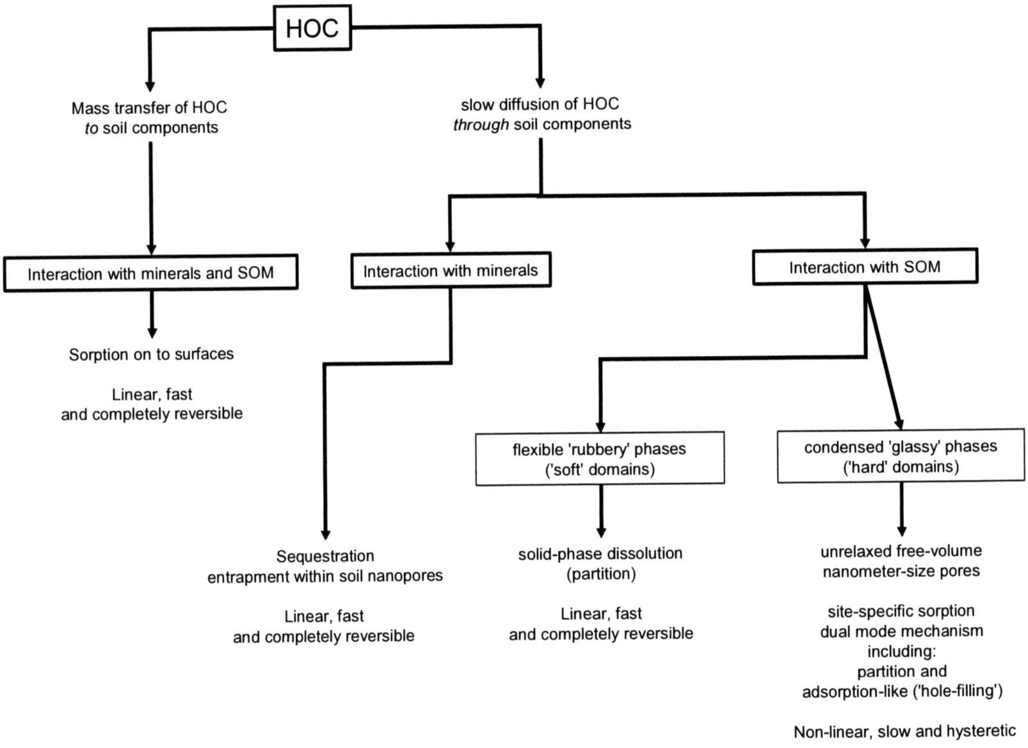

Figure 1. Hydrophobic organic contaminant (HOC) interactions with soil and sediment media.

A characteristic feature of the glassy phase is the presence of unrelaxed free-volume in the form of internal nanometer-size pores, in which site-specific sorption can take place (White and Pignatello, 1999). It has been shown that sorption of hydrophobic organic compounds (HOCs) in rubbery domains occurs by solid-phase dissolution (partition) and is linear, fast and completely reversible (White and Pignatello, 1999). While sorption in glassy domains occurs by a dual mode mechanism that includes both partition and adsorption-like ('hole-filling') interactions and is non-linear, slow and hysteretic (White and Pignatello, 1999; Xing and Pignatello, 1997; Xing et al., 1996). Although organic contaminants are still present in the soil, sometimes at relatively high concentrations, compound sequestration has shown to result in decreased (bio)availability/(bio)accessibility, as, for example, illustrated through a decreased extractability by non-exhaustive extractions (Hatzinger and Alexander, 1995; Johnson et al., 2002), decrease in extent and rate of microbial mineralisation (Hatzinger and Alexander, 1995), earthworm accumulation (Johnson et al., 2002), and decreased toxicity (Alexander et al., 2002). While related to the physical and chemical phenomena that govern extractability these latter observations reflect compound bioavailability or bioaccessibility.

In addition to, and occurring in parallel with, changes in compound partitioning with time are changes in residual compound concentration. These residual concentrations being dependent upon inherent compound degradability, the physicochemical properties of the compounds (Mackay and Callcott, 1998) and the (bio)availability/(bio)accessibility of the compound to be degraded or to leach or to volatilise. It is generally accepted that the degree of persistence of PAHs in soil increases with increasing molecular mass (and number of aromatic rings) of the PAHs. The 'lighter' PAHs (2 to 3 rings) are more susceptible to volatilisation and other abiotic loss processes (hydrolysis, reduction, oxidation and photolysis) than the 'heavier' PAHs. Similarly, lighter PAHs are more susceptible to biodegradation than the heavier PAHs, which are inherently recalcitrant to microbial degradation (Cerniglia, 1984). Indeed, as molecular mass increases, compound hydrophobicity, measured as a partition coefficient between octanol/water (K_{ow}), can increase over several orders of magnitude (Figure 2). K_{ow} (usually expressed as Log K_{ow}) is the ratio, at equilibrium, of the concentration of a compound in octanol to the concentration in water. As a result, heavier PAHs partition more strongly with the soil solid phase (in particular soil organic matter) and thereby become less (bio)available/(bio)accessible.

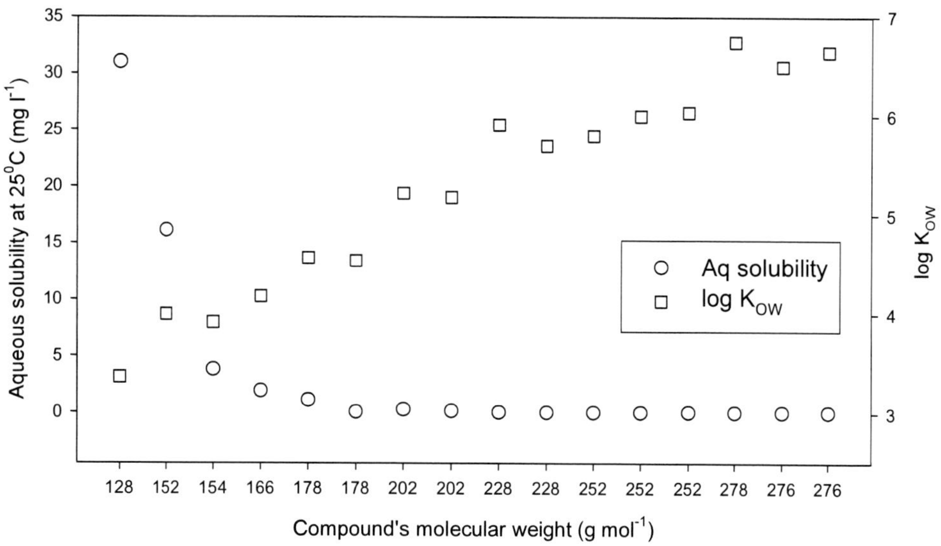

Figure 2. Commensurate decrease in aqueous solubility and increase in K_{ow} with increasing molecular weight (adapted from data in Table 1).

Natural communities of prokaryotes (e.g. bacteria) and eukaryotes (e.g. fungi) possess great physiological versatility. They can metabolise and, in many cases, mineralise (i.e. convert an organic compound to inorganic products such as CO_2 and water) many of the environmental organic pollutants of current concern, including synthetic compounds. However, for degradation to take place contaminants need to be partitioned in the soil in such a way that they are still accessible to the microbes. Assuming these residues are accessible the most significant pathway for bacterial PAH biodegradation is by hydroxylation (Juhasz and Naidu, 2000). Many different species of bacteria are able to utilise 2 to 4 ring PAHs as a carbon and energy source, resulting in complete mineralisation of the compounds (see Juhasz and Naidu (2000) for an extensive list of references). To date, no bacteria have been found that can utilise heavier PAHs as sole growth substrate. Instead, the bacteria require an alternative substrate to induce the synthesis of enzymes capable of transforming these high molecular weight PAHs. This type of PAH degradation is known as 'cometabolism' (Kanaly et al., 1997; Kanaly and Bartha, 1999). In addition to bacteria, many ligninolytic and non-ligninolytic fungi possess the ability to cometabolically oxidise PAHs (see Juhasz and Naidu (2000) for an extensive list of references).

3. Extraction Methodologies in the Context of PAH Partitioning

Semple et al. (2004) recently published an A-page feature in Environmental Science and Technology in response to a 'persuasive' feature by Ehlers and Luthy (2003). Both articles were concerned with the need to include assessment of contaminant bioavailability in the risk assessment of contaminated soils and sediments. The Semple et al. (2004) feature specifically considered the semantics of the words '*bioavailability*' and '*bioaccessibility*'. The following definitions were proposed: '*bioavailable*' compound is that which is freely available to cross an organism's cellular membrane from the medium the organism inhabits at a given time' while '*bioaccessible*' compound is that which is available to cross an organism's cellular membrane from the environment, if the organism has access to the chemical; however the chemical may be either physically removed from the organism or only bioavailable after a period of time (i.e. bioaccessible includes what is actually bioavailable and potentially bioavailable) (Figure 3). Semple et al. (2004) concluded their feature with the following: 'our proposed definitions raise a fundamental methodological question: can the bioavailable portion of substance *x* to species *y* actually be measured, and if so, how?'

Obviously there are a myriad of biological assays that can be used to assess contaminant availability for potential bioaccumulation, biodegradability or toxicity. In the case of bioaccumulation: analysing tissue, from, for example, earthworms from contaminated locations or by exposing laboratory-reared specimens to soil samples from the field are assessment options (Jager et al., 2005). In the case of assessing biodegradation potential of soil-associated contaminants: inoculation of soil with cultures of catabolically active microorganisms might be used in, combination with optimisation of nutrient conditions, to establish the bioavailable amounts of contaminants present in soil samples (Stokes et al., 2005). With respect to toxicity of available contaminants: mortality, fecundity, growth, and juvenile maturation, represent a variety of possible tests that can be employed (Booth et al., 2005).

In addition to these 'direct' measures of bioavailability using biological receptors, a myriad of extraction methodologies to quantify 'bioaccessible/bioavailable' contaminant concentrations in soils and sediments have been proposed (and are reviewed in Section 6). The extraction methods presented in this review have been broadly divided into two categories: exhaustive techniques (Section 4) and non-exhaustive techniques (Section 6).

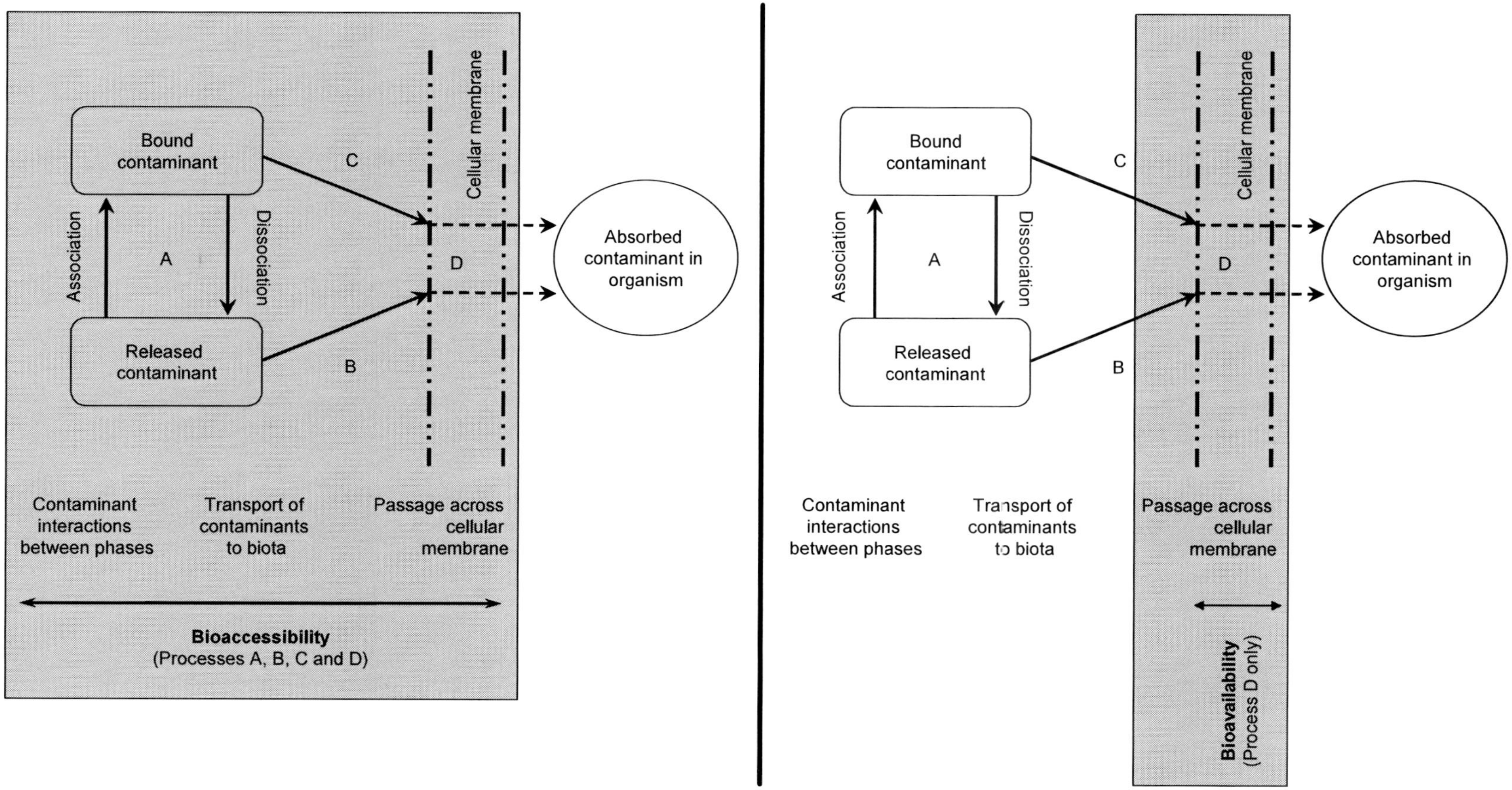

Figure 3: A schematic representation distinguishing the terms bioavailable- and bioaccessible-contaminant in the context of soils/sediments-organisms interaction (adapted from Semple et al., 2004).

4. Exhaustive Extractions

An exhaustive extraction, by definition, should liberate as much analyte as possible from a given matrix. Thus, exhaustive extraction methodologies are often considered to provide an evaluation of 'total' contaminant concentrations present. These exhaustive extraction techniques are based on the release of compounds following interactions with a selected solvent. The most commonly used methodologies are presented in this section and include Soxhlet extraction, shake extraction, alkaline saponification, ultrasonic extraction, supercritical fluid extraction (SFE), subcritical water extraction, microvave-assisted extraction (MAE) and pressurised liquid extraction (PLE).

4.1. Soxhlet Extraction

During a Soxhlet extraction, solvent is continuously refluxed through the system for several hours (typically between 6 to 48 h). First the liquid solvent, usually between 200 and 300mL, is evaporated from a round-bottomed flask, condensed back to a liquid, which is allowed to fall into a permeable vessel containing the sample. The solvent of choice must be one in which the contaminants dissolve readily. Non-polar organic contaminants such as PAHs have low aqueous solubilities but strong affinities for soil organic matter. The solvent must therefore have the ability to displace the contaminants from the matrix to which they are sorbed and into solution. In so doing, as the solvent passes through the sample, the dissolved contaminants are removed. The solvent and analyte drop back into the distillation flask where the solvent is re-evaporated and recycled. The boiling point of the contaminants should always be much higher than that of the organic solvent to ensure that the solvent remains pure throughout the extraction and that the contaminants remain trapped in the flask. The solid matrix to be extracted should be permeable to the extracting solvent. Should the chosen solvent be immiscible in water, the moist sample can be air-dried or a drying agent, typically anhydrous sodium sulphate (Na_2SO_4), can be mixed with the matrix, in order to remove the water, thereby allowing intimate contact between the solvent and the contaminant. A drying agent such as Na_2SO_4 has the advantage over oven or air-drying of minimising the loss of the more volatile compounds. Crushing the sample into a fine homogeneous matrix will aid the extraction process. The organic solvents used for Soxhlet are essentially the same as those used for shake extraction (Table 2). No studies, however, appear to exist in which the efficiencies of extraction solvents using Standard Reference Materials (SRM) for Soxhlet extraction have been tested. However, the USEPA recommend either acetone/hexane (1:1, v/v) or DCM/acetone (1:1, v/v).

The USEPA, Soxhlet extraction (Method 3540) is 'applicable to the isolation and concentration of water-insoluble and slightly water-soluble organics in preparation for a variety of chromatographic procedures' (USEPA Method 3540C). Prior to analysis, the extract solution often requires a concentration step. This is achieved by either using a rotary evaporator or a nitrogen blowdown technique. In light of this concentration step, the importance of an initially pure extraction solvent has to be recognised. Soxhlet extraction is very efficient and is generally used as a benchmark technique for total PAH concentrations. For instance, the U.S National Institute for Standards and Technology (NIST) provide Standard Reference Materials (SRMs) with certified PAH concentrations (e.g. SRM 1649a,

1650a and 1944) based on Soxhlet extractions. Soxhlet, however, is not as selective as many other extraction techniques. Indeed, the extract solutions often contain a significant amount of coextractives, non-target matrix components, such as other organic compounds of similar solubilities. This results in very dark and highly turbid extracts (likened by Hawthorne et al. (2000a) to the colour of black coffee). These matrix organics need to be removed from the extract solution prior to its analysis (typically by gas chromatography/mass spectrometry (GC/MS) or high precision liquid chromatography (HPLC)), as failing to do so would most certainly result in sample interferences obscuring instrumental responses. Various clean-up techniques for PAHs have been used (Codina et al., 1994; Guerin, 1999; Sun et al., 1998; USEPA Method 3600C, 1996; USEPA Method 3620b, 1996; USEPA Method 3630c, 1996; USEPA Method 3640a, 1994; USEPA Method 3650b, 1996): with gel permeation chromatography clean-up, solid-phase extraction clean-up and acid-base partitioning clean-up the most commonly applied.

Table 2. List of common solvents used in organic extractions, (1) (BDH, 1990).

Solvent	Boiling Point, °C[(1)] At normal temperature and pressure	Solubility in water[(1)] at 20°C
Acetone	56	Miscible in all proportions
Acetonitrile	80	Miscible in all proportions
Benzene	80	Slightly soluble (1.8g/L)
Butan-1-ol	118	Soluble (77g/L) at 20°C)
Dichloromethane, DCM	40	Slightly soluble (20g/L at 20°C)
Ethanol	78	Miscible in all proportions
Methanol	65	Miscible in all proportions
n-hexane	69	Practically insoluble (0.14g/L)
N-methyl-2-pyrrolidone	202	Miscible in all proportions
2-propanol	82	Miscible in all proportions
Toluene	111	Practically insoluble (0.5g/L)
Water	100	Miscible in all proportions

4.2. Shake Extraction

Simple shake extractions using organic solvents have typically been considered as mild extractants. Such extractions have commonly been proposed as chemical means through which sequestration can be evaluated (Alexander and Alexander, 2000; Hatzinger and Alexander, 1995; Nam et al., 1998). However, there are also reports where organic solvents used in simple shake extractions have overestimated bioavailability (McLeod and Semple, 2000; Northcott and Jones, 2001a; Reid et al., 2000) or indeed performed as well as Soxhlet extractions (Khodadoust et al., 2000; Noordkamp et al., 1997; Northcott and Jones, 2001a; Reid et al., 2000), (Table 3).

Shake extractions are simple and rely upon placing the contaminated sample along with the extraction solvent in a sealed vessel. The vessels are placed on a shaker typically at room temperature. The mixture is then centrifuged or filtered and depending on the colour and turbidity of the extract, a clean-up step may be undertaken. The efficiency of the solvent in removing the contaminants from the sample is dependent upon the solubility of the analyte in the solvent. Aqueous solubility of the solvent can also be an important factor if moist samples are to be analysed. A list of solvents typically found associated with organic solvent extractions is given in Table 2. More often than not, mixed solvents are used rather than a single solvent. Indeed, under certain conditions, a particular solvent might be most effective in removing the low molecular weight while less effective in removing the high molecular weight PAHs. In which case, this solvent combined with a complementary one showing good recoveries for the high molecular weight PAHs, could potentially produce an ideal mixed solvent. Another scenario where mixed solvents might be used is if water is present in the sample to be extracted. If so, a pair of miscible solvents would be used, where one has a high aqueous solubility and the other not. In shake extractions, the soil:solvent ratios can range from 1:100 to 1:2 (g:mL). Optimum extraction times need to be established from extraction kinetics data, which are specific to each experiment depending on the nature of the sample, the analyte, the choice of solvent and the sample:solvent ratio. Various studies report extraction times ranging from 1 to 72 hours. Sequential solvent extractions have also been documented (Khodadoust et al., 2000). These techniques process the sample in multiple stages. Several combinations exist, but each time a sample is extracted a first time with a particular solvent (pure or mixed), centrifuged, the supernatant is removed and the same sample is then extracted in a similar manner with fresh solvent. The solvents may be changed between each different stage.

Table 3. Comparison of PAH recoveries from Soxhlet and from shake extraction.

Reference	Soil type	Solvent	PAHs	Soxhlet	Shake
(Noordkamp et al., 1997)	Sandy loam, Industrial soil	Acetone	Total	$mg\ kg^{-1}_{dm}$ 108 ± 15	$mg\ kg^{-1}_{dm}$ 108 ± 8
(Khodadoust et al., 2000)	Sandy loam, MGP	Mixed (5% pentanol, 10% water, 85% ethanol)		$mg\ kg^{-1}_{dm}$	$mg\ kg^{-1}_{dm}$
			naphthalene	2400	3500
			2-methyl-naphthalene	1900	2000
			fluorene	580	540
			phenanthrene	1500	1500
			fluoranthene	780	740
			pyrene	900	820
			benzo(b+k)flouranthene	440	380
			benzo(a)pyrene	320	280
			indeno(1,2,3-cd)pyrene	200	140
			benzo(g,h,i)perylene	180	160

4.3. Saponification

Organic solvents (usually methanol) can be used in combination with bases (potassium hydroxide, sodium hydroxide or tetra methyl ammonium hydroxide) to extract soil-associated organic contaminants. This technique is known as alkaline saponification. The saponification step has been shown to significantly increase the recovery of PAHs in soil and sediment (Eschenbach et al., 1994; McLeod and Semple, 2000; Noordkamp et al., 1997; Northcott and Jones, 2001a), (Table 4).

Table 4. Recoveries of PAHs from soil by solvent extraction, Soxhlet and saponification.

Reference	PAHs	Amount extracted		
		Solvent shake	**Soxhlet**	**Saponification**
(Noordkamp et al., 1997)		(methanol) $mg\ kg^{-1}{}_{dm}$	No data	(KOH/methanol) $mg\ kg^{-1}{}_{dm}$
	total	55 ± 1		76 ± 14
(Northcott and Jones, 2001a)		(butanol) %	(DCM) %	(KOH/methanol) %
	phenanthrene	91.3 ± 2.1	88.9 ± 0.5	94.7 ± 0.3
	pyrene	87.0 ± 2.7	82.4 ± 1.0	90.7 ± 0.4
	benzo(a)pyrene	68.5 ± 2.6	60.3 ± 1.4	77.3 ± 0.4

Eschenbach et al. (1994) observed that the saponification step usually extracted 20-30% more than the simple solvent extraction. It has been suggested that alkaline hydrolysis cleaves the ester bonds, which form part of the macromolecular structure of humic polymers, resulting in an increase in the availability of the contaminants. Addition of a saponification element to an organic solvent extraction enhances recoveries by releasing the PAHs no longer accessible to organic solvents on account of them being trapped in the soil matrix. As with Soxhlet extraction, the extract may require cleaning up to avoid interferences during subsequent analyses. In a study by Noordkamp et al. (1997), although the comparisons were not made directly, tabulated results suggest that Soxhlet and shake extractions with acetone or acetone: water (1:4, v/v) performed better than methanolic saponification at room temperature (Table 5).

Table 5. Comparison of methods in terms of total PAH recovery from industrially contaminated soil. Data from Noordkamp et al. (1997).

Solvent used	PAH recovery, $mg\ kg^{-1}{}_{dm}$			
	Solvent extraction	**Soxhlet extraction**	**Saponification at room T°**	**Saponification (+ microwave)**
Acetone/water (4:1)	104 ± 16	-	-	-
Acetone	108 ± 8	108 ± 15	-	-
KOH/methanol	-	-	76 ± 14	91 ± 27
TMAH/methanol	-	-	70 ± 14	103 ± 19

The recoveries matched those of Soxhlet and organic solvent extraction when the saponification extraction was carried out in a microwave oven at 70°C (Table 4). The saponification technique has the advantage of not being affected by sample water content. Indeed, Heemken et al. (1997), reported recoveries of PAHs from undried (56 wt%) suspended particulate matter (SPM) as 97% ± 2.0% RSD of the dried SPM recoveries.

4.4. Ultrasonic Extraction

Sonication has developed from the more traditional liquid partitioning extraction techniques, where organic solvent was added to a sample and subjected to mechanical shaking. The shaking was subsequently replaced by high frequency sound energy (ultrasound). The vibrations help release the trapped contaminants by breaking up the soil aggregates and disrupting the matrix, thereby increasing the surface area of the soil sample that is in contact with the solvent. The energy can be administered to the sample in two different ways. An ultrasonic probe can be inserted into the sample/solvent mixture or the mixture contained in appropriate glassware can be immersed in an ultrasonic bath. Although the former method delivers more intense energy, the probe can be abraded when used for the extraction of soil or sediment samples. Gradual erosion of the probe would lead to changes in the energy output and therefore in the extraction efficiencies over time. The ultrasonic bath, although less intense, has the advantage of applying more consistent energies. As with the aforementioned techniques, the ultrasonic approach relies upon intimate contact between the organic solvent and the sample matrix. Consequently the choice of solvent (Table 2) is crucial to the efficiency of the extraction. Extractions described in the literature typically use 10-50 mL of solvent for 0.5-15 g samples, which are sonicated for 0.5-1 h. Extraction time and amount of solvent do, however, vary significantly amongst authors. As is the case with Soxhlet, the extracts may require clean-up step prior to analysis. Recoveries of PAHs from sonication are influenced by sample texture (Brilis and Marsden, 1990), the method being more efficient for free-flowing samples. Provided a suitable solvent is chosen, such as for example acetone or 2-propanol, the water content of a moist sample does not affect the efficiency of ultrasonic extraction. Ultrasonic extraction, or USEPA Method 3550, is regarded as being less rigorous than other so-called exhaustive techniques (USEPA Method 3550b, 1996). In some cases, sonication was found to be as successful as Soxhlet (Almaric and Mouvet, 1997; Hollender et al., 2003; Sun et al., 1998), (Table 6), or when intensive extraction (8hrs) was carried out (2179 and 2073 mg kg^{-1} PAH after sonication and Soxhlet extraction, respectively, (Guerin, 1999)), however in other studies, the method consistently under performed (65 ± 4 and 108 ±

Table 6. Comparable extraction efficiencies between sonication and Soxhlet extraction. Data from Almaric and Mouvet (1997).

Phenanthrene concentration, mg $kg^{-1}{}_{dm}$	Mean recoveries, % ± % RSD	
	Sonication	Soxhlet
0.2	54.3 ± 10.9	72.3 ± 38.1
1.0	96.9 ± 8.2	80.2 ± 15.0
10.0	83.7 ± 13.3	77.5 ± 12.5

15 mg kg^{-1} for sonication and Soxhlet PAH recoveries, respectively, (Heemken et al., 1997), (Mzoughi et al., 2002; Noordkamp et al., 1997)).

4.5. Supercritical Fluid Extraction (SFE)

Supercritical fluid extraction makes use of a supercritical fluid in place of an organic solvent to extract the contaminants. Many supercritical fluids can be used, for example N_2O, H_2O, CO_2. CO_2 is preferred due to its conveniently low critical pressure and temperature (73.8 bar, 31°C), low reactivity, non-toxicity, non-flammability and low cost. Supercritical fluids typically have much greater molecular diffusivities (10^{-4} *vs.* 10^{-5} cm^2 s^{-1}) and lower viscosities (10^{-4} *vs.* 10^{-3} N s m^{-2}) than liquid solvents, properties that improve the rate of mass transfer of a solute from the matrix surface to the bulk extraction fluid (Benner, 2000). In other words, one would expect, given the inherent properties of supercritical fluids, a much faster removal of analytes from samples. Under optimised conditions, SFE can be completed in less than 1h, much less than the more conventional organic solvent extractions. Enhancements of the solvating power of a supercritical fluid can be achieved by either adjusting the pressure (Langenfeld et al., 1993) and/or the temperature (Benner, 1998; Langenfeld et al., 1993; Yu et al., 1990). Langenfeld et al. (1993) found that raising the extraction temperature from 50°C to 200°C significantly increased PAH recoveries from urban air particulates but not from highly contaminated soil (Table 7). Increases in pressure from 350 atm to 650 atm, at 50°C, had no effect on extraction efficiencies (Table 7). The efficiency of the technique has been shown to be quite variable. It is generally accepted that for a given matrix, recovery decreases with increasing PAH molecular weight (e.g. RSM HS-4, (Reimer and Suarez, 1995), soils from municipal waste disposal site and coal tar contaminated soil, (Hollender et al., 1997), various SRMs, (Benner, 1998), MGP soil, (Hawthorne et al., 2000b)).

Table 7. Effect of temperature and pressure on SFE efficiency. The values are percent recoveries relative to certified concentrations. The certified concentrations are based on Soxhlet (SRM 1649) and EPA Methods 3540 and 3550 (Soxhlet and Sonication) (Soil). Data from Langenfeld et al. (1993).

PAHs	% recovery ± % RSD					
	SRM 1649			'US EPA certified' contaminated soil		
	350 atm 50°C	350 atm 200°C	650 atm 50°C	350 atm 50°C	350 atm 200°C	650 atm 50°C
Phenanthrene	35 ± 7	80 ± 12	34 ± 23	127 ± 6	126 ± 4	135 ± 8
Fluoranthene	49 ± 9	142 ± 10	49 ± 22	142 ± 6	137 ± 6	146 ± 9
Pyrene	35 ± 7	105 ± 9	37 ± 23	135 ± 7	126 ± 5	137 ± 7
Benzo(a)anthracene	46 ± 6	102 ± 6	42 ± 11	107 ± 12	96 ± 4	98 ± 15
Chrysene	49 ± 4	93 ± 7	47 ± 9	104 ± 6	97 ± 4	98 ± 12
Benzo(a)pyrene	23 ± 20	69 ± 8	29 ± 8	78 ± 23	88 ± 13	65 ± 8

The extraction efficiency has been shown to increase with the addition of a modifier (Hollender et al., 1997), usually methanol although many types can be used but at the expense of extracting more matrix components, thereby adding the need for a clean-up step. The presence of small amounts of water (up to 10%) in the sample has been shown to increase the extraction recoveries of PAHs. It has been suggested that when the water content is low, the soil will preferentially adsorb the water, releasing the weakly adsorbed PAHs. Too much water, however, would saturate the CO_2, rendering it more polar and therefore less attractive to PAHs (Laitinen et al., 1994). A study by Benner (1998) showed that although in certain cases the addition of a modifier could significantly enhance recoveries, not all modifiers performed as well. Hollender et al. (1997) also noted this and added that modifier volume needed to be considered also. The aim of the modifier is to interact with the analyte causing it to rapidly desorb into the supercritical fluid and outcompete the analyte by occupying the matrix sites thus preventing the contaminants from readsorbing back on to those sites. The modifier can also modify the matrix itself thereby providing access to internal sites (Librando et al., 2004). It follows, as was suggested by Yang et al. (1995) that the interactions of the modifier with the matrix are more important than the solubility of the PAHs in the modifier. Furthermore, the SFE operating conditions need to be tailored to each soil sample. Indeed, conditions that yield quantitative recoveries of a pollutant from one sample matrix may not be adequate to quantitatively extract the same pollutant from a different sample. Benner (1998) used the same extraction conditions (40.5 MPa CO_2 at 150°C, wet samples with 10% DCM) on two aquatic sediments (SRM 1941a, 1944). The extraction of SRM 1944 yielded quantitative recoveries ($\geq$ 90% of certified concentrations) of all PAHs (except perylene at 89% ± 9%). However, for SRM 1941a, recoveries were much lower especially for high molecular weight PAHs, with recoveries ranging from 80% ± 8% for benzo(a)anthracene to 48% ± 6% for benzo(g,h,i)perylene.

4.6. Subcritical Water Extraction

At ambient temperature and pressure, water is very polar (dielectric constant, $\varepsilon \sim 80$) and therefore is an unsuitable solvent for the extraction of nonpolar organic compounds such as PAHs. However, under supercritical conditions (22.1 MPa, 374°C), the dielectric constant of water is much lower ($\varepsilon = \sim$5-15). This explains the good PAH recoveries obtained using water SFE (Hawthorne et al., 1994). However, these extreme operating conditions preclude the method from being used for routine extractions. Various studies have demonstrated that quantitative recoveries of PAHs can be obtained under much milder conditions (250°C, 5 MPa), using subcritical water. The efficiency of this technique primarily depends on the temperature (Hageman et al., 1996; Hawthorne et al., 1994) and to a far lesser extent on the pressure, at which the extraction is carried out. Miller and Hawthorne (1998) showed how the solubilities of PAHs in water could be significantly increased simply by increasing the temperature. As long as the pressure is high enough to maintain the extractant in the liquid state, variations of pressure thereafter were shown to have little effect on recoveries (Hawthorne et al., 1994). Under optimised conditions, subcritical (or superheated) water has quantitatively recovered PAHs from different matrices (Hawthorne et al., 1994; Yang et al., 1997), from certified highly contaminated soil and air particulates (up to 87% and up to 12% more efficient than Soxhlet in soil and air particulates respectively, (Hawthorne et al., 1994)) to sludges (recoveries ranged from 92% for benzo(g,h,i)perylene to 99% for phenanthrene, (Yang et al., 1997)). Subcritical water extractions generally show good reproducibility (small

RSD) and can be performed in as little as 15 minutes. On account of water being the extractant, samples do not require the time consuming step of drying prior to being processed.

4.7. Microwave Assisted Extraction (MAE)

Microwave-assisted extraction uses microwave technology to heat organic solvent while in contact with the sample. Two types of systems exist. The solvent can either be heated up in an open vessel (atmospheric-MAE) or in a sealed (pressurised-MAE) vessel. The latter allows rapid heating of the solvent to temperatures well above its boiling point thus reducing the overall extraction time from hours to minutes. Operating conditions can vary from study to study, although the extraction time is often 5-20 min with a further 40-50 min generally required for cooling, centrifugation and extract concentration. Extra care should be taken during the filtration step to avoid losses of the more volatile PAHs (e.g. naphthalene), (Lopez-Avila et al., 1994). MAE (closed vessel system) can extract up to 12 samples simultaneously. MAE typically requires between 30-50 mL of solvent, almost 10 times less than Soxhlet. The amount of sample and solvent used should be in such proportions as to allow complete submersion of the sample to prevent electrical arcing from occurring (Barnabas et al., 1995). The choice of solvent for MAE is essential. Since MAE relies on microwave radiation, the solvent must be capable of absorbing this type of energy and converting it to heat energy. Microwave energy is a high frequency (typically 2.5 GHz) electromagnetic radiation, which causes the permanent dipoles present in the analyte and solvent to align themselves with the electric field at a rate of about 4.9×10^9 times per second (Barnabas et al., 1995), resulting in rapid heating. Thus, for heating to occur, the solvent must be polar (e.g. methanol, acetone, DCM) or must be a mixture of a non-polar and polar solvent (e.g. hexane:acetone). Moisture content of a sample does not seem to affect recoveries, although low extraction yields might be expected if a water immiscible solvent is used. Pressurised-MAE, atmospheric-MAE, SFE, PLE and Soxhlet extraction efficiencies were all compared for the extraction of PAHs from native contaminated soils, with total PAH recoveries of 1578, 1492, 1544, 1537 and 1623 mg kg^{-1} respectively (Saim et al., 1997). Lopez-Avila et al. (1994) compared recoveries from different standard materials (marine sediments and soil) and concluded that method performance was a function of the matrix. Barnabas et al. (1995) applied a statistical technique (Central Composite Design, CCD) to assess the optimum MAE operating conditions (temperature, extraction time, volume). They assigned upper and lower limits for these variables of 120 and 40°C, 20 and 5 min, 30 and 50 mL, respectively. They found that the PAHs recoveries from the various extractions in the CCD were independent of the operating parameters. MAE (120°C, 20 min, 40 mL of DCM) was shown to be dependent on analyte concentration, as demonstrated by the poor reproducibility obtained for the extraction of low contaminated soil in comparison to a highly contaminated soil. Barnabas et al. (1995) found large RSD values for the individual PAHs ranging from 4.1% ($n = 4$, naphthalene) to 60.9% ($n = 4$, indeno(1,2,3-cd)pyrene) with an average of the RSD for individual PAHs of 37.4%. Soxhlet, however, showed even lower reproducibility (average RSD for individual PAHs of 43.7%). The RSD value for MAE of a highly contaminated soil for the total PAHs extracted was 2.4% ($n = 6$).

4.8. Pressurised Liquid Extraction (PLE)

Pressurised liquid extraction, also known as accelerated solvent extraction (ASE®), uses conventional liquid organic solvents at elevated pressure and temperature to extract organic pollutants from solid samples. Standard operating conditions proposed by the USEPA (Method 3545) (100°C, 14MPa, 5 minutes extraction time + 5 minutes equilibration time) have led to quantitative recoveries of PAHs from a range of different matrices (Heemken et al., 1997; Richter et al., 1996; Schantz et al., 1997). Under such conditions, the method benefits from reduced extraction time compared to all of the previously mentioned techniques and solvent volume (20-30mL). The samples should ideally be dry and finely ground (particle size less than 0.5 mm) (Dionex, 1998). PLE is an exhaustive, non-selective extraction methodology, which can extract many non-target compounds along with the desired analytes. A clean-up step prior to analysis is therefore recommended to remove the unwanted substances, which may interfere with the determination and quantification of the target analytes. Saim et al. (1998) investigated the dependence of PLE operating variables (pressure, temperature, extraction time) on the recovery of 16 PAHs from contaminated soils using a central composite design. In terms of total PAH recovery, PLE efficiencies were not significantly different ($p < 0.05$), suggesting that PLE is independent of pressure, temperature and extraction time, within the operating limits used in the CCD (i.e. 1000 to 2400 psi, 40 to 200°C, 2 to 16 min, respectively). They also evaluated the effect of different extraction solvents, using constant operating conditions (2100 psi, 100°C, 5 min extraction time + 5 min equilibration time) on recovery and found that as long as a polar organic solvent was used, the recoveries of PAHs were independent of the choice of solvent.

5. Appraisal of 'Exhaustive' Techniques

The ideal extraction method should be effective, cheap, rapid, environmentally friendly (i.e. uses a minimal amount of organic solvent) and be user friendly (Figure 4). Based on these criteria, none of the extraction techniques presented here stands out as ideal on all accounts.

Extraction techniques need to be reproducible and as complete as possible. The quality of different methodologies can be established by comparing their yields against certified values from various Standard Reference Materials (SRM) that reflect different conditions (i.e. range of matrices, range of concentrations). Heemken et al. (1997) compared extraction yields of PAHs from sediment material HS-6 and suspended particulate matter (SPM) for Soxhlet, alkaline saponification, ultrasonication, SFE and PLE. Their results showed that all methods beside ultrasonication gave similar recoveries with good RSDs and were comparable to the certified values. Ultrasonication, in agreement with the USEPA (Method 3550b) description, yielded somewhat lower recoveries than the other methods, though powders or finely ground samples were more efficiently extracted. Efficiencies of MAE (Saim et al., 1997) and subcritical water extraction (Hawthorne et al., 1994; Yang et al., 1997) have also been proven.

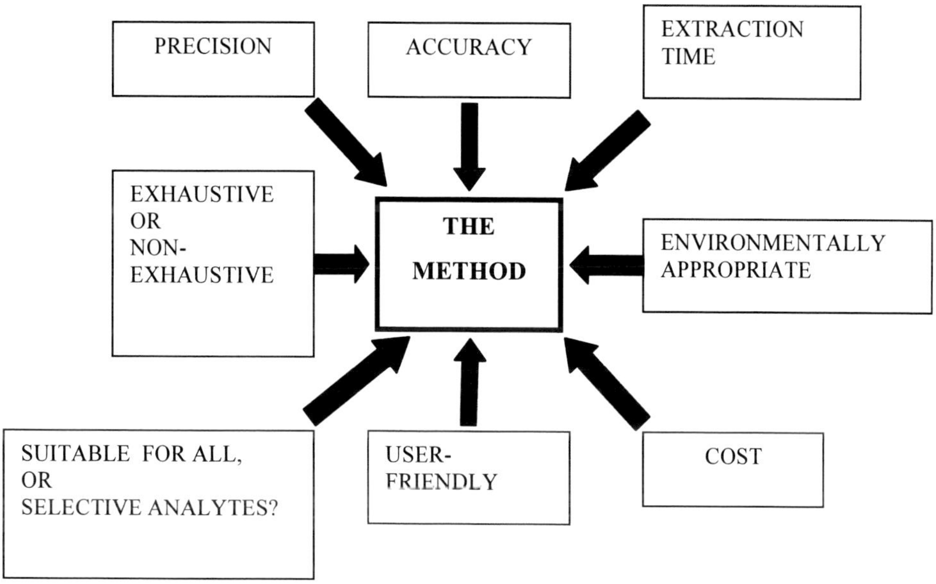

Figure 4: General criteria used in selecting the most appropriate extraction technique

Soxhlet is considered as the benchmark technique. It is highly efficient and highly standardised. Unfortunately, it is extremely time consuming particularly when time for drying, extraction, concentration and clean-up are allowed for. Furthermore, Soxhlet is also solvent intensive. Not only is this arguably non-environmentally friendly but it has also been noted that when large volumes of solvent are used, the risk of losses of the more volatile compounds during the concentration step is increased. Soxtec, an automated Soxhlet extraction method is available (USEPA Method 3541). Although designed for polychlorinated biphenyls (PCBs), this method is also applicable to other semi volatile organic compounds and has been shown to perform as well as Soxhlet in shorter time periods. Shake extractions also require large volumes of solvent and long extraction times. However, this method is less standardised compared to Soxhlet since kinetic data is first required to determine optimal extraction time. Saponification is also time consuming and solvent intensive but has the added advantage of not requiring a drying step, thus saving some time where compared to Soxhlet. Since alkaline saponification relies on cleavage of ester bonds within the organic matter matrix, this method lends itself to the extraction of samples with high organic matter content. Ultrasonication can also cope with moist samples and has an extraction time of around 1.5 h (much less than the aforementioned techniques). Unfortunately, as mentioned earlier, the technique has been shown to be less efficient than the others discussed here. Nonetheless, many laboratories prefer ultrasonication for its convenience and ease of use.

Due to growing environmental concern regarding solvent production and disposal, techniques that are less solvent intensive are preferred. There are no advantages in using large volumes of solvent: the expense of purchasing large quantities of pure organic solvents, the cost of safely disposing of contaminated solvent waste, the hazards associated with such substances (e.g. flammable, toxic and carcinogenic properties) and the potential impacts on the environment (air pollution, water and soil contamination, harm wildlife). In light of this, several other techniques such as SFE, subcritical water, PLE and MAE have been developed.

SFE has been intensively studied and in most cases provides a good alternative to the more traditional solvent extractions. However, traditional techniques for the extraction of PAHs from sediment and soil samples are well established, whereas method development for SFE is laborious. Furthermore, PAHs are best extracted using modified CO_2, thus not eliminating the usage of solvent, although very much reducing it. Subcritical water is very environmentally friendly in that it does not use any organic solvent during the extraction, and only a few millilitres as collection solvent. Hawthorne et al. (2000b) have shown that coupling subcritical water with solid-phase extraction (SPE) enabled quantitative recoveries and collection of contaminants, in good agreement with the certified values, while reducing the need for any organic solvent. MAE provides a rapid throughput by extracting up to 12 samples simultaneously in about 20 min (i.e. less than 2 min per sample) but still uses 30-50 mL of solvent. MAE is not yet standardised for PAHs, which implies many operating conditions (temperature, extraction time, choice of solvent) need to be varied in order to find the optimum working conditions. PLE is a very complete method, with reports of better recoveries than Soxhlet for the higher molecular weight PAHs (Schantz et al., 1997). It is, however, not the perfect method since although being fast and efficient, it requires organic solvent (though very little), a drying and clean-up step and is probably the most expensive technique to operate. These methods do provide a faster throughput and consume less organic solvent but are generally more costly and less user friendly, requiring expertise to operate.

6. Non-exhaustive Extractions

By definition 'non-exhaustive' extraction methods do not seek to liberate the entire burden of contaminant from a given matrix. Rather these methodologies seek to establish insight into contaminant partitioning. In many circumstances these non-exhaustive methodologies aim to establish an estimate of contaminant bioavailability/bioaccessibility (see Sections 2 and 3). Over the last decade, it has become apparent that total contaminant concentrations determined by exhaustive extraction overestimate the contaminant fraction available to soil biota (Hatzinger and Alexander, 1995; Reid et al., 2000), thus overestimating the potential risk to humans and other living organisms. Recent publications (Ehlers and Luthy, 2003; Semple et al., 2004) have highlighted the importance of including bioavailability assessments when conducting risk assessments of contaminated soils. Several 'non-exhaustive' extraction techniques have been developed in recent years and the following are discussed in this section: solid-phase extraction (SPE), cyclodextrin extraction (CD), surfactant extraction and persulfate oxidation. Solvent extraction, supercritical fluid extraction (SFE) and (super)heated (or subcritical) water extraction are also revisited in this section.

6.1. Aqueous Based Extractions

Solid phase extraction, cyclodextrin extraction and surfactant extraction are all considered in the aqueous based extraction section. These methods are proposed to mimic the mass transfer mechanism of a sorbed contaminant to a microbial degrader. It is generally accepted that before a molecule can be taken up by microorganisms, it must partition into the aqueous phase (Bosma et al., 1997). In the following three methods, just as for biodegradation, the key transfer is that to the aqueous phase.

The main driving force in these extraction methods is the ability to maintain a high hydrophobic organic contaminant (HOC) concentration gradient between the soil/sediment and the aqueous phase (Bosma et al., 1997). The additions of solid sorbents, cyclodextrin molecules or surfactants, with which the HOCs readily associate, provide a sink for readily desorbed HOCs.

6.1.1. Solid Phase Extraction

Solid phase extraction (SPE) is an extraction technique, which uses a solid sorbent to extract HOCs from the aqueous phase. Different sorbents can be used such as resin beads (e.g Tenax TA or XAD) or membrane disks (e.g C18 disks). Fibres coated with a hydrophobic polymer (e.g. polydimethylsiloxane) are used for solid phase microextraction (SPME).

Many SPE studies have been performed using Tenax TA. Tenax is a solid porous polymer based on 2,6-diphenyl-ρ-phenylene oxide, for which PAHs have an affinity similar to the affinity for organic carbon (Cornelissen et al., 1997). Recent studies have adapted their methodology from Cornelissen et al. (1997). Briefly, extractions are carried out in a separation funnel, to which soil or sediment, water and previously conditioned Tenax beads are added. To prevent biodegradation during the extraction, mercuric chloride ($HgCl_2$) can also be added. To maintain the concentration gradient between the soil/sediment and the aqueous phase, fresh Tenax is added throughout the experiment. The separation funnel is continuously shaken to ensure the slurry and the Tenax beads are well mixed. At selected times, the Tenax is separated from the soil/sediment suspension. Tenax is easily separated since the soil/sediment suspension sinks to the bottom of the separation funnel while the Tenax beads float to the top and adhere to the glass wall of the funnel. The Tenax beads can in turn be extracted by shaking with hexane, which is subsequently analysed by HPLC (Cornelissen et al., 1997). Alternatively the slurry can be centrifuged and the soil/sediment pellet extracted with acetone, which is then analysed by HPLC (Cuypers et al., 2002). Typically the removal of readily bioavailable PAHs takes several hours (low molecular weight) to several days (high molecular weight) (Cuypers, 2001). Tenax has been shown to successfully extract the rapidly desorbing PAHs from soils and sediments. Cornelissen et al. (1998) showed that the extent of possible PAH (microbial) degradation could be roughly predicted from the initial rapidly desorbing fraction. Cuypers et al. (2002) showed that the residual 3- and 4-ring PAH concentrations after a 264h SPE (33% and 58% for the PAH contaminated Assendelft sediment and 20% and 34% for the PAH and mineral oil contaminated Petroleum Harbour sediment, respectively) were comparable to the residual concentrations after 21d of biodegradation (35% and 58% for Assendelft and 20% and 40% for Petroleum Harbour, respectively). For 5,6-ring PAHs, the extent of biodegradation was successfully predicted for the Assendelft sediment (residual concentrations of 97% and 98% for SPE and biodegradation, respectively), while for the Petroleum Harbour sediment, SPE overestimated the extent of biodegradation (residual concentrations of 78% and 100% for SPE and biodegradation, respectively). Although it was suggested that biodegradation in the latter sediment was limited by microbial factors, thus not all the 'bioavailable' compounds had been successfully degraded by the microorganisms, a phenomenon also described by others (Cornelissen et al., 1998; Cuypers et al., 2000; Lei et al., 2004). Cuypers et al. (2001) demonstrated that Tenax could also be used to predict the extent of total petroleum hydrocarbons (TPH) microbial degradation (in an optimised slurry reactor). Comparison of the residual concentrations after SPE with the residual concentrations after biodegradation showed that the extent of TPH biodegradation could be well predicted by a short (168h) SPE. Similarly to the results found in other studies (Cornelissen et al., 1998; Cuypers et al., 2000; Cuypers et al., 2002), it was found that the method slightly underestimated the degradation of readily bioavailable hydrocarbons (possibly due to the shorter extraction time (168h) compared to the biodegradation time (84d)), whereas it slightly overestimated the degradation of poorly biodegradable hydrocarbons. Overall, Tenax has proved to be a successful method for the prediction of PAHs and TPHs bioavailability in soils and sediments.

Amberlite XAD resins have been used to extract PCBs in sediments (Carroll et al., 1994) and PAHs in soils (Northcott and Jones, 2001b). More recently, styrenedivinylbenzene XAD-2 resin has been used to assess the bioavailability of PAHs in field-contaminated sediments (Lei et al., 2004). Using a XAD-2 to sediment ratio (w/w) of 4, as to benefit from the faster kinetics without compromising the ease of separation of the resins from the sediment, Lei et al. (2004) found that the residual concentration in the sediment after 16 weeks of XAD-2-assisted desorption of 2-, 3- and 4-ring PAHs correlated well with the residual PAH concentrations after 24 weeks of biodegradation (slope = 1.04, r^2 = 0.875). For 5- and 6-ring PAHs, no appreciable biodegradation was observed, yet considerable levels of desorption were measured. The authors suggested this could be due to microbial factors.

Other SPE methods are briefly mentioned but have not been discussed in detail; primarily because manuscripts do not correlate PAH extraction from soil/sediment with data from microbial bioassays, precluding discussion as to their potential usage as bioavailability predictors.

Octadecyl-modified silica disks (C18) have been used to extract DDT, DDE and DDD (Tang et al., 1999), PAHs and PCBs (Krauss and Wilcke, 2001; Tang et al., 2002) from soil/sediment solutions. In these studies, extraction with C18 disks was related to earthworm uptake or benthic uptake (Lake et al., 1996).

SPME uses a fused silica fibre with a hydrophobic coating, allowing hydrophobic contaminants to partition from the aqueous phase. SPME has been used to extract a range of contaminants (TNT, PCB, BTEX) from the aqueous phase in a non-depletive manner (Arthur and Pawliszyn, 1990; Conder et al., 2003; Mayer et al., 2000; Sijm et al., 2000; Vaes et al., 1996; Zygmunt et al., 2001). None of these studies, however, have correlated SPME residues with data from bioasssays so as to predict bioavailability to specific organisms.

6.1.2. Cyclodextrin Extraction

Cyclodextrins are cyclic oligosaccharides formed from the enzymatic degradation of starch by bacteria (Wang and Brusseau, 1993). The unique property of these compounds is that they are hydrophilic and also contain a hydrophobic organic cavity within their molecular structure. The high aqueous solubilities of the molecules stems from the array of hydroxyl functional groups on the exterior of the molecule, while the size of the apolar cavity (6.2 Å for hydroxypropyl-β-cyclodextrin (HPCD)), allows the formation of a 1:1 water soluble inclusion complex between the cyclodextrin molecule and a low polarity organic compound (Shieh and Hedges, 1996). With larger compounds, β-cyclodextrin has been shown to form 1:2 inclusion complexes, where two macrocycles accommodate one guest (Cescutti et al., 1996; Wang and Brusseau, 1995). Alternatively, the larger γ-cyclodextrin (internal diameter of 7.9 Å) can also be used.

It has been proposed that contaminant mass transfer, from the solid phase to the aqueous phase, governs microbial bioavailability (Bosma and Harms, 1996; Bosma et al., 1997; Carmichael et al., 1997; Stucki and Alexander, 1987; Yeom and Ghosh, 1998), in particular the size of the rapidly desorbable fraction (Cornelissen et al., 1998). This idea stems from the observation that substrate uptake by microorganisms is more extensive in the aqueous phase (Miller and Alexander, 1991; Ogram et al., 1985).

Reid and co-workers (Reid et al., 1999; Reid et al., 1998; Reid et al., 2000) proposed and validated the use of cyclodextrin extraction for the prediction of PAH bioavailability. They hypothesised that an aqueous solution of HPCD would extract labile soil-associated PAHs while the strongly bound or sequestered contaminants would not readily transfer to the aqueous phase and thus would not be extracted. This extraction technique therefore closely mimics the key transfer to the aqueous phase, a mass transfer mechanism, which dictates microbial bioavailability.

Contaminated soil samples (1.5 g) are extracted using an aqueous solution of 60mM HPCD (25 mL). The extractions are carried out in centrifuge tubes that are continuously shaken for 20h then centrifuged to separate the supernatant. The supernatant can then be analysed or alternatively the centrifuged pellet can be extracted and analysed for PAHs by HPLC.

Reid and co-workers (Reid et al., 1999; Reid et al., 1998; Reid et al., 2000) showed that for soils spiked with $^{12}C/^{14}C$-phenanthrene as the only contaminant, the amount of HPCD-extractable ^{14}C-phenanthrene was strongly correlated to the bioavailable fraction (slope = 0.977, r^2=0.964) as determined by the extent of mineralised ^{14}C-phenanthrene by catabolically active phenanthrene degraders.

Semple et al. (2003) further investigated the HPCD extraction method by using soil containing a range of PAHs from a former coke works site. They also found a strong correlation (r^2 = 0.98) between the HPCD-extractable and the microbially degraded fractions. Cuypers et al. (2002) also found a good agreement between residual concentrations of 3-, 4- and 5,6-ring PAHs after HPCD extraction and after biodegradation of the Assendelft sediment, which contained 122 mg kg^{-1} PAHs (16 EPA).

The use of HPCD to predict microbial bioavailability in mixed contaminant sytems (Cuypers et al., 2002, (Petroleum Harbour sediment); Swindell and Reid, 2003 (soil with a range of diesel concentrations) is not as effective as in single contaminant systems (Reid et al., 2000) or multiple PAH contaminant systems (Cuypers et al., 2002, (Assendelft sediment); Semple et al., 2003). In Cuypers et al. (2002), where Petroleum Harbour sediment was heavily contaminated in PAHs (2036 mg kg^{-1}) and mineral oil (13600 mg kg^{-1}), HPCD extraction overestimated the extent of 5,6-ring PAH biodegradation, while correctly predicted the extent of biodegradation of the 3- and 4-ring PAHs. In Swindell and Reid (2003), phenanthrene biodegradation was overestimated by HPCD in all treatments aged less than 225d. However, for aged residues (after 225d), HPCD predicted phenanthrene biodegradation in treatments containing 20, 200 and 2000 mg kg^{-1} of diesel, while overestimated biodegradation in the 20000 mg kg^{-1} of diesel treatment. It was suggested that diesel could have acted as a co-extractant thus enhancing the extraction efficiency, while also negatively impacting the degraders thereby reducing mineralisation.

6.1.3. Surfactant Extraction

Surfactants are amphiphilic molecules (i.e. they consist of a hydrophobic and a hydrophilic moiety). Beyond a certain concentration (the critical micelle concentration, CMC), the surfactant molecules form aggregates, known as micelles. The surfactants cluster together so as to form a hydrophobic core and a hydrophilic outer surface. The presence of micelles increases the apparent solubility of HOCs as these compounds preferentially partition into the hydrophobic core of the water-soluble micelles. This partitioning into the micelles leads to a high HOC concentration gradient between the soil/sediment and the aqueous phase, thus promoting further desorption of soil-associated compounds.

Surfactant sorption to the soil/sediment and the effect of surfactant concentration on HOC extraction efficiency are two parameters, which need to be studied to ensure an optimised surfactant extraction (see Cuypers et al. (2002) for details on these preparatory steps). Soil/sediment is then added to a solution of $HgCl_2$ and surfactant (concentration based on the optimisation experiments), which is mixed on a rotary tumbler until the samples are equilibrated. The suspensions are then centrifuged and the HOC concentration of the liquid and/or solid phase subsequently determined.

Volkering et al. (1998) used a surfactant extraction (Triton X-100) to predict soil-associated mineral oil microbial bioavailability. Although the study was not complete, preliminary results were promising. The residual oil concentration after surfactant extraction (500 mg kg^{-1}) was comparable to the residual oil concentration after 10 weeks bench-scale bioremediation (450 mg kg^{-1}). Although these are of the same range, the GC-chromatogram for their results showed slightly higher concentrations of mineral oil (especially for linear alkanes C12-C18) after surfactant extraction than after 10 weeks biodegradation.

So far, no other study has successfully predicted the microbial bioavailability of soil-associated HOCs using surfactants. Cuypers et al. (2002) tested the method for PAH bioavailability prediction. They found that residual PAH concentrations after surfactant extraction (Triton X-100) were considerably lower than the residual PAH concentrations after biodegradation, particularly for the 5,6-ring PAHs. Microorganisms removed approximately 65%, 40% and less than 5% of 3-ring, 4-ring and 5,6-ring PAHs respectively. Triton X-100, however, removed approximately 70-80% PAHs irrespective of their molecular weight.

The results from Cuypers et al. (2002) are in agreement with most of the literature on surfactants. These have been extensively studied for their capacity to enhance the bioavailability of HOCs (Grasso et al., 2001; Jordan and Cunningham, 1999; Mulligan et al., 2001; Tiehm et al., 1997; Volkering et al., 1997; Yeom et al., 1996). The mechanisms responsible for the enhancement have been reviewed by Jordan and Cunningham (1999) and Volkering et al. (1997).

6.2. Persulfate Oxidation

Persulfate oxidation is a low temperature wet oxidation method developed by Cuypers and co-workers (Cuypers et al., 2001; Cuypers et al., 2000) for the prediction of PAH microbial bioavailability. The method extends from work by Weber et al. (1992) and Young and Weber (1995), who showed that persulfate oxidation only oxidised amorphous (soft) organic matter (OM), the residual OM showing increased organic carbon partition coefficient, K_{oc} values. In partitioning models, soil/sediment organic matter (SOM) has often been described as a homogeneous amorphous phase with a linear sorption isotherm (Chiou et al., 1983). Recent studies show that SOM is much more heterogeneous, displaying a range of organic fractions with near linear to non-linear sorption isotherms (Pignatello, 1998; Weber et al., 1999; Weber et al., 1991, 1992; Xing and Pignatello, 1997). To accommodate for this non-ideal sorption behaviour, various conceptual models have been developed (Chiou and Kile, 1998; Chiou et al., 2000; Weber et al., 1999; Xing et al., 1996). Here we refer to an amorphous (soft) SOM domain to which readily bioavailable HOCs are sorbed and a condensed (hard) SOM domain where the HOCs are recalcitrant.

The method is based on the selective chemical oxidation by persulfate of the amorphous SOM, with negligeable effect to the condensed SOM, so as to only remove the bioavailable contaminants. The methodology is described in Cuypers et al. (1999b). Briefly, soil or sediment is mixed with $K_2S_2O_8$ and demineralised water to obtain a persulfate ($S_2O_8^{2-}$) to OM ratio of 12 g:g and an aqueous persulfate concentration of 0.0357 g mL^{-1}. The slurry is shaken for 3h at 70°C and subsequently filtered. Solids are then extracted and analysed by conventional methods (e.g. HPLC).

Cuypers and co-workers (Cuypers et al., 2001; Cuypers et al., 1999a; Cuypers et al., 2000) tested persulfate oxidation as a method to predict bioavailability by comparing residual contaminant concentrations after oxidation with residual concentrations after 21 or 84d of catabolically induced biodegradation. Under the optimised conditions described earlier, residual concentrations of PAHs (sum of 16 EPA PAHs, 2,3-, 4- and 5,6-ring PAH groups) strongly correlated with residual concentrations after biodegradation (Table 8).

Table 8. Regression parameters for the correlations between residual concentrations after persulfate oxidation versus residual concentrations after biodegradation. Data from Cuypers et al. (2000).

PAHs	Regresion parameters	
	slope	r^2
16 EPA	0.97	0.95
2, 3 ring	1.02	0.91
4 ring	0.99	0.95
5, 6 ring	0.91	0.97

Table 9. Residual concentrations of TPH after persulfate oxidation (70 °C, 3h) and after biodegradation. Data from Cuypers et al. (2001).

Soils and sediments (initial concentrations ± st dev), mg kg^{-1}	Residual concentrations after persulfate oxidation, mg kg^{-1}	Residual concentrations after biodegradation, mg kg^{-1}
Petroleum Harbour (13637 ± 134)	12380 ± 961	6719 ± 561
Arcadis Petroleum (8416 ± 503)	5940 ± 346	1307 ± 63
Van Velde Buren (34096 ± 2682)	34232 ± 2633	6182 ± 88

The method was also tested to predict the bioavailability of TPH. Persulfate oxidation significantly underestimated the bioavailable fraction of TPH for the three soils/sediments tested (Table 9). The authors of the study concluded persulfate oxidation was unfit for the prediction of TPH bioavailability (Cuypers et al., 2001).

6.3. Other Methods

The following three methods (organic solvent extraction, supercritical fluid extraction and (super)heated water extraction) have already been described as exhaustive extraction methods. However, there are reports where these methods have been used in a non-exhaustive way and results correlated to microbial degradation.

6.3.1. Organic Solvent Extraction

Alexander and co-workers have studied solvent extractions and their efficiency to predict bioavailability of PAHs and/or pesticides to microorganisms and/or earthworms (Alexander and Alexander, 2000; Chung and Alexander, 1998; Kelsey et al., 1997; Liste and Alexander, 2002; Nam et al., 1998; Tang and Alexander, 1999; Tang et al., 2002). In all these studies, they have found that the extent of PAHs removed by a mild extraction with organic solvents decreases with increasing ageing and that these changes in extractability parallel similar changes in bioavailability.

The methodology is slightly different from when solvents are used as exhaustive extractants. Here, soil (~ 5 g) is placed in a centrifuge tube to which 20 mL of solvent is added. The tubes are vigorously shaken on a vortex mixer for 8s, then centrifuged for 10 min at 12000 *g*. A sample of supernatant is filtered and analysed by HPLC (Liste and Alexander, 2002).

Liste and Alexander (2002) have obtained very good correlations between bioavailability to microorganisms and extractability using *n*-butanol for low molecular weight PAHs (phenanthrene and pyrene) in spiked Lima loam soil (r^2= 0.971 and 0.994, respectively). In this same study, extractability when expressed as a percentage of extraction or bioavailability at t = 0 and plotted against bioavailability gave a slope of 0.952 and 0.998 for phenanthrene and pyrene respectively, suggesting that the extraction with *n*-butanol was able to directly predict (in a relative sense) the bioavailable amount. However, in a different study (Alexander and Alexander, 2000), using high molecular weight PAHs (dimethylbenz(a)anthracene (DMBA), benzo(a)pyrene (BaP) and phenylanthracene (PA)), the extent of recovery using *n*-butanol from the Lima loam soil and the extent bioavailable (for mutagenesis using bacterial genotoxicity assays) were significantly different: 63.3%, 77.8% and 56.0% recovered for DMBA, BaP and PA respectively against 30.7%, 21.5% and 24.2% bioavailable. In this study, correlations between the percent bioavailable of DMBA, BaP and PA with the percent extracted from 6 different soils by *n*-butanol gave r^2 values of 0.757, 0.478 and 0.620, respectively (based on r values of 0.870, 0.691 and 0.787, respectively (Alexander and Alexander, 2000)), and slopes of 0.351, 0.296 and 0.296, respectively (based on values of % recovered by *n*-butanol and % bioavailable (Alexander and Alexander, 2000)). In a study using 16 different soils, Chung and Alexander (1998) found that the 'declines in bioavailability in the soils were not highly correlated with the decreases in extractability' using ethanol as an extractant for PAHs.

6.3.2. Supercritical Fluid Extraction (SFE)

Sequential SFE with pure CO_2 has been investigated by Loibner and co-workers (Loibner et al., 1997; Loibner et al., 1998; Szolar et al., 2004) and Hawthorne and co-workers (Bjorklund et al., 1999; Bjorklund et al., 2000; Hawthorne et al., 1999; Hawthorne and Grabanski, 2000; Pilorz et al., 1999). The method has been used to characterise the behaviour of PAHs and PCBs in soil. By varying the extraction parameters (CO_2 pressure and temperature) to provide sequentially stronger extraction conditions, the selected contaminants will be extracted according to their availability. Hawthorne and Grabanski (2000) defined four different 'sites' from which contaminants will desorb: fast, moderate, slow and very slow. By applying sequentially stronger SFE conditions, contaminants associated to these sites are selectively extracted.

Hawthorne and Grabanski (2000) showed that in general, the percent of each group of PAHs removed using the 'fast' SFE fraction best agreed with the percent of each group of PAHs removed during a 1 year large-scale field bioremediation of a PAH impacted soil from a manufactured gas plant (MGP) site. Based on the similarity of the curves obtained for SFE-extractability (as a function of OM% and ageing time) and those for biodegradability and toxicity, Loibner et al. (1998) concluded that SF-extractability may prove to be a powerful tool for predicting toxicity as well as biodegradation success for PAH contaminated soil. However these results are to be regarded with caution as similar curves do not necessarily imply that the same PAH molecules are involved.

More recently, Hawthorne et al. (2001) used SFE profiles fitted with a simple two-site (fast and slow) model to obtain the fraction of each PAH that was rapidly released (F). These F values agreed with the actual removal of PAHs obtained after 343d of field remediation, especially for the lower molecular weight PAHs (Table 9). However, SFE had a tendency to overestimate the release of middle to high molecular weight PAHs (fluoranthene to benzo(ghi)perylene). The curve-fit approach was generally unable to yield F values for the high molecular weight PAHs due to their slow extraction behaviour. This method of curve fitting was also applied to 15 different soil samples collected from an oil gas and coal gas MGP (Hawthorne et al., 2002). In this study, F values for coal gas site samples obtained with SFE agreed well ($r^2 = 0.87$, slope = 0.93). However, SFE overestimated the release of PAHs from the oil gas site samples.

6.3.3. (Super)0Heated Water Extraction

The use of (super)heated water extraction as a means of predicting bioavailability has not been extensively studied. Direct correlations between percent extractable by this method and percent bioavailable, determined from bioassay data, have not been carried out. However, Johnson and Weber (2001), successfully correlated predicted rates of long-term release based on the (super)heated water extraction method with experimental desorption data carried out at 25°C over a period of 250-640d. Desorption rate constants were computed by fitting a three parameter, biphasic first order model to desorption data of phenanthrene obtained by (super)heated extraction. Three different temperatures (75°, 100° and 150°C) were used, which enabled the computation of slowly desorbing rate constants (K_s) as a function of temperature (T). By using the Arrhenius equation, plotting ln K_s as a function of T^{-1}, the apparent activation energies (E_a) could be calculated from the respective slopes ($-E_a/R$) of the resulting straight lines. From these apparent activation energies, the rate constants at any given temperature can be extrapolated thus enabling desorption behaviours at any desired ambient temperature to be predicted.

In this particular study, the 25°C predictions matched the measured data (measured at 25°C) to within 3%, 10% and 20% respectively for the Lachine shale, Chelsea soil and Wagner soil.

Table 10. Comparison of the fractions of individual PAHs removed after 343d of bioremediation with the fast released fractions (F) obtained from curve fitting of the SFE data from day 0. [a] nd, not determined. [b]---, curve fitting did not yield a suitable F value. Data from Hawthorne et al. (2001).

	Fraction removed by bioremediation rel. to initial concentration (± Standard Deviation) A	F values determined by SFE (with CO_2, 200 min, 200 bar, 50°C) (Confidence intervals, 95%) B	A/B
Naphthalene	0.83 (0.1)	0.86 (0.85-0.88)	0.97
1-methylnaphthalene	0.93 (0.1)	0.97 (0.96-0.98)	0.96
2-methylnaphthalene	0.95 (0.1)	0.94 (0.93-0.95)	1.01
Acenaphthylene	0.77 (0.2)	0.72 (0.71-0.73)	1.07
Acenaphthene	0.93 (0.1)	nd[a]	nd[a]
Fluorene	0.91 (0.2)	0.97 (0.95-0.99)	0.94
Dibenzothiophene	0.81 (0.1)	0.96 (0.93-0.98)	0.84
Phenanthrene	0.95 (0.2)	0.97 (0.93-0.99)	0.98
Anthracene	0.87 (0.1)	0.86 (0.85-0.88)	1.01
Fluoranthene	0.69 (0.1)	0.89 (0.85-0.92)	0.78
Pyrene	0.58 (0.04)	0.92 (0.82-0.97)	0.63
Benz[*a*]anthracene	0.38 (0.01)	0.80 (0.71-0.88)	0.48
Chrysene	0.34 (0.01)	0.75 (0.67-0.82)	0.45
Benzo[*b*+*k*]fluoranthene	0.34 (0.1)	---[b]	nd[a]
Benzo[*e*]pyrene	0.23 (0.03)	0.44 (0.42-0.46)	0.52
Benzo[*a*]pyrene	0.16 (0.02)	0.36 (0.31-0.50)	0.44
Indeno[*1,2,3-cd*]pyrene	0.26 (0.1)	0.12 (0.10-0.13)	2.17
Benzo[*ghi*]perylene	0.04 (0.01)	0.07 (0.06-0.08)	0.57

7. Appraisal of Non-exhaustive Extractions

As already stressed the assessment of bioavailability/bioaccessibility by a non-exhaustive chemical method can never be all-encompassing. Therefore, a single method to predict bioavailability/bioaccessibility to the soil biota is impossible. Instead, it is imperative to select an appropriate organism to which bioavailability is assessed. Based on this limitation, an ideal non-exhaustive extraction method should be able to access the entire labile pool of the soil-associated compounds and to correlate with uptake or degradation by a specific organism or consortium of organisms. In addition, the method should be simple to use and rapid.

The development of non-exhaustive techniques to assess bioavailability is a relatively recent quest. Consequently, to date, none of the selected methods described in this review could be described as ideal. Some, however, appear more promising than others.

The use of Triton X-100 surfactant has been deemed unsuitable (too rigorous) for the prediction of microbial bioavailability of PAHs (Cuypers et al., 2002). However, it was successful in assessing mineral oil bioavailability (Volkering et al., 1998). Generally, current research using surfactants is directed towards the enhancement of contaminant bioavailability, not it's prediction (Makkar and Rockne, 2003; Zhao et al., 2005).

To date, the use of (super)heated water extraction as a means of predicting bioavailability has not been extensively studied. Direct correlations between percent extractable by this method and percent bioavailable, determined from bioassay data, have not been carried out. Instead, the method was successfully correlated to desorption rates (Johnson and Weber, 2001). Nonetheless, more research is needed to further validate this method and assess its suitability in predicting microbial bioavailability.

Residual PAH concentrations of light to moderate molecular weight PAHs, after a mild SFE method, were in good agreement with those obtained after 1 year of bioremediation (Hawthorne and Grabanski, 2000; Hawthorne et al., 2001, 2002). The studies were carried out using soils and sediments from manufactured gas plants, representing a range of PAH concentrations and matrices. However, the method had a tendency to overestimate the release of 4-ring PAHs and was unable to predict the release of the high molecular weight PAHs (Hawthorne et al., 2001).

Persulfate oxidation can provide a fast and efficient prediction of the amount of residual PAH contamination that can be expected after microbial degradation under optimised conditions. So far the method has been tested on historically contaminated soils and sediments containing a range of PAHs and in some cases high TPH concentrations (Cuypers et al., 2000). The method, however, is unfit for the prediction of TPH bioavailability (Cuypers et al., 2001).

The solvent shake extraction has so far shown mixed performances (Alexander and Alexander, 2000; Chung and Alexander, 1998; Liste and Alexander, 2002). Further work is needed to understand why the method can predict bioavailability in certain cases and not in others (e.g. different soil type, different PAHs). Indeed optimal extraction conditions (solvent of choice, extent and vigour of agitation) vary with pollutants, type of target organism and soil type. The use of organic solvents may interact with the SOM and consequently affect the bioavailability of compounds by competition for sorption sites and by swelling (Cuypers, 2001).

SPE (using Tenax TA or XAD-2) and HPCD methods have both been applied successfully to the prediction of PAH bioavailability in single and multiple contaminant systems involving a range of PAHs for both laboratory and field contaminated soils and sediments (Cuypers et al., 2002; Lei et al., 2004; Reid et al., 2000; Semple et al., 2003). These two methods (SPE and HPCD), based on mass transfer, are more conceptually relevant to the assessment of microbial bioavailability than the other methods mentioned above. Indeed, both methods mimic the contaminant mass transfer, from the solid phase to the aqueous phase, a key transfer for microbial uptake. However, unlike surfactants, Tenax TA and HPCD do not interact with the soil or sediment matrix (Cuypers et al., 2002).

8. Conclusions

This manuscript has reviewed methodologies applicable to the extraction of polycyclic aromatic hydrocarbons (PAHs) associated with soil and sediment matrices. It has drawn the distinction between exhaustive extraction techniques that seek to liberate as much analyte as possible from a given matrix and non-exhaustive extraction techniques that strive to reveal how a compound is partitioning within a matrix. Given that exhaustive extraction techniques vary in their efficacy (and as such the term 'exhaustive' is only nominal) the reality is that all of the techniques reviewed here are on a spectrum of exhaustiveness.

Many non-exhaustive techniques offer secondary advantages; in particular a significant reduction in organic solvent usage. Due to growing environmental concern regarding solvent production and disposal, techniques that are less solvent intensive are likely to be embraced in the future. The development of non-exhaustive techniques, that are solvent minimal and that reflect bioavailability/bioaccessibility fractions of compound, has been a relatively recent quest. To date, none of the selected methods described in this review could be described as ideal. Some, however, appear more promising than others with respect to matching chemical extractability with bioavailability/bioaccessibility of compound to specified receptors. Where methods have been developed with a key transfer process that governs bioavailability/bioaccessibility in mind (such as SPE and HPCD) the most promising results have been obtained.

There is now a need for those who assess and regulate contaminated land to accept that extraction methodologies that reflect compound partitioning and in particular bioavailability/bioaccessibility fractions of compound with respect to pivotal receptors (e.g. earthworms or microorganisms) are arguably the most appropriate measures of environmental significance. Indeed, as contaminated land is regulated more and more based on the risk it poses there is a need for extraction methodologies that reflect bioavailability/bioaccessibility to be accepted, in preference to exhaustive extraction techniques, in order that significant pollutant linkages (source-pathway-receptor) can be established in the most appropriate way.

References

Alexander, M., 1995. How toxic are toxic-chemicals in soil. *Environmental Science and Technology* **29**: 2713-2717.

Alexander, R.R., Alexander, M., 2000. Bioavailability of genotoxic compounds in soils. *Environmental Science and Technology* **34**: 1589-1593.

Alexander, R.R., Tang, J.X., Alexander, M., 2002. Genotoxicity is unrelated to total concentration of priority carcinogenic polycyclic aromatic hydrocarbons in soils undergoing biological treatment. *Journal of Environmental Quality* **31**: 150-154.

Almaric, L., Mouvet, C., 1997. Comparison of SFE, soxhlet and sonication for the extraction of organic contaminants from sandstone. *International Journal of Environmental Analytical Chemistry* **68**: 171-186.

Arthur, C.L., Pawliszyn, J., 1990. Solid-phase microextraction with thermal-desorption using fused-silica optical fibers. *Analytical Chemistry* **62**: 2145-2148.

Barnabas, I.J., Dean, J.R., Fowlis, I.A., Owen, S.P., 1995. Extraction of polycyclic aromatic hydrocarbons from highly contaminated soils using microwave energy. *Analyst* **120**: 1887-1904.

BDH, 1990. Hazards data sheet.

Benner, Jr.B.A., 1998. Summarizing the effectiveness of supercritical fluid extraction of polycyclic aromatic hydrocarbons from natural environmental samples. *Analytical Chemistry* **70**: 4594-4601.

Benner, Jr.B.A., 2000. Supercritical fluid extraction of organics in environmental analysis. Encyclopedia of analytical chemistry, applications, theory and instrumentation: John Wiley and Sons Ltd. p 3143-3172.

Bjorklund, E., Bowadt, S., Mathiasson, L., Hawthorne, S.B., 1999. Determining PAH sorption desorption behavior on sediments using selective supercritical fluid extraction. 1. Desorption from historically contaminated samples. *Environmental Science and Technology* **33**: 2193-2203.

Bjorklund, E., Nilsson, T., Bowadt, S., Pilorz, K. Mathiasson, L., Hawthorne, S.B., 2000. Introducing selective supercritical fluid extraction as a new tool for determining sorption/desorption behaviour and bioavailability of persistent organic pollutants in sediment. *Journal of Biochemical and BiophysicalMethods.* **43**: 295-311.

Booth, L.H., Heppelthwaite, V.J., O'Halloran, K., 2005. Effects-based assays in the earthworm *Aporrectodea caliginosa* - Their utilisation for evaluation of contaminated sites before and after remediation. *Journal of Soils and Sediments*. **5**: 87-94.

Bosma, T.N.P., Harms, H., 1996. Bioavailability of organic pollutants. *EAWAG News* **40E**: 28-31.

Bosma, T.N.P., Middeldorp, P.J.M., Schraa, G., Zehnder, A.J.B., 1997. Mass transfer limitation of biotransformation: Quantifying bioavailability. *Environmental Science and Technology* **31**: 248-252.

Breedveld, G.D., Sparrevik, M., 2000. Nutrient-limited biodegradation of PAH in various soil strata at a creosote contaminated site. *Biodegradation* **11**: 391-399.

Brilis, G.M., Marsden, P.J., 1990. Comparative evaluation of soxhlet and sonication extraction in the determination of polynuclear aromatic hydrocarbons in soil. *Chemosphere* **21**: 91-98.

Carmichael, L.M., Christman, R.F., Pfaender, F.K., 1997. Desorption and mineralization kinetics of phenanthrene and chrysene in contaminated soils. *Environmental Science and Technology* **31**: 126-132.

Carroll, K.M., Harkness, M.R., Bracco, A.A., Balcarcel, R.R., 1994. Application of a permeant polymer diffusional model to the desorption of polychlorinated-biphenyls from Hudson river sediments. *Environmental Science and Technology* **28**: 253-258.

Cerniglia, C.E., 1984. Microbial metabolism of polycyclic aromatic hydrocarbons. *Advances in Applied Microbiology* **30**: 31-71.

Cescutti, P., Garozzo, D., Rizzo, R., 1996. Study of the inclusion complexes of aromatic molecules with cyclodextrins using ion spray mass spectrometry. *Carbohydrate Research* **290**: 105-115.

Chiou, C.T., Kile, D.E., 1998. Deviations from sorption linearity on soils of polar and nonpolar organic compounds at low relative concentrations. *Environmental Science and Technology* **32**: 338-343.

Chiou, C.T., Kile, D.E., Rutherford, D.W., Sheng, G.Y., Boyd, S.A., 2000. Sorption of selected organic compounds from water to a peat soil and its humic-acid and humin fractions: Potential sources of the sorption nonlinearity. *Environmental Science and Technology* **34**: 1254-1258.

Chiou, C.T., Porter, P.E., Schmedding, D.W., 1983. Partition equilibria of non-ionic organic-compounds between soil organic-matter and water. *Environmental Science and Technology* **17**: 227-231.

Chung, N.H., Alexander, M., 1998. Differences in sequestration and bioavailability of organic compounds aged in dissimilar soils. *Environmental Science and Technology* **32**: 855-860.

Codina, G., Vaquero, M.T., Comellas, L., Broto-Puig, F., 1994. Comparison of various extraction and clean-up methods for the determination of polycyclic aromatic hydrocarbons in sewage sludge-amended soils. *Journal of Chromatography A* **673**:21-29.

Conder, J.M., La Point, T.W., Lotufo, G.R., Steevens, J.A., 2003. Nondestructive, minimal-disturbance, direct-burial solid-phase microextraction fiber tecnique for measuring TNT in sediment. *Environmental Science and Technology* **37**: 1625-1632.

Cornelissen, G., Rigterink, H., Ferdinandy, M.M.A., Van Noort, P.C.M., 1998. Rapidly desorbing fractions of PAHs in contaminated sediments as a predictor of the extent of bioremediation. *Environmental Science and Technology* **32**: 966-970.

Cornelissen, G., vanNoort, P.C.M., Govers, H.A.J., 1997. Desorption kinetics of chlorobenzenes, polycyclic aromatic hydrocarbons, and polychlorinated biphenyls: Sediment extraction with Tenax (R) and effects of contact time and solute hydrophobicity. *Environmental Toxicology and Chemistry* **16**: 1351-1357.

Cuypers, C., 2001. Bioavailability of polycyclic aromatic hydrocarbons in soils and sediments: Prediction of bioavailability and characterization of organic matter domains. [PhD-thesis]. Wageningen: The Netherlands: Wageningen University. 161 p.

Cuypers, C., Clemens, R., Grotenhuis, T., Rulkens, W., 2001. Prediction of petroleum hydrocarbon bioavailability in contaminated soils and sediments. *Soil and Sediment Contamination* **10**: 459-482.

Cuypers, C., Grotenhuis, J.T.C., Rulkens, W.H., 1999a. Development of a persulfate oxidation method for the prediction of pah bioavailability in soils and sediments. In: Bioremediation Technologies for Polycyclic Aromatic Hydrocarbon Compounds, Proceedings 5th In Situ and On-Site Bioremediation Symposium Battelle Press, Columbus, OH.

Cuypers, C., Grotenhuis, T., Joziasse, J., Rulkens, W., 2000. Rapid persulfate oxidation predicts PAH bioavailability in soils and sediments. *Environmental Science and Technology* **34**: 2057-2063.

Cuypers, C., Pancras, T., Grotenhuis, T., Rulkens, W., 2002. The estimation of PAH bioavailability in contaminated sediments using hydroxypropyl-beta-cyclodextrin and Triton X-100 extraction techniques. *Chemosphere* **46**: 1235-1245.

Cuypers, M.P., Grotenhuis, J.T.C., Rulkens, W.H., 1999b. Prediction of PAH bioavailability in soils and sediments by persulfate oxidation. Bioremediation technologies for polycyclic aromatic hydrocarbon compounds. p 241-246.

Dionex, 1998. Methods optimization in accelerated solvent extraction (ASE). *Technical notes* **108**:

Ehlers, L.J., Luthy, R.G., 2003. Contaminant bioavailability in soil and sediment. *Environmental Science and Technology* **37**: 295A-302A.

Eschenbach, A., Kastner, M., Bierl, R., Schaefer, G., Mahro, B., 1994. Evaluation of a new effective method to extract pahs from soil samples. *Chemosphere* **28**: 683-692.

Grasso, D., Subramaniam, K., Pignatello, J.J., Yang, Y., Ratte, D., 2001. Micellar desorption of polynuclear aromatic hydrocarbons from contaminated soil. *Colloids and Surfaces a-Physicochemical and Engineering Aspects* **194**: 65-74.

Guerin, T.F., 1999. The extraction of aged polycyclic aromatic hydrocarbon (PAH) residues from a clay soil using sonication and a soxhlet procedure: A comparative study. *Journal of Environmental Monitoring* **1**: 63-67.

Guerin, W.F., Boyd, S.A., 1992. Differential bioavailability of soil-sorbed naphthalene to 2 bacterial species. *Applied and Environmental Microbiology* **58**: 1142-1152.

Hageman, K.J., Mazeas, L., Grabanski, C.B., Miller, D.J., Hawthorne, S.B., 1996. Coupled subcritical water extraction with solid-phase microextraction for determining semivolatile organics in environmental solids. *Analytical Chemistry* **68**: 3892-3898.

Hatzinger, P.B., Alexander, M., 1995. Effect of aging of chemicals in soil on their biodegradability and extractability. *Environmental Science and Technology* **29**: 537-545.

Hatzinger, P.B., Alexander, M., 1997. Biodegradation of organic compounds sequestered in organic solids or in nanopores within silica particles. *Environmental Toxicology and Chemistry* **16**: 2215-2221.

Hawthorne, S.B., Bjorklund, E., Bowadt, S., Mathiasson, L., 1999. Determining PCB sorption/desorption behavior on sediments using selective supercritical fluid extraction. 3. Sorption from water. *Environmental Science and Technology* **33**: 3152-3159.

Hawthorne, S.B., Grabanski, C.B., 2000. Correlating selective supercritical fluid extraction with bioremediation behavior of PAHs in a field treatment plot. *Environmental Science and Technology* **34**: 4103-4110.

Hawthorne, S.B., Grabanski, C.B., Martin, E., Miller, D.J., 2000a. Comparisons of soxhlet extraction, pressurized liquid extraction, supercritical fluid extraction and subcritical water extraction for environmental solids: Recovery, selectivity and effects on sample matrix. *Journal of Chromatography A* **892**: 421-433.

Hawthorne, S.B., Poppendieck, D.G., Grabanski, C.B., Loehr, R.C., 2001. PAH release during water desorption, supercritical carbon dioxide extraction, and field bioremediation. *Environmental Science and Technology* **35**: 4577-4583.

Hawthorne, S.B., Poppendieck, D.G., Grabanski, C.B., Loehr, R.C., 2002. Comparing PAH availability from manufactured gas plant soils and sediments with chemical and biological tests. 1. PAH release during water desorption and supercritical carbon dioxide extraction. *Environmental Science and Technology* **36**: 4795-4803.

Hawthorne, S.B., Trembley, C.L., Moniot, C.L., Grabanski, C.B., Miller, D.J., 2000b. Static subcritical water extraction with simultaneous solid-phase extraction for determining polycyclic aromatic hydrocarbons on environmental solids. *Journal of Chromatography A* **886**: 237-244.

Hawthorne, S.B., Yang, Y., Miller, D.J., 1994. Extraction of organic pollutants from environmental solids with sub- and supercritical water. *Analytical Chemistry* **66**: 2912-2020.

Heemken, O.P., Theobald, N., Wenclawiak, B.W., 1997. Comparison of ASE and SFE with soxhlet, sonication and methanolic saponification extractions for the determination of organic micropollutants in marine particulate matter. *Analytical Chemistry* **69**:2171-2180.

Hollender, J., Koch, B., Lutermann, C., Dott, W., 2003. Efficiency of different methods and solvents for the extraction of polycyclic aromatic hydrocarbons from soils. *International Journal of Environmental Analytical Chemistry* **83**: 21-32.

Hollender, J., Schneine, J., Dott, W., Heinzel, M., Hagemann, H.W., Götz, G.K.E., 1997. Extraction of polycyclic aromatic hydrocarbons from polluted soils with binary and ternary supercritical phases. *Journal of Chromatography A* **776**: 233-43.

Huang, W.L., Weber, W.J., 1997. A distributed reactivity model for sorption by soils and sediments .10. Relationships between desorption, hysteresis, and the chemical characteristics of organic domains. *Environmental Science and Technology* **31**: 2562-2569.

Jagerm T., Van der Wal, L., Fleuren, R.H.L.J., Barendregt ,A., Hermens, J.L.M., 2005. Bioaccumulation of organic chemicals in contaminated soils: Evaluation of bioassays with earthworms. *Environmental Science and Technology* **39**: 293-298.

Johnson, M.D., Weber, W.J., 2001. Rapid prediction of long-term rates of contaminant desorption from soils and sediments. *Environmental Science and Technology* **35**:427-433.

Johnson, D.L., Jones, K.C., Langdon, C.J., Piearce, T.G., Semple, K.T., 2002. Temporal changes in earthworm availability and extractability of polycyclic aromatic hydrocarbons in soil. *Soil Biology and Biochemistry* **34**: 1363-1370.

Jordan, R.N., Cunningham, A.B., 1999. Surfactant enhanced bioremediation: A review of the effects of surfactants on the bioavailability of hydrophobic organic chemicals in soils. In: Baveye, P., Block, J.-C., Goncharuk, V.V., editor. Bioavailability of organic xenobiotics in the environment, NATO ASI series 2, environment 64. Dordrecht: Kluwer Academic Publishers. p 463-496.

Juhasz, A.L., Naidu, R., 2000. Bioremediation of high molecular weight polycyclic aromatic hydrocarbons: A review of the microbial degradation of benzo(a)pyrene. *International Biodeterioration and Biodegradation* **45**: 57-88.

Kanaly, R., Bartha, R., Fogel, S., Findlay, M., 1997. Biodegradation of C-14 benzo(a)pyrene added in crude oil to uncontaminated soil. *Applied and Environmental Microbiology* **63**: 4511-4515.

Kanaly, R.A., Bartha, R., 1999. Cometabolic mineralization of benzo(a)pyrene caused by hydrocarbon additions to soil. *Environmental Toxicology and Chemistry* **18**: 2186-2190.

Kastner, M., Breuer-Jammali, M., Mahro, B., 1998. Impact of inoculation protocols, salinity, and pH on the degradation of polycyclic aromatic hydrocarbons (PAHs) and survival of PAH-degrading bacteria introduced into soil. *Applied and Environmental Microbiology* **64**: 359-362.

Kelsey, J.W., Kottler, B.D., Alexander, M., 1997. Selective chemical extractants to predict bioavailability of soil-aged organic chemicals. *Environmental Science and Technology* **31**: 214-217.

Khodadoust, A.P., Bagchi, R., Suidan, M.T., Brenner, R.C., Sellers, N.G., 2000. Removal of PAHs from highly contaminated soils found at prior manufactured gas operations. *Journal of Hazardous Materials* **B80**: 159-174.

Krauss, M., Wilcke, W., 2001. Biomimetic extraction of PAHs and PCBs from soil with octadecyl-modified silica disks to predict their availability to earthworms. *Environmental Science and Technology* **35**: 3931-3935.

Laitinen, A., Michaux, A., Aaltonen, O., 1994. Soil cleaning carbon dioxide extraction: A review. *Environmental Technology* **15**: 715-727.

Lake, J.L., McKinney, R., Osterman, F.A., Lake, C.A., 1996. C-18-coated silica particles as a surrogate for benthic uptake of hydrophobic compounds from bedded sediment. *Environmental Toxicology and Chemistry* **15**: 2284-2289.

Langenfeld, J.J., Hawthorne, S.B., Miller, D.J., Pawliszyn, J., 1993. Effects of temperature and pressure on SFE efficiencies of polycyclic aromatic hydrocarbons and polychlorinated biphenyls. *Analytical Chemistry* **65**: 338-344.

Lei, L., Suidan, M.T., Khodadoust, A.P., Tabak, H.H., 2004. Assessing the bioavailability of PAHs in field-contaminated sediment using XAD-2 assisted desorption. *Environmental Science and Technology* **38**: 1786-1793.

Librando, V., Hutzinger, O., Tringali, G., Aresta, M., 2004. Supercritical fluid extraction of polycyclic aromatic hydrocarbons from marine sediments and soil samples. *Chemosphere* **54**: 1189-1197.

Liste, H.H., Alexander, M., 2002. Butanol extraction to predict bioavailability of PAHs in soil. *Chemosphere* 46: 1011-1017.

Loibner, A.P., Gartner, M., Schlegl, M., Hautzenberger, I., Braun, R., 1997. PAHs: Rapid estimation of their bioavailability in soil. *In situ and on-site bioremediation*, vol 5. P 617-622.

Loibner, A.P., Szolar, O., Schlegl, M., Gartner, M., Braun, R., 1998. Bioavailability of PAHs in soil and ecotoxicological considerations. Contaminated soil '98, vols 1 and 2. P 797-799.

Lopez-Avila, V., Young, R., Beckert, W.F., 1994. Microwave-assisted extraction of organic compounds from standard reference soils and sediments. *Analytical Chemistry* **66**: 1097-1106.

Mackay, D., Callcott, D., 1998. Partitioning and physical chemical properties of PAHs. In: Nielson, A.H., editor. The handbook of environmental chemistry Vol.3 Part I, PAHs and related compounds. Berlin Heidelberg: Springer-Verlag.Mader, B.T., UweGoss, K., Eisenreich, S.J., 1997. Sorption of nonionic, hydrophobic organic chemicals to mineral surfaces. *Environmental Science and Technology* **31**: 1079-1086.

Mayer, P., Vaes, W.H.J., Wijnker, F., Legierse, K., Kraaij, R.H., Tolls, J., Hermens, J.L.M., 2000. Sensing dissolved sediment porewater concentrations of persistent and bioaccumulative pollutants using disposable solid-phase microextraction fibers. *Environmental Science and Technology* **34**: 5177-5183.

McGinley, P.M., Katz, L.E., Weber, W.J., 1993. A distributed reactivity model for sorption by soils and sediments .2. Multicomponent systems and competitive effects. *Environmental Science and Technology* **27**: 1524-1531.

McGinley, P.M., Katz, L.E., Weber, W.J., 1996. Competitive sorption and displacement of hydrophobic organic contaminants in saturated subsurface soil systems. *Water Resources Research* **32**: 3571-3577.

Makkar, R.S., Rockne, K.J., 2003. Comparison of synthetic surfactants and biosurfactants in enhancing biodegradation of polycyclic aromatic hydrocarbons. *Environmental Toxicology and Chemistry* **22**: 2280-2292.

McLeod, C.J.A., Semple, K.T., 2000. Influence of contact time on extractability and degradation of pyrene in soils. *Environmental Science and Technology* **34**: 4952-4957.

Miller, D.J., Hawthorne, S.B., 1998. Method for determining the solubilities of hydrophobic organics in subcritical water. *Analytical Chemistry* **70**: 1618-1621.

Miller, M.E., Alexander, M., 1991. Kinetics of bacterial-degradation of benzylamine in a montmorillonite suspension. *Environmental Science and Technology* **25**: 240-245.

Mulligan, C.N., Yong, R.N., Gibbs, B.F., 2001. Surfactant-enhanced remediation of contaminated soil: A review. *Engineering Geology* **60**: 371-380.

Mzoughi, N., Hellal, F., Dachraoui, M., Villeneuve, J.P., Cattini, C., de Mora, S.J., El Abed, A., 2002. Methodology of extraction of polycyclic aromatic hydrocarbons. Application to sediment from the Bizerte lagoon (Tunisia). *Comptes Rendus Geoscience* **334**: 893-901.

Nam, K., Alexander, M., 1998. Role of nanoporosity and hydrophobicity in sequestration and bioavailability: Tests with model solids. *Environmental Science and Technology* **32**: 71-74.

Nam, K., Chung, N., Alexander, M., 1998. Relationship between organic matter content of soil and the sequestration of phenanthrene. *Environmental Science and Technology* **32**: 3785-3788.

Noordkamp, E.R., Grotenhuis, J.T.C., Rulkens, W.H., 1997. Selection of an efficient extraction method for the determination of polycyclic aromatic hydrocarbons in contaminated soil and sediment. *Chemosphere* **35**: 1907-1917.

Northcott, G.L., Jones, K.C., 2001a. Partitioning extractability and formation of nonextractable PAH residues in soil. 1. Compound differences in ageing and sequestration. *Environmental Science and Technology* **35**: 1103-1110.

Northcott, G.L., Jones, K.C., 2001b. Partitioning, extractability, and formation of nonextractable PAH residues in soil. 2. Effects on compound dissolution behavior. *Environmental Science and Technology* **35**: 1111-1117.

Ogram, A.V., Jessup, R.E., Ou, L.T., Rao, P.S.C., 1985. Effects of sorption on biological degradation rates of (2,4- dichlorophenoxy)acetic acid in soils. *Applied and Environmental Microbiology* **49**: 582-587.

Piatt, J.J., Brusseau, M.L., 1998. Rate-limited sorption of hydrophobic organic compounds by soils with well-characterized organic matter. *Environmental Science and Technology* **32**: 1604-1608.

Pignatello, J.J., 1998. Soil organic matter as a nanoporous sorbent of organic pollutants. *Advances in Colloid and Interface Science* **77**: 445-467.

Pilorz, K., Bjorklund, E., Bowadt, S., Mathiasson, L., Hawthorne, S.B., 1999. Determining PCB sorption desorption behavior on sediments using selective supercritical fluid extraction. 1. Describing PCB extraction with simple diffusion models. *Environmental Science and Technology* **33**: 2204-2212.

Reid, B.J., Jones, K.C., Semple, K.T., 1999. Can bioavailability of PAHs be assessed by a chemical means? In: *Bioremediation Technologies for Polycyclic Aromatic Hydrocarbon Compounds,* Proceedings 5th In Situ and On-Site Bioremediation Symposium Battelle Press, Columbus.

Reid, B.J., Semple, K.T., Jones, K.C., 1998. Prediction of bioavailability of persistent organic pollutants by a novel extraction technique. *Contaminated soil* '98, vols 1 and 2. P 889-890.

Reid, B.J., Stokes, J.D., Jones, K.C., Semple, K.T., 2000. Nonexhaustive cyclodextrin based extraction for the evaluation of PAH bioavailability. *Environmental Science and Technology* **34**: 3174-3179.

Reimer, G., Suarez, A., 1995. Comparison of supercritical fluid extraction and soxhlet extraction for the analysis of native polycyclic aromatic hydrocarbons in soil. *Journal of Chromatography A* **699**: 253-263.

Richter, B.E., Jones, B.A., Ezzell, J.L., Porter, N.L., Avdalovic, N., Pohl, C., 1996. Accelerated solvent extraction: A technique for sample preparation. *Analytical Chemistry* **68**: 1033-1039.

Saim, N., Dean, J.R., Abdullah, M.P., Zakaria, Z., 1997. Extraction of polycyclic aromatic hydrocarbons from contaminated soil using soxhlet extraction, pressurised and atmospheric microwave-assisted extraction, supercritical fluid extraction and accelerated solvent extraction. *Journal of Chromatography A* **791**: 361-366.

Saim, N., Dean, J.R., Abdullah, M.P., Zakaria, Z., 1998. An experimental approach for the determination of polycyclic aromatic hydrocarbons from highly contaminated soil using accelerated solvent extraction. *Analytical Chemistry* **70**: 420-424.

Schantz, M.M., Nichols, J.J., Wise, S.A., 1997. Evaluation of pressurized fluid extraction for the extraction of environmental matrix reference material. *Analytical Chemistry* **69**: 4210-4219.

Semple, K.T., Doick, K.J., Jones, K.C., Burauel, P., Craven, A., Harms, H., 2004. Defining bioavailability and bioaccessibility of contaminated soil and sediment is complicated. *Environmental Science and Technology* **38**: 228A-231A.

Semple, K.T., Stokes, J.D., Clasper, P.J., Wilkinson, A., Reid, B.J., 2003. Cyclodextrins: A chemical mimic of PAH bioavailability in soils? In: In Situ and On-Site Bioremediation Symposium. Orlando, FL. Battelle Press, Columbus, OH.

Shieh, W.J., Hedges, A.R., 1996. Properties and applications of cyclodextrins. *Journal of Macromolecular Science-Pure and Applied Chemistry* **A33**: 673-683.

Sijm, D., Kraaij, R., Belfroid, A., 2000. Bioavailability in soil or sediment: Exposure of different organisms and approaches to study it. *Environmental Pollution* **108**: 113-119.

Smith, V.H., Graham, D.W., Cleland, D.D., 1998. Application of resource-ratio theory to hydrocarbon biodegradation. *Environmental Science and Technology* **32**: 3386-3395.

Stucki, G., Alexander, M., 1987. Role of dissolution rate and solubility in biodegradation of aromatic-compounds. *Applied and Environmental Microbiology* **53**: 292-297.

Sun, F., Littlejohn, D., Gibson, M.D., 1998. Ultrasonication extraction and solid-phase extraction clean-up for determination of USEPA 16 priority pollutant PAHs in soils by reversed-phase liquid chromatography with ultraviolet absorption detection. *Analytica Chimica Acta* **364**: 1-11.

Swindell, A.L., Reid, B.J., 2003. Predicting soil-associated contaminant bioavailability in single and multiple contaminant systems. In: In Situ and On Site Bioremediation. Orlando, USA. Battelle Press, Columbus, OH.

Szolar, O.H.J., Rost, H., Hirmann, D., Hasinger, M., Braun, R., Loibner, A.P., 2004. Sequential supercritical fluid extraction (SSFE) for estimating the availability of high molecular weight polycyclic aromatic hydrocarbons in historically polluted soils. *Journal of Environmental Quality* **33**: 80-88.

Tang, J.X., Alexander, M., 1999. Mild extractability and bioavailability of polycyclic aromatic hydrocarbons in soil. *Environmental Toxicology and Chemistry* **18**: 2711-2714.

Tang, J.X., Liste, H.H., Alexander, M., 2002. Chemical assays of availability to earthworms of polycyclic aromatic hydrocarbons in soil. *Chemosphere* **48**: 35-42.

Tang, J.X., Robertson, B.K., Alexander, M., 1999. Chemical-extraction methods to estimate bioavailability of DDT, DDE, and DDD in soil. *Environmental Science and Technology* **33**: 4346-4351.

Tiehm, A., Stieber, M., Werner, P., Frimmel, F.H., 1997. Surfactant-enhanced mobilization and biodegradation of polycyclic aromatic hydrocarbons in manufactured gas plant soil. *Environmental Science and Technology* **31**: 2570-2576.

USEPA Method 3540C, 1996. Methods for evaluating solid waste, physical/chemical methods, SW-846 manual. Washington DC, USA: U.S. government printing office.

USEPA Method 3550B, 1996. Methods for evaluating solid waste, physical/chemical methods, SW-846 manual. 3rd ed. Washington DC, USA: U.S. government printing office.

USEPA Method 3600C, 1996. Methods for evaluating solid waste, physical/chemical methods, SW-846 manual. 3rd ed. Washington DC, USA: U.S. government printing office.

USEPA Method 3620B, 1996. Methods for evaluating solid waste, physical/chemical methods, SW-846 manual. 3rd ed. Washington DC, USA: U.S. government printing office.

USEPA Method 3630C, 1996. Methods for evaluating solid waste, physical/chemical methods, SW-846 manual. 3rd ed. Washington DC, USA: U.S. government printing office.

USEPA Method 3640A, 1994. Methods for evaluating solid waste, physica/chemical methods, SW-846 manual. 3rd ed. Washington DC, USA: U.S. government printing office.

USEPA Method 3650B, 1996. Methods for evaluating solid waste, physical/chemical methods, SW-846 manual. 3rd ed. Washington DC, USA: U.S. government printing office.

Vaes, W.H.J., Hamwijk, C., Ramos, E.U., Verhaar, H.J.M., Hermens, J.L.M., 1996. Partitioning of organic chemicals to polyacrylate-coated solid phase microextraction fibers: Kinetic behavior and quantitative structure-property relationships. *Analytical Chemistry* **68**: 4458-4462.

Volkering, F., Breure, A.M., Rulkens, W.H., 1997. Microbiological aspects of surfactant use for biological soil remediation. *Biodegradation* **8**: 401-417.

Volkering, F., Quist, J.J., van Velsen, A.F.M., Thomassen, P.H.G., Olijve, M., 1998. A rapid method for predicting the residual concentration after biological treatment of oil-polluted soil. Contaminated soil '98, vols 1 and 2. p 251-259.

Wang, X.J., Brusseau, M.L., 1993. Solubilization of some low-polarity organic-compounds by hydroxypropyl-beta-cyclodextrin. *Environmental Science and Technology* **27**: 2821-2825.

Wang, X.J., Brusseau, M.L., 1995. Cyclopentanol-enhanced solubilization of polycyclic aromatic- hydrocarbons by cyclodextrins. *Environmental Science and Technology* **29**: 2346-2351.

Weber, W.J., Huang, W.L., LeBoeuf, E.J., 1999. Geosorbent organic matter and its relationship to the binding and sequestration of organic contaminants. *Colloids and Surfaces a-Physicochemical and Engineering Aspects* **151**: 167-179.

Weber, W.J., McGinley, P.M., Katz, L.E., 1991. Sorption phenomena in subsurface systems - concepts, models and effects on contaminant fate and transport. *Water Research* **25**: 499-528.

Weber, W.J., McGinley, P.M., Katz, L.E., 1992. A distributed reactivity model for sorption by soils and sediments .1. Conceptual basis and equilibrium assessments. *Environmental Science and Technology* **26**: 1955-1962.

White, J.C., Hunter, M., Pignatello, J.J., Alexander, M., 1999. Increase in bioavailability of aged phenanthrene in soils by competitive displacement with pyrene. *Environmental Toxicology and Chemistry* **18**: 1728-1732.

White, J.C., Pignatello, J.J., 1999. Influence of bisolute competition on the desorption kinetics of polycyclic aromatic hydrocarbons in soil. *Environmental Science and Technology* **33**: 4292-4298.

WHO. 1998. Environmental health criteria 202. Polycyclic aromatic hydrocarbons, selected non-heterocyclic. Geneva: International programme on chemical safety, United Nations Environment Programme and World Health Organisation.

Wild, S.R., Jones, K.C., 1995. Polynuclear aromatic hydrocarbons in the United Kingdom environment: A preliminary source inventory and budget. *Environmental Pollution* **88**: 91-108.

Xing, B.S., Pignatello, J.J., 1997. Dual-mode sorption of low-polarity compounds in glassy poly(vinyl chloride) and soil organic matter. *Environmental Science and Technology* **31**: 792-799.

Xing, B.S., Pignatello, J.J., Gigliotti, B., 1996. Competitive sorption between atrazine and other organic compounds in soils and model sorbents. *Environmental Science and Technology* **30**: 2432-2440.

Yang, Y., Gharaibeh, A., Hawthorne, S.B., Miller, D.J., 1995. Combined temperature modifier effects on supercritical CO_2 extraction efficiencies of polycyclic aromatic hydrocarbons from environmental samples. *Analytical Chemistry* **67**: 641-646.

Yang, Y., Hawthorne, S.B., Miller, D.J., 1997. Class selective extraction of polar, moderately polar and nonpolar organics from hydrocarbon wastes using subcritical water. *Environmental Science and Technology* **31**: 430-437.

Yeom, I.T., Ghosh, M.M., 1998. Mass transfer limitation in PAH-contaminated soil remediation. *Water Science and Technology* **37**: 111-118.

Yeom, I.T., Ghosh, M.M., Cox, C.D., Ahn, K.H., 1996. Dissolution of polycyclic aromatic hydrocarbons from weathered contaminated soil. *Water Science and Technology* **34**: 335-342.

Young, T.M., Weber, W.J., 1995. A distributed reactivity model for sorption by soils and sediments .3. Effects of diagenetic processes on sorption energetics. *Environmental Science and Technology* **29**: 92-97.

Yu, X.B., Wang, X.P., Bartha, R., Rosen, J.D., 1990. Supercritical fluid extraction of coal tar contaminated soil. *Environmental Science and Technology* **27**: 1732-1738.

Zhao, B.W., Zhu, L.Z., Li, W., Chen, B.L., 2005. Solubilization and biodegradation of phenanthrene in mixed anionic-nonionic surfactant solutions. *Chemosphere* **58**: 33-40.

Zygmunt, B., Jastrzebska, A., Namiesnik, J., 2001. Solid phase microextraction - a convenient tool for the determination of organic pollutants in environmental matrices. *Critical Reviews in Analytical Chemistry* **31**: 1-18.

In: Environmental Pollution: New Research
Editor: Rachel H. Plattenberg, pp. 263-293 ISBN 1-60021-285-9
© 2007 Nova Science Publishers, Inc.

Chapter 8

Modelling Tidal and Residual Circulation in the Alborán Sea: Applications to Oil and Chemical Spill Dispersion

Raúl Periáñez
Dpt Física Aplicada 1, EUITA, Universidad de Sevilla
Crta Utrera km 1, 41013-Sevilla, Spain

Abstract

The Strait of Gibraltar and the Alborán Sea are the only connection between the Atlantic Ocean and the Mediterranean Sea. Intense shipping activities occur in the area, including massive transport of chemicals and oil. Thus, it is relevant to have a dispersion model that can be used in an emergency situation after an accident, to help the decision-making process. Such dispersion models require a deep knowledge of the physical oceanography of the region of interest, with detailed simulations of tides and residual (average) circulation. In this work, a particle-tracking dispersion model that can be used to simulate the dispersion of oil, radioactive and chemical substances in the system Strait of Gibraltar-Alborán Sea is described. Tides are simulated using a barotropic model and for the average circulation a reduced-gravity model is applied. This model is able to reproduce the main features of the Alborán circulation (the well known western Alborán gyre). The dispersion model is run off-line, using previously computed tidal and residual currents. The contamination patch is simulated by a number of particles whose individual paths are computed, diffusion being modelled using a Monte Carlo method. Contaminant concentrations may be obtained from the density of particles. Specific processes for each contaminant have to be included (for instance radioactive decay or buoyancy and biodegradation of oil) and they are discussed in detail. Several examples of dispersion computations for several contaminants and under different wind and circulation conditions are presented. It is interesting to note that, under certain conditions, contaminants are trapped in the gyre.

1. Introduction

Numerical models for simulating pollutant dispersion are nowadays being developed since they can be used for decision-making purposes after releases of contaminants into the marine environment. In particular, particle-tracking methods are well suited for problems in which high contamination gradients are involved, since they do not introduce numerical diffusion. Also, they can give very fast answers, specially if the hydrodynamic calculations are made off-line and tidal analysis and computed residuals are used to reconstruct water movements, which avoids the CFL (Courant-Friederichs-Lewy) stability limitations in the dispersion calculations. Thus, particle-tracking models are very useful predictive tools that can be used for assessing contamination after accidental or deliberate releases. Particle-tracking models have been used to simulate the dispersion of passive tracers [1, 2], radionuclides [3, 4, 5], oil spills [6, 7] and even contaminated milk [8] in coastal waters.

The area of the Strait of Gibraltar has a high ecological value, being essential in marine and aerial migratory processes. As the only connection between the Mediterranean Sea and the Atlantic Ocean, marine turtles and mammals (dolphin, porpoise, sperm whale, killer whale) travel through the Strait and the Alborán Sea. Also, the region has a high tourist interest, since many kilometers of sand beaches in the area of the Strait and the Costa del Sol (Málaga) attract thousands of tourists each year. Finally, there are also some important towns. A release of contamination in the area as a consequence of an accident (or a deliberate release) can lead to relevant ecological and economic impacts.

Shipping activities in the area of interest are very intense, again due to the fact that it is the only connection between the Atlantic and the Mediterranean. There is a traffic over 70000 merchant vessels per year, 30 % of them declaring hazardous cargos. Traffic of oil tankers is about 5000 vessels per year. Transit of nuclear submarines and of vessels transporting radioactive waste must also be considered. Only in the Strait of Gibraltar there are over 12000 vessels (mostly passenger ferries) crossing between the north and south coasts. Connections between the towns of Málaga and Almería (Spain) and the north African coast must be added (see figure 1 for locations).

It is usual to have adverse meteorological conditions in the Strait, with more than 54 % of days of moderate to poor visibility and 13 % of days with persistent fog conditions. Winds must be added, with frequent east and west gales. East winds (*levantes*) blow an average of 165 days per year, predominantly from April to October, with an average speed of the order of 50 km/h. Maximum speed reaches 125 km/h. Gusts of winds can remain up to 7-10 days. West winds (*ponientes*) blow an average of 60 days per year, from November to March predominantly. Minimum and maximum speeds are 30 and 90 km/h. West winds are not as persistent as *levantes*, lasting for some 12-36 hours.

The particular conditions in the area (intense traffic and adverse meteorology) make navigation difficult. Indeed, 81 accidents have occurred in the Strait in the last years, with 14 collisions and 16 groundings. For instance, in 1990 there was a collision between the oil tanker Hesperus and the chemical tanker Sea Spirit. More recently, a ferry and a gas tanker collided off Ceuta. Due to the particular water circulation patterns in the area, a spill occurring in the Strait is introduced into the Alborán Sea basin in a time scale ranging from some hours to some days [9]. Thus, the area of the Alborán Sea may be affected by an accident occurring not only inside the Alborán Sea itself, but in the Strait of Gibraltar as well. The objective of this work consists of describing a spill model that covers the Strait of Gibraltar and the Alborán Sea that can be applied to radioactive and chemical contaminants and to oil.

The hydrodynamic is solved in advance. A 2D depth-averaged barotropic model is used to obtain tidal currents in the region. A reduced gravity model is applied to obtain

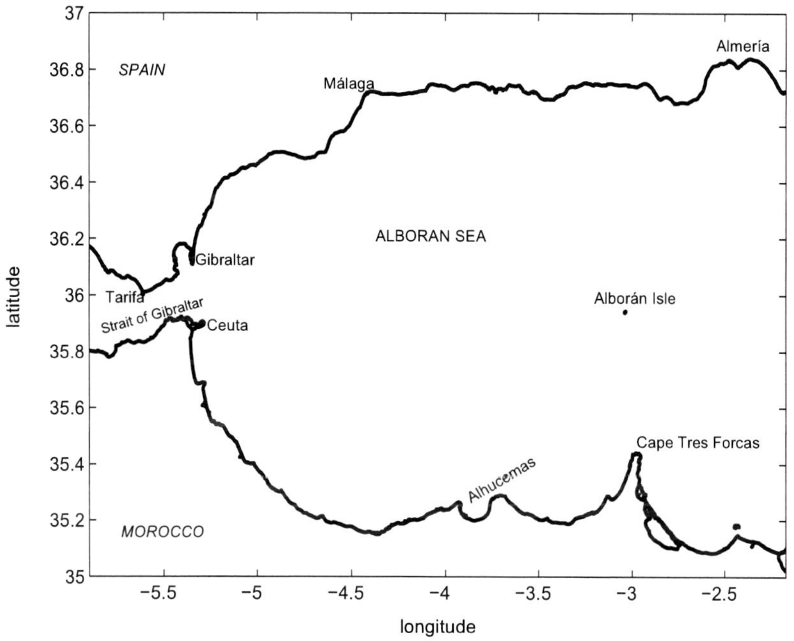

Figure 1. Map of the Strait of Gibraltar and Alborán Sea.

the average water circulation in the upper water layer, since only spills occurring at the surface will be considered by the moment. Tidal and average circulation are stored in files that will be read by the dispersion code to compute water current at any time and position. Dispersion is solved using a particle tracking method. Thus, the spill is simulated by a number of particles, each of them equivalent to a number of units (for instance kg or Bq), whose paths are followed in time. Specific processes for each contaminant are included in the model using Monte Carlo techniques (radioactive decay, oil evaporation and biodegradation, buoyancy etc). Contaminant concentrations may be obtained at the desired time from the density of particles per water volume unit.

Some characteristics of the physical oceanography of the region are briefly described in the next section. The hydrodynamic and dispersion models are next presented. Finally some examples of calculations are given.

2. Physical Oceanography of the Area

The water circulation in the Strait of Gibraltar is characterized by a surface inflow of Atlantic water and a deep outflow of more dense Mediterranean water. Exchanged flows are [10] of the order of 0.7 Sv (1 Sv=10^6 m^3/s) with a net inflow into the Mediterranean Sea of about 0.05 Sv. This net inflow compensates the excess of evaporation over precipitation and river supply in the Mediterranean. This description of the exchange as a simple two-layer system flowing in opposite directions is a good first approximation [9]. The surface circulation in the Alborán Sea is as follows. The Atlantic water penetrates the Strait of Gibraltar and reaches mean velocities of the order of 0.6 m/s. This water forms a jet that enters the Alborán Sea in a east-northeast direction. The jet flows along the Spanish coast and curves to the south. Part of it flows to the west, incorporating to an anticyclonic gyre, while the remainder flows to the African coast between Cape Tres Forcas and Alborán Isle [11].

The Alborán gyre is an almost permanent feature that fills the western Alborán Sea between the Strait of Gibraltar and the Alborán Isle, thus extending some 200 km in the west-east direction. The gyre extends towards the south, reaching the coast of Morocco [11, 12]. The vertical scale of the gyre is about 200 m. It has also been suggested [13] that a coastal feature (Cape Tres Forcas) acts as a barrier to the gyre growth.

With respect to the tides in the area, an important feature of the tidal flow in the Strait is that it can be considered, as a first approach, as barotropic. Indeed, 93 % of the variance of current velocity in the semidiurnal band has a barotropic character in the Strait [14]. Tsimplis and Bryden [10] have pointed out that tidal currents are barotropic and larger than the mean inflow or outflow. The semidiurnal tide dominates ADCP records (acoustic doppler current profiler) in the Strait, obscuring the expected two-layer character of the mean flow. The tidal signal is so strong that it reverses the currents near the bottom for a part of each tidal cycle. As a consequence, 2D depth-averaged models have already been applied to simulate surface tides in the Strait [15]. Tsimplis et al. [16] have even used a 2D barotropic model for simulating tides in the whole Mediterranean Sea. The success of these models indicates that the baroclinic component is of secondary importance. However, it is also known that the surface manifestation of baroclinic tidal currents is usually small.

In the case of the main tidal constituent, M_2, tide amplitude at the Atlantic entrance of the Strait is about 0.8 m. This amplitude decreases towards the east. Thus, at the Mediterranean entrance is only about 0.3 m. The amplitude of the tide is reduced more in the Alborán Sea, reaching 0.09 m at the east limit of our domain (see map in figure 1). The associated currents decrease in a similar way: from tidal currents of the order of 1 m/s in the Strait to currents of a few cm/s in the Alborán Sea basin. Similar behaviour is observed for the S_2 constituent.

From this analysis it seems evident that tides are relevant, for the transport of pollutants in water, in the area of the Strait of Gibraltar. However, once that pollutants enter the Alborán Sea the main mechanism is the residual (average) circulation. Contaminants may remain trapped in the Alborán gyre, and an appropriate description of this average circulation is thus required for a realistic calculation of contaminant dispersion.

Tides are described by means of a 2D depth-averaged model, since as commented above, it is a good approximation. A reduced-gravity model has been used for the average circulation. The reduced-gravity model is the simplest model able to simulate the major features of circulation above the pycnocline. Essentially it is a two-layer model in which the lower layer is infinitely deep and at rest. Thus, the interface between the two layers may deform without any resulting motion in the lower layer. Of course, this kind of model cannot simulate circulation of the deep layer and thus, dispersion of contaminants in this layer cannot be solved. However, as a first approach to the problem, we will study only contaminant releases occurring at the surface. Since the pycnocline acts as a barrier for vertical mixing, pollutants will remain in the surface layer and the reduced gravity model is thus enough for our purposes. Models are described in more detail below.

3. Description of Models

The hydrodynamic models (barotropic and reduced-gravity models) are described first. Next, the dispersion particle tracking model is presented.

3.1. 2D Depth-Averaged Barotropic Model

As commented before, this model is used to simulate tides in the domain of interest. The depth-averaged hydrodynamic equations may be written as:

$$\frac{\partial z}{\partial t} + \frac{\partial}{\partial x}(Hu) + \frac{\partial}{\partial y}(Hv) = 0 \tag{1}$$

$$\frac{\partial u}{\partial t} + u\frac{\partial u}{\partial x} + v\frac{\partial u}{\partial y} + g\frac{\partial z}{\partial x} - \Omega v + \frac{\tau_u}{\rho H} = A\left(\frac{\partial^2 u}{\partial x^2} + \frac{\partial^2 u}{\partial y^2}\right) \tag{2}$$

$$\frac{\partial v}{\partial t} + u\frac{\partial v}{\partial x} + v\frac{\partial v}{\partial y} + g\frac{\partial z}{\partial y} + \Omega u + \frac{\tau_v}{\rho H} = A\left(\frac{\partial^2 v}{\partial x^2} + \frac{\partial^2 v}{\partial y^2}\right) \tag{3}$$

where u and v are the depth averaged water velocities along the x and y axis, D is the depth of water below the mean sea level, z is the displacement of the water surface above the mean sea level measured upwards, $H = D + z$ is the total water depth, Ω is the Coriolis parameter ($\Omega = 2w \sin\beta$, where w is the earth rotational angular velocity and β is latitude), g is acceleration due to gravity, ρ is water density and A is the horizontal eddy viscosity. τ_u and τ_v are friction stresses that are written in terms of a quadratic law:

$$\begin{aligned} \tau_u &= k\rho u\sqrt{u^2+v^2} \\ \tau_v &= k\rho v\sqrt{u^2+v^2} \end{aligned} \tag{4}$$

where k is the bed friction coefficient.

The solution of these equations provides the water currents at each point in the model domain and for each time step. Currents are treated through standard tidal analysis and tidal constants are stored in files that will be read by the dispersion code to calculate the advective transport of particles. The model includes the two main tidal constituents, M_2 and S_2. Thus, the hydrodynamic equations are solved for each constituent and tidal analysis is also carried out for each constituent separately.

Some open boundary conditions must be provided to solve the hydrodynamic equations. Surface elevations are specified, from observations, along open boundaries of the computational domain. A radiation condition is applied to the water velocity component that is normal to the open boundary:

$$\frac{\partial \phi}{\partial t} = c\frac{\partial \phi}{\partial n} \tag{5}$$

where ϕ is the current component normal to the boundary, in the direction n, and c is a phase speed calculated as in [17]. Water flux across a land boundary is set to zero as usual.

Equations are solved using a explicit finite difference scheme [18, 20]. The computational domain (figure 2) extends from $35°00'$ N to $36°56'$ N and from $5°54'$ W to $2°10'$ W. Resolution of the grid is 2 minutes, that is equivalent to $\Delta x = 2972$ m and $\Delta y = 3341$ m.

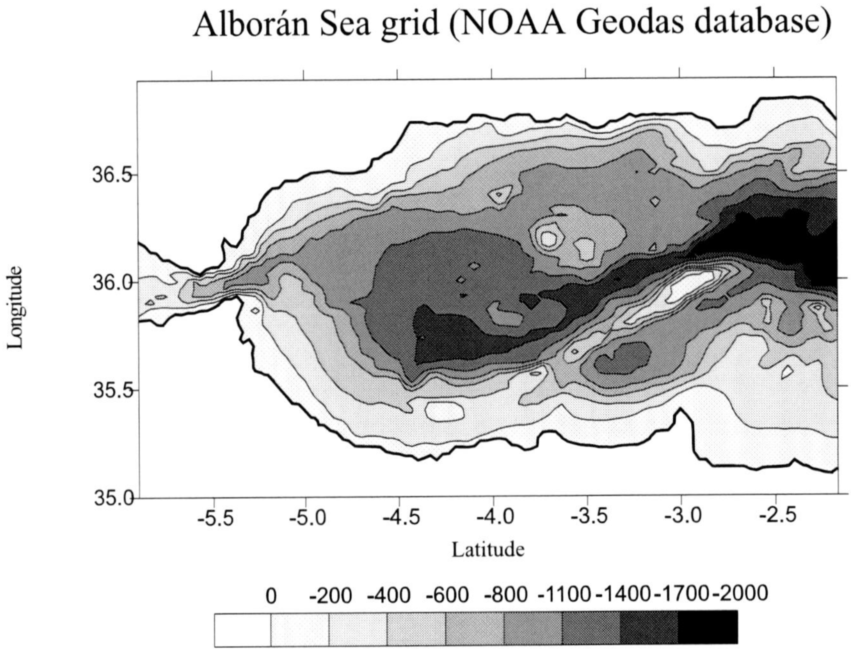

Figure 2. Map of the computational domain showing water depths.

Water depths were downloaded from the NOAA Geodas database and are shown in figure 2. Time step, limited by the CFL condition is $\Delta t = 10$ s. Once a stable periodic solution is achieved, tidal analysis is carried out to determine tidal constants that are used by the particle-tracking code.

3.2. Reduced-Gravity Model

The reduced-gravity model is, as commented above, the simplest model able to simulate the major features of circulation above the pycnocline. Essentially it is a two-layer model in which the lower layer is infinitely deep and at rest. Thus, the interface between the two layers may deform without any resulting motion in the lower layer. The term reduced gravity is derived from a balance of vertical forces in which the downward force due to acceleration of gravity is reduced by the upward buoyancy force. In spite of this relative simplicity, reduced gravity models are used at present to study different problems in oceanography [28, 29, 30, 31].

Equations are expressed in the following way [32]:

$$\frac{\partial h}{\partial t} + \frac{\partial hu}{\partial x} + \frac{\partial hv}{\partial y} = 0 \tag{6}$$

$$\frac{\partial u}{\partial t} + u\frac{\partial u}{\partial x} + v\frac{\partial u}{\partial y} + g'\frac{\partial h}{\partial x} - \Omega v + \frac{\tau_u}{\rho_1 h} = A\left(\frac{\partial^2 u}{\partial x^2} + \frac{\partial^2 u}{\partial y^2}\right) \tag{7}$$

$$\frac{\partial v}{\partial t} + u\frac{\partial v}{\partial x} + v\frac{\partial v}{\partial y} + g'\frac{\partial h}{\partial y} + \Omega u + \frac{\tau_v}{\rho_1 h} = A\left(\frac{\partial^2 v}{\partial x^2} + \frac{\partial^2 v}{\partial y^2}\right) \tag{8}$$

where h is the thickness of the upper layer and g' is the reduced gravity, written as:

$$g' = g\frac{\Delta\rho}{\rho_2} \tag{9}$$

In this equation $\Delta\rho$ is the density difference between the lower and upper layer and ρ_1 and ρ_2 are the densities of the upper and lower layers respectively. Friction between both water layers is described by a quadratic law, as in the barotropic model:

$$\begin{aligned} \tau_u &= k\rho_1 u\sqrt{u^2+v^2} \\ \tau_v &= k\rho_1 v\sqrt{u^2+v^2} \end{aligned} \tag{10}$$

where k is the friction coefficient. These equations are very similar to those described before for the barotropic motion. Here the upper layer thickness replaces the surface elevation in the continuity equation and in the gravity term of the momentum equations. Also, the reduced gravity replaces g in these equations.

The same numerical scheme as in the barotropic model was used to solve the reduced gravity equations. One restriction of layered models is that when the topography intersects the interface the solution becomes inestable [13]. To avoid this, the topography of the Alborán Sea was restricted to a minimum of 350 m. The downloaded bathymetry was artificially modified to describe the Alborán Isle as approximately 20×10 km^2 (the island disappears if minimum depth is restricted to 350 m due to the limited resolution of the downloaded data). These dimensions are a good estimate of the island size at 200 m depth. Such modification was also carried out in a previous study [13] to include the shallow topography associated with the island.

The CFL stability condition is less restrictive in this case than in the barotropic model since now g is replaced by the reduced gravity g'. Thus time step for the explicit integration is increased to $\Delta t = 800$ s. The model is integrated specifying water currents along the open western boundary until a stable solution is achieved. The computed steady current and upper layer thickness over the domain are stored in files that will be read by the dispersion code. Radiation boundary conditions are used along the open eastern boundary. Some authors [12] have found that to reproduce the gyre it is essential to use no-slip conditions along the coast. No-slip conditions consists of setting both the normal and tangential flows along the coastline to zero. However, slip (only the normal current is set to zero along the coast) and no-slip conditions have been tested in this work.

3.3. Particle-Tracking Dispersion Model

In particle-tracking models a contamination release is simulated by a number of particles, each one of them equivalent to a number of units (for instance kg or Bq). The path followed by each particle is then calculated. Contaminant concentrations may be calculated at the desired time from the density of particles per water volume unit. Processes specific for each contaminant (for instance radioactive decay or oil biodegradation) have to be described adequately. The different processes will be treated separately.

3.3.1. Advection and Diffusion

Advection is computed solving for each particle the equation:

$$\frac{d\vec{r}}{dt} = \vec{q} \tag{11}$$

where $\vec{r}$ is the position vector of the particle and $\vec{q}$ is the current vector solved in components u and v. This equation is approximated by:

$$\frac{\vec{r}(t + \Delta t) - \vec{r}(t)}{\Delta t} = \vec{q}(t) \tag{12}$$

where Δt is the time step used in the model. It can be observed that equation 12 constitutes a first order explicit approximation to the advection equation, since the water current changes continuously as the particle moves. However, for real sea dispersion calculations a first order accuracy scheme is adequate. When simulating the movement of drogues in an estuarine environment, Elliott and Clarke [19] found no improvements of results when a second order accuracy scheme was used. Indeed, the effects of turbulence will mask any small errors in the advection scheme. Thus, in practice, it is enough to obtain current components, $u(t)$ and $v(t)$, at the grid cell where the particle is located and use them directly into equation 12. Nevertheless, more complex interpolation algorithms may be applied as well [3].

The particle-tracking model is three-dimensional, but the hydrodynamic calculations provide depth-averaged currents. In the main body of water above the logarithmic layer, the flow gradually increases in a manner which may be represented as [21]:

$$u_{z'} = u_s \left(\frac{D - z'}{D} \right)^{1/m} \tag{13}$$

where $u_{z'}$ is the current speed at a level z' below the sea surface and u_s is the surface flow. From observations, it has been deduced that m ranges between 5 and 7. The surface current can be deduced from the depth averaged one [21]:

$$u_s = \frac{m+1}{m} \overline{u} \tag{14}$$

where $\overline{u}$ is the depth averaged current. Thus, components u and v of the current at any depth can be obtained from their depth-averaged values (provided by the hydrodynamic model) applying equations 13 and 14. This current profile has already been used [22] before. The total water depth is replaced by the thickness of the upper water layer in the case of the residual current, provided by the reduced gravity model. Actual water depth, however, is used to obtain the profiles of tidal currents, since they are obtained from a barotropic model in which the complete water column is considered. An example of a current profile obtained from these equations may be seen in figure 3.

Wind is typically included in particle-tracking models assuming that the surface wind-induced current is 3 % of the wind speed measured 10 m above the sea surface [6, 21]. This current decreases logarithmically to zero at a depth z_1. This depth is assumed to be 20 m

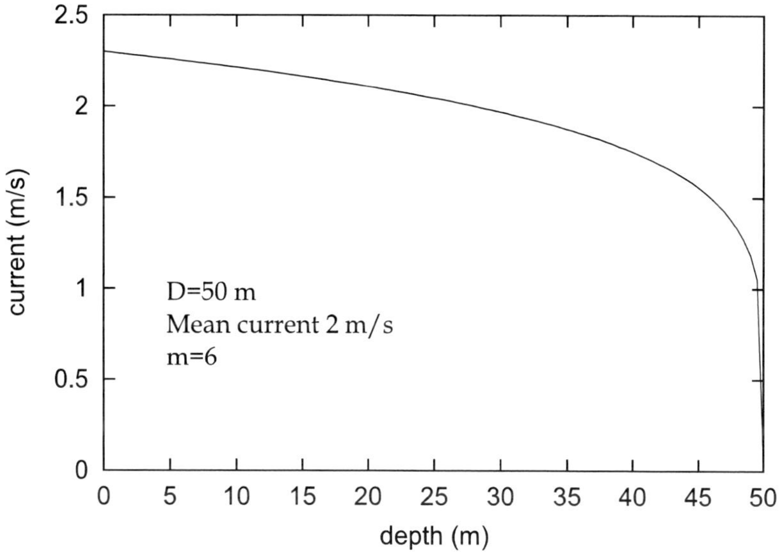

Figure 3. Current profile for a mean current of 2 m/s and a water depth of 50 m.

[23]. The wind induced current at any depth z' below the surface is written as [21]:

$$u_{z'} = \begin{cases} u_0 - \frac{u^*}{\kappa} \ln\left(\frac{z'}{z_0}\right) & \text{if} \quad z' < z_1 \\ 0 & \text{if} \quad z' \geq z_1 \end{cases} \tag{15}$$

where u_0 is the surface wind-induced current, $\kappa = 0.4$ is the von Karman constant, u^* is a friction velocity and z_0 is the sea surface roughness length, which has values between 0.5 and 1.5 mm. It has been obtained [21] that the friction velocity can be estimated as

$$u^* = 0.0012W \tag{16}$$

for a wide range of conditions, where W is wind speed 10 m above the sea surface. From these equations, the wind effect on the advection of particles can be calculated. Of course, the current profile is solved in the u and v components. An example of the current profile produced by a 20 m/s wind may be seen in figure 4.

Three dimensional diffusion is simulated using a random walk method. It has been shown [6] that it is a simulator of Fickian diffusion provided that the maximum size of the horizontal step given by the particle, D_h, is:

$$D_h = \sqrt{12K_h\Delta t} \tag{17}$$

in the direction $\theta = 2\pi RAN$, where RAN is a random number between 0 and 1. This equation gives the maximum size of the step. In practice, it is multiplied by RAN to obtain the real size at a given time and for a given particle. Similarly, the maximum size of the vertical step is:

$$D_v = \sqrt{2K_v\Delta t} \tag{18}$$

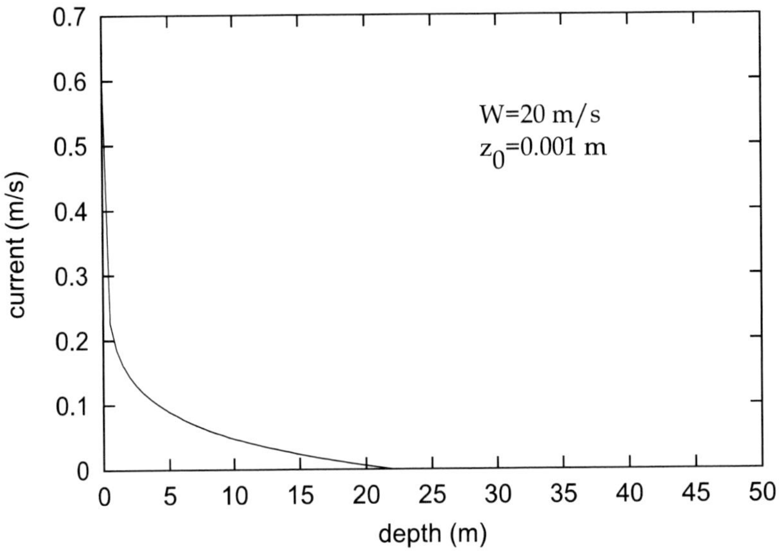

Figure 4. Current profile for a wind speed of 20 m/s and a water depth of 50 m.

given either towards the sea surface or the sea bottom. K_h and K_v are the horizontal and vertical diffusion coefficients respectively. While there is no stability criterion equivalent to the CFL condition in the particle tracking calculations, it is wise to ensure that each particle does not move through a distance that exceeds the grid spacing during each time step.

Some boundary conditions must be specified when computing advection and diffusion of particles. In the case of an open boundary, particles that cross it are removed from the computation. In practice, a label (for instance 1) is assigned initially to each particle. If a particle must be removed from computations at a given time (because has moved out of the domain or has decayed, see below) the label is changed to a new value (say 0 for instance). The label of each particle is checked each time step before computing advective and diffusive transport. If the label is zero such calculations are not carried out. Of course, these particles are not considered for calculating contaminant concentrations from the density of particles (see below).

In addition, as with a finite difference model, flux of particles through land boundaries, sea surface and the interface between the two water layers is not allowed. If a given particle steps onto land when computing diffusion, this movement is not allowed and new random numbers are generated to repeat the particle displacement. In practice, less than five repetitions are required to obtain a valid particle displacement even in the case of a complicated topography.

Values for the diffusion coefficients have to be provided. The horizontal diffusion coefficient value depends on the horizontal grid spacing, following [24, 3]

$$K_h = 0.2055 \times 10^{-3} \Delta x^{1.15} \tag{19}$$

The present grid resolution gives $K_h = 2.0$ m^2/s. For the vertical diffusion coefficient a typical value of 0.001 m^2/s is used [8, 3, 24, 25].

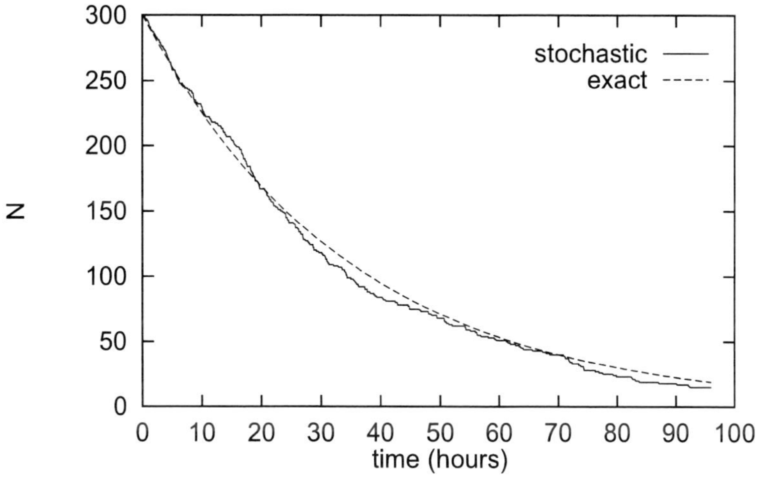

Figure 5. Solution of the radioactive decay equation with the stochastic method and exact solution

The adsorption of pollutants by suspended and bottom sediments can also be simulated with a particle-tracking model [4, 20]. However, these processes are neglected in the present study since suspended matter concentrations are very low in the area, typically 0.1-0.5 mg/L [26]. Also, since the pycnocline is acting as a barrier for vertical mixing, interactions of pollutants with bed sediments can be neglected.

3.3.2. Radioactive Decay

Consider the radioactive decay equation:

$$\frac{\partial C}{\partial t} = -\lambda C \tag{20}$$

where λ is the radioactive decay constant. This equation can be treated using a stochastic method if it is assumed that the probability p of removal of a particle at each time step is [6, 4, 27]:

$$p = 1 - e^{-\lambda \Delta t} \tag{21}$$

In practice, a random number is generated for each particle at each time step. If $RAN \leq p$ then the particle is removed from the computation. The solution of the radioactive decay equation with this method is shown in figure 5 for a radionuclide with a half-life of 1 day and using 300 particles. It can be seen that the stochastic solution mimics the exact solution rather well. The agreement is getting better as the number of particles used in the calculation increases. Typical particle number in real models is of the order of several thousands.

3.3.3. Oil Processes

Particle-tracking techniques have been widely applied to simulate the dispersion of oil spills in the sea [6, 7, 23, 33, 34]. As in the case of a chemical or radioactive spill, the motion of individual oil droplets, the sum of which constitute the oil spill, is followed. Oil droplets are advected by the currents induced by tides, winds and residual circulation. They are also affected by three dimensional diffusion induced by water turbulence, as described above. In addition, droplets have a size distribution so that larger ones tend to remain in the water surface and move in the direction of wind. Smaller droplets mix downwards because of turbulence and shear diffusion results in a patch elongated in the current direction. The model also includes the effects of surface evaporation of oil and decomposition within the water column (for instance because of biodegradation).

The buoyancy force depends on the density and size of droplets. The rise velocity, w, can be described as [6]:

$$w = \frac{gd^2(1 - \rho_0/\rho)}{18\nu} \tag{22}$$

for small droplets with diameter $d \leq d_c$ (laminar motion). In this equation ρ and ρ_0 are the densities of water and oil respectively and ν is the water kinematic viscosity. For large droplets with $d > d_c$ (turbulent motion) the rise velocity is:

$$w = \left(\frac{8}{3}gd(1 - \rho_0/\rho)\right)^{1/2} \tag{23}$$

The critical diameter, d_c, is given by the expression

$$d_c = \frac{9.52\nu^{2/3}}{g^{1/3}(1 - \rho_0/\rho)^{1/3}} \tag{24}$$

that is deduced matching the Reynolds numbers at which the transition from laminar to turbulent flow occurs.

The diameter of each oil droplet in the simulation is assigned randomly between a minimum and maximum diameter, $d_{min} - d_{max}$.

The effects of oil evaporation and decomposition are treated in a similar way as radioactive decay. Thus, the probability of removal of particle in a time step is given by equation 21. The decay constant is related to the e-folding time by $\lambda = 1/T_e$. Different e-folding times are used for evaporation, T_{ev}, and decomposition, T_{de}. Additionally, only particles within a depth $z_{ev} = 0.25$ m below the surface can be evaporated, whereas droplets at any depth can experience decomposition [6]. The age of each oil droplet is also calculated each time step (the age is the time elapsed since a given particle was released). Oil droplets are not allowed to decay after a specified age. This simulates the rapid evaporation that occurs after the release, while ensuring that some droplets remain in the computation as old and weathered subsurface oil [6]. Typical values for these parameters will be given below.

If during a computation an oil droplet reaches the coastline, it is considered *beached*. Thus, the droplet stays in the coast without moving any more. In practice this is achieved by assigning a special label to particles that are beached.

3.4. Calculation of Contaminant Concentrations

The output of a particle tracking model is the position of each particle. Concentrations can be obtained by counting the density of particles (number of particles per water volume unit):

$$C = N \times R \tag{25}$$

where C is concentration, N is density of particles and R is the number of units (for instance Bq) that is equivalent to each particle:

$$R = \frac{I}{NP} \tag{26}$$

where I is the total contaminant input and NP is the number of particles used in the simulation. Of course, computed concentrations correspond to average values over the mesh $(\Delta x \times \Delta y \times h)$, where h is a given value for water depth. In the case of a chemical spill this depth is fixed as the thickness of the upper water layer, since the pycnocline acts as a barrier for further vertical mixing and all particles will remain in this layer. In the case of an oil spill, however, concentrations are given in units/m^2.

3.5. Computational Scheme

The dispersion model runs off-line using standard tidal analysis. Tides can be represented as a number of harmonics in the form of an equation

$$z(t) = \sum_{i=1}^{n} h_i \cos(w_i t - g_i) \tag{27}$$

where the amplitude and phase of each harmonic, h_i and g_i, are denoted as tidal constants and n is the number of constituents (harmonics) that are considered. Tidal analysis consists of determining, by a standard fitting algorithm, the tidal constants for each constituent included in the model and for each grid cell. These tidal constants have to be calculated for both components of the flow, u and v, and for water elevation z. Tidal constants, for each tide constituent, are calculated by the same hydrodynamic code and are stored in files that will be read by the dispersion model. Once that tidal constants are known, computation of flow and water elevation at any cell and time just involves the evaluation and addition of a few cosine terms. This is very fast, and simultaneously, the dispersion model is not limited by the CFL condition. It has to be clearly pointed out that tidal analysis is carried out running the hydrodynamic model for each constituent separately. This technique is usually applied in rapid response Lagrangian models, although has also been used in finite difference dispersion models [35]. It has the clear advantage of joining the strengths of the off-line mode with a temporal resolution that is high enough to solve tidal processes. Such strengths are that the stability conditions imposed by the transport equation are much less restrictive than the CFL condition due to the hydrodynamic equations. Thus, a larger time step can be used. This, together with the fact that hydrodynamic calculations have all been carried out in advance, and thus only the transport equation is solved each time step, implies that computation is much faster.

The reduced gravity model provides the residual current in the domain, that is stored in another file. Nevertheless, residual currents are controlled by the water inflow through the Strait of Gibraltar that, as will be discussed in some more detail below, presents some variability. Thus, a factor that acts as a modulator of the residual current amplitude must be introduced. If 1 is used, the residual current for the average water inflow through the Strait is used in the calculations. These average currents may be amplified or reduced by specifying values for the modulator larger or smaller than 1 respectively.

Table 1. Information required by the model to be introduced by the user.

Release point coordinates
Select instantaneous/continuous release option
Wind speed
Wind direction
Release date (day,month,year)
Release time, UTC, (hours,minutes)
Residual current modulator
Simulation time (days)
Magnitude of the release in the corresponding units
Contaminant decay costant (radioactive)
Oil spill additional information:
Oil density
Droplet minimum and maximum sizes
e-folding times (evaporation and decomposition)

The total current at any time and position in the domain is calculated by the dispersion code as the residual plus the wind and tidal currents (produced by the two main constituents as commented before) obtained from the corresponding tidal constants. Date and time of the discharge (and duration in the case of continuous releases) must be specified since the fate of the release will depend on the tidal state when it took place. Thus, the appropriate phase of each tidal constituent at $t = 0$ must be specified. The values used in this model correspond to the origin of time being January 1, 2003 at 0:15 hours Greenwich time.

As commented before, there is no stability criterion equivalent to the CFL condition in the particle tracking calculations, although it is wise to ensure that each particle does not move through a distance that exceeds the grid spacing during each time step. Thus, time step was fixed as $\Delta t = 300$ s.

The dispersion code automatically reads 6 files which contain the topography of the Strait, amplitudes and phases of the two tidal constituents included in the model (M_2 and S_2), and the residual circulation. The user must also include some information for each simulation, related to the release characteristics. This information is summarized in table 1. Wind speed and direction are included into a file that is read by the dispersion code. Variable wind fields (uniform wind over the domain but changing in time) may be specified for each simulation.

The model provides several output files. It gives the coordinates of each particle at several times during the simulation. Thus, snap shots of particles can be drown to study the evolution of the discharge along time. In particular, 12 snaps shots at constant intervals along the simulation are provided. Another file contains a map of the final contaminant concentration over the domain computed from the density of particles per water volume unit. Optionally, the time evolution of contaminant concentration at desired points over the domain can be obtained.

A simplified flow diagram of the whole computation process is shown in figure 6. The

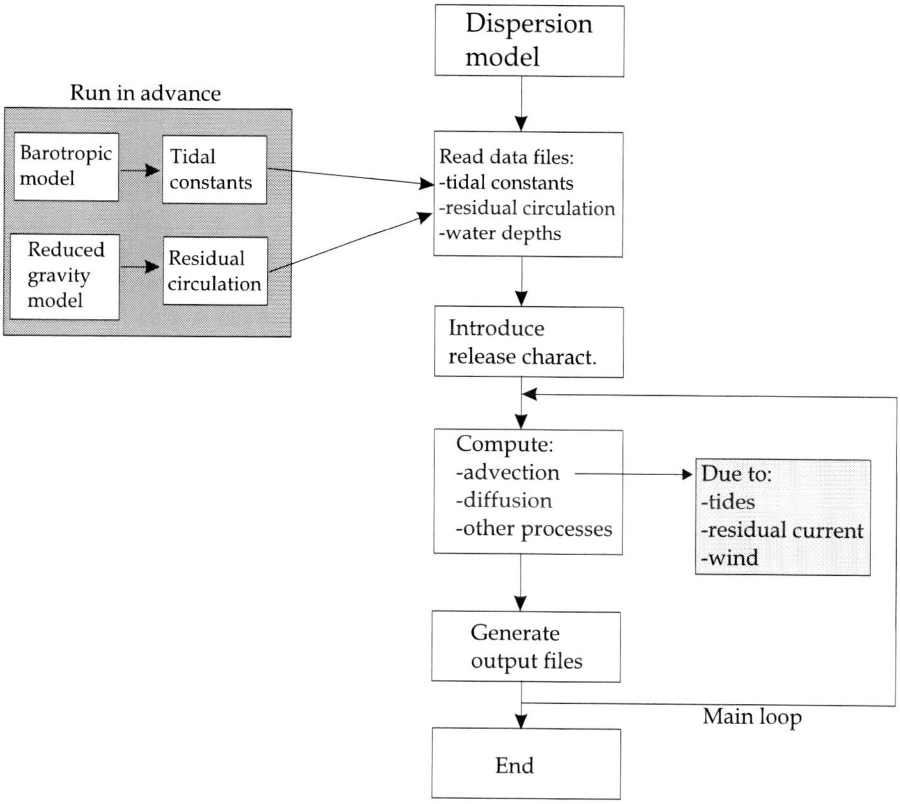

Figure 6. Flow diagram of the computation process in the model.

hydrodynamics is solved in advance and files generated are used by the dispersion model to calculate the currents at any position of the domain for each time step.

When a continuous release is simulated the number of particles that are being tracked increases linearly during the computation until the time at which the release ceases is reached. At this time (and for longer times if the computation extends beyond the end time of the release) there are NP particles in the simulation (some of which may have decayed or moved out of the region of interest) and therefore the required accuracy is achieved at the end of the simulation. At intermediate times, for example at 10 % of the way through the computation, there will only be $NP/10$ particles in the simulation and therefore the accuracy of the intermediate results will be lower than that obtained at the end of the simulation. In such circumstances the model user must decide on the simulation time at which the highest accuracy results are required. If the interest is in short time scales after the release, the model should not be run to simulate a long continuous release. However, there is no computational reason why the technique cannot be applied to releases that continue over decades and algorithms have been devised that can mimic both sporadic and long term spill scenarios.

Table 2. Observed and computed amplitudes (A, cm) and phases (g, deg) of tidal elevations at several locations indicated in figure 1.

	M_2				S_2			
Station	A_{obs}	g_{obs}	A_{comp}	g_{comp}	A_{obs}	g_{obs}	A_{comp}	g_{comp}
Tarifa	42	57	41	45	14	85	16	79
Ceuta	30	50	32	52	11	76	12	86
Málaga	17	59	18	42	7	72	8	77
Alhucemas	18	58	18	56	7	80	7	89
Almería	9	51	9	49	4	78	4	77
Gibraltar	30	46	29	41	11	72	12	77

4. Model Results

Results from the hydrodynamic models (barotropic and reduced gravity) are described and discussed first. Then some examples of dispersion calculations are presented.

4.1. Hydrodynamic Calculations

Computed values of the amplitude and phase of the water surface elevation produced by the M_2 and S_2 tides have been compared with observed values [16, 36] at different locations in the Strait of Gibraltar and the Alborán Sea. Results are given in table 2. Locations may be seen in the map in figure 1. It can be observed that there is a good agreement between the measured and computed tidal constants for both constituents.

A tidal chart for the M_2 tide is shown in figure 7. The same color joins points with the same tide amplitude (corange chart). The amplitude decreases quickly in the Strait of Gibraltar, from some 0.75 m in the Atlantic entrance to 0.30 in the section of Gibraltar. There is a further, although slower, amplitude reduction in the Alborán Sea, reaching only some 0.10 m at the eastern boundary. The phase of the tide is shown in figure 8, where points with the same tidal phase are joined (cotidal chart). It can be seen that corange lines are oriented in a south-north direction, while cotidal lines are essentially in a northeast direction, in agreement with observations and the earlier computations in [16].

The spatial distribution of the M_2 current amplitude is presented in figure 9 as an example. This distribution is in good agreement with that presented in [16]. The current amplitude in most of the Alborán Sea is some 0.03 m/s. Only in some specific areas larger currents are obtained. This is the case of the south-north section going from Cape Tres Forcas to Alborán Isle and the Spanish coast. Currents of the order of 0.06 to 0.09 m/s are computed close to the Cape, around Alborán Isle and close to the Spanish coast. This is in good agreement with the earlier computations presented in [16]. Current amplitude increases as approaching Gibraltar Strait, where, as expected, maximum values are obtained. Indeed, currents of the order of 0.8 m/s are computed in some locations. For clarity reasons, however, the scale maximum in figure 9 is limited to 0.2 m/s.

In the case of the S_2 tide, results show a similar decrease in the tide amplitude, from

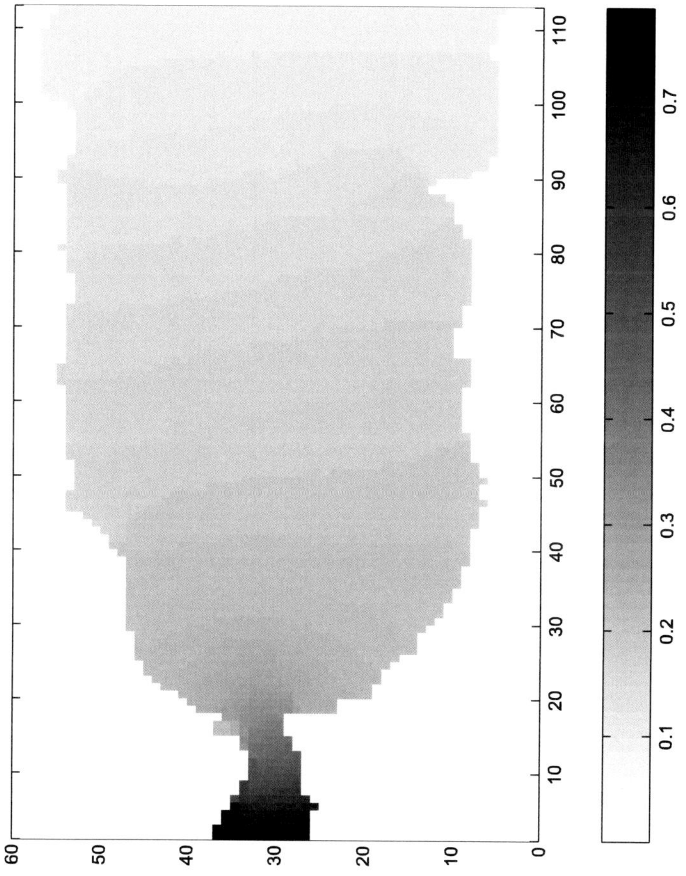

Figure 7. Corange chart for the M_2 tide. Amplitude of the tide is given in m. Numbers in the axes correspond to the grid cell number.

some 0.25 m at the Atlantic entrance of the Strait to some 0.04 m at the eastern open boundary of the domain. Current amplitudes decrease from maximum values of the order of 0.25 m/s in the Strait of Gibraltar to currents weaker than 1 cm/s in the Alborán Sea.

It can be said that, in general, the 2D depth-averaged barotropic model is giving a realistic description of tides in the Strait of Gibraltar and Alborán Sea. The average circulation provided by the reduced gravity model is discussed next.

The value of the reduced gravity has been fixed as 0.01 m/s^2, following previous work [37] and horizontal viscosity is 30 m^2/s. A stable solution is obtained with this viscosity setting.

A computed steady current distribution obtained from the reduced gravity model with slip boundary conditions along the coast can be seen in figure 10. The jet of Atlantic water entering through the Strait of Gibraltar is clearly observed. This jet flows towards the east along the Spanish coast and partially curves to the south before reaching Alborán Island.

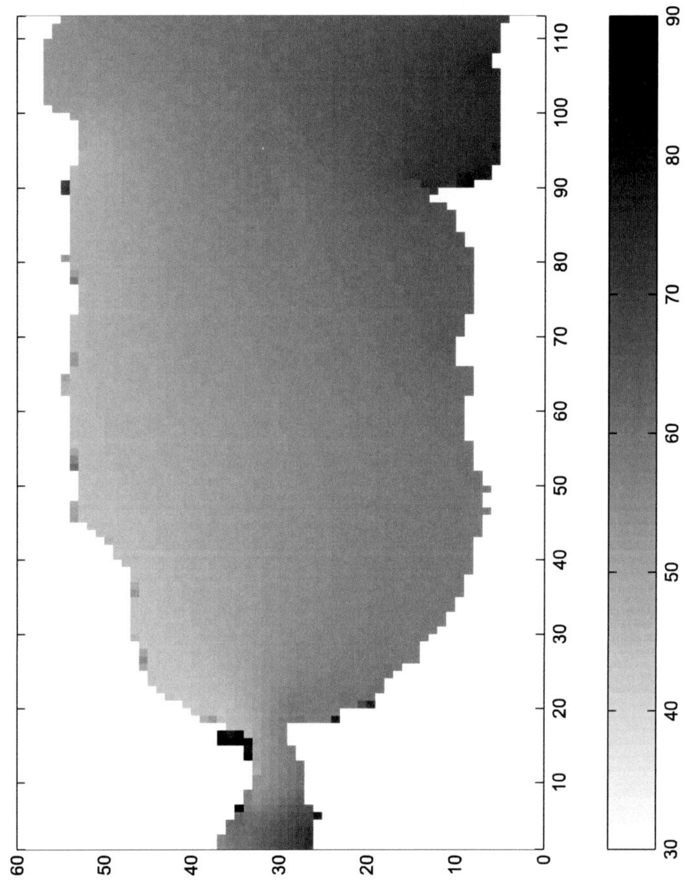

Figure 8. Cotidal chart for the M_2 tide. Phase of the surface elevation is given in degrees. Numbers in the axes correspond to the grid cell number.

Part of this flow continues to the east between Cap Tres Forcas and Alborán Island and the remaining rotates towards the west, flowing along the African coast. A gyre of anticyclonic circulation is thus completed. This circulation is in qualitative agreement with that deduced from a satellite thermal image, that is shown in figure 11 (from [37]). It can be clearly observed in this photograph that warmer water, of Atlantic origin, is accumulated in the centre of the gyre.

Horizontal dimension of the gyre is of the order of 180 km, in agreement with previous studies [12, 13]. This circulation pattern is almost always present [11, 12], although the gyre size may change from the entire western Alborán basin to less than one-half that size [42].

It is interesting to observe that a small gyre of cyclonic circulation is obtained by the Spanish coast, north of the incoming jet. This gyre has been obtained with other models

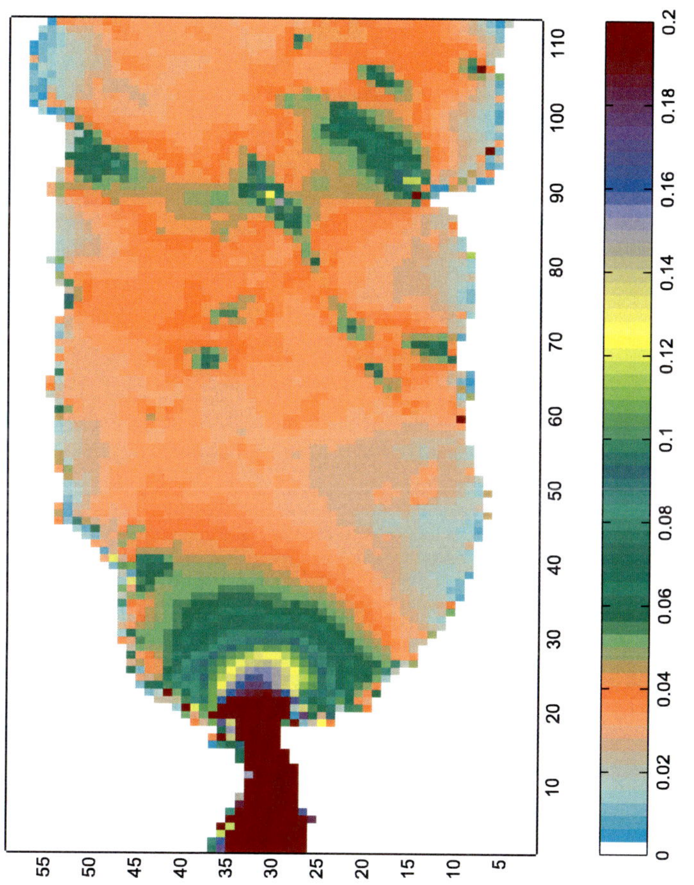

Figure 9. Current amplitudes (m/s) for the M_2 tide. Numbers in the axes correspond to the grid cell number.

[13] and has also been observed in the area [41].

The computed steady inflow through the Strait of Gibraltar is 0.86 Sv. A review on historic estimations of the Atlantic inflow may be seen in [38]. First estimations gave results of about 1.2 Sv. A more realistic value was proposed to be 0.72 Sv [38]. More recently [10], the inflow has been revised to 0.66 ± 0.47 Sv, where the large error is due to seasonal variability. Thus, our result is giving a good estimation of the average transport of Atlantic water into the Alborán Sea. Other transport estimations, by means on numerical models, are 0.53 Sv [39] and 0.72 Sv [43].

The amplitude of the seasonal signal in the inflow is about 0.2 Sv, following calculations in [39]. Maximum transports generally occur in April and minimum transports in August, although there is not a full agreement between different authors in this point. To account for the variability in the water inflow through the Strait of Gibraltar, the residual current modulator was introduced. The default value, 1, corresponds to the average inflow 0.86 Sv, as calculated by the reduced gravity model. Nevertheless, this is a first approximation. A further development of the model will include the construction of a database of residuals. Thus, a specific residual current distribution over the domain will be used in dispersion calculations for each specific inflow magnitude. Such database will include unusual circulation patterns (the collapse of the gyre). These events appear to occur when the inflow through the Strait of Gibraltar is minimum [39]. The numerical calculations in [37] suggest that the collapse of the western Alborán gyre is related to a westward advection of the gyre, that is caused by an enhanced transport towards the Atlantic in the deep water layer. When the lower layer outflow exceeds the upper layer inflow (because a decrease in the inflow through Gibraltar and/or an increase in the deep flow through the Strait of Sicily) there is a westward propagation and the gyre disappears [37]. Other authors [40] have found that the gyre is more stable during summer months, while in winter the prevailing situation is a coastal circulation (the incoming Atlantic water curves to the south as enters the Alborán Sea from Gibraltar Strait and flows to the east along the African coast).

The computed thickness of the upper water layer is shown in figure 12. It can be clearly seen that there is a depression of the interface in the centre of the gyre. This is in agreement with previous calculations carried out with a reduced gravity model [12, 37]. The high pressure area makes water to rotate clockwise around it.

It has finally been observed that results with slip and no-slip boundary conditions along the coast are essentially the same. Thus, it is not essential to use no-slip conditions to produce the gyre, as was stated in [12]. Nevertheless, results are sensitive to the friction coefficient. Results presented here correspond to a friction coefficient equal to 0.001. If it is increased to 0.005, the small cyclonic gyre located by the Spanish coast disappears. The western Alborán gyre disappears if friction is increased further.

4.2. Dispersion Calculations

Some examples of results obtained with the dispersion model are presented in this section. Chemical spills will be considered first. Some experiments with instantaneous discharges have been carried out. It has been assumed that releases occurred in all cases in January 1, 2005 at 0:00 hours UTC (just as an example, there is not any particular reason to select this date and time). No wind is considered and the decay constant is set to zero (stable chemical pollutant). The total pollutant discharged is 1×10^{12} units. Finally, the current amplitude modulator is 1 (average inflow through the Strait of Gibraltar).

In the first experiment, the fate of a release occurring at the centre of the Strait of Gibraltar [grid cell (10,30)] has been simulated. In this case the patch is introduced in the Alborán Sea by the Atlantic water jet and moves rapidly towards the east, in a trajectory that is initially parallel to the Spanish coast. Then curves to the south and the patch is divided by Alborán Island. The time required by the patch to reach the east boundary of the domain is of some 50 days, which implies an average velocity of the order of 9 cm/s. Three of the 12 snapshots produced by the model are shown in figure 13 as an example.

If the same experiment is repeated at a short temporal scale, simulating only the first 4 days, then the snapshots produced by the model show how the patch moves in the Strait forward and backwards because of tidal oscillations. These movements cannot be appreciated

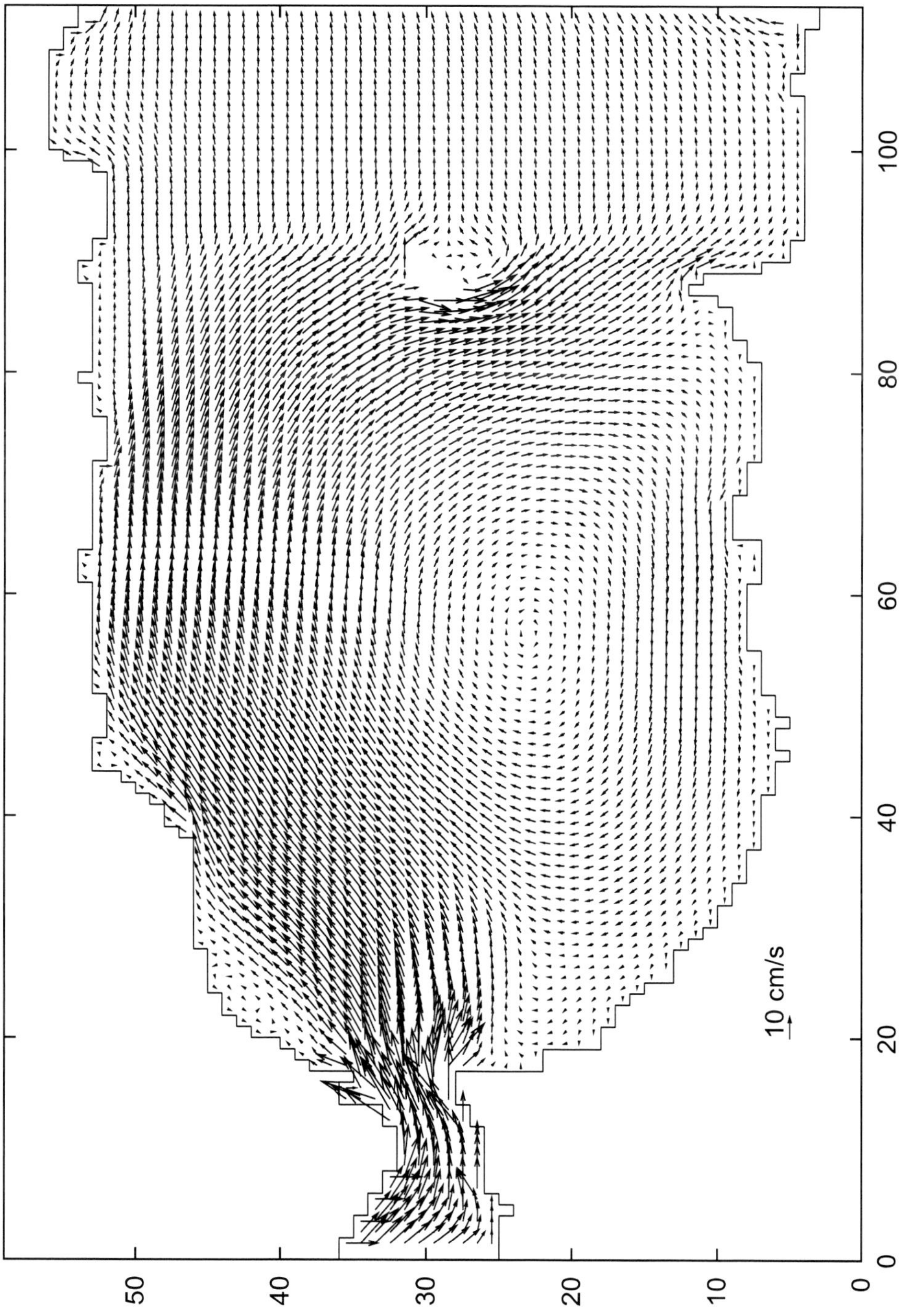

Figure 10. Steady currents computed with the reduced gravity model.

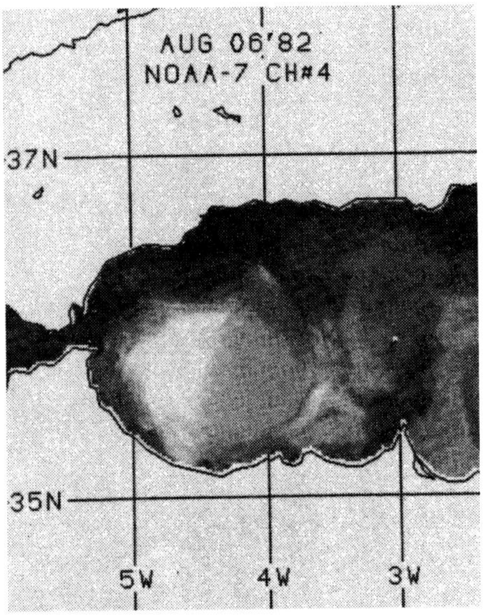

Figure 11. Classic image of the western Alborán gyre as seen by the thermal sensor of the NOAA 7 satellite. The darker tones refer to colder water, whereas lighter tones refer to warmer water [37].

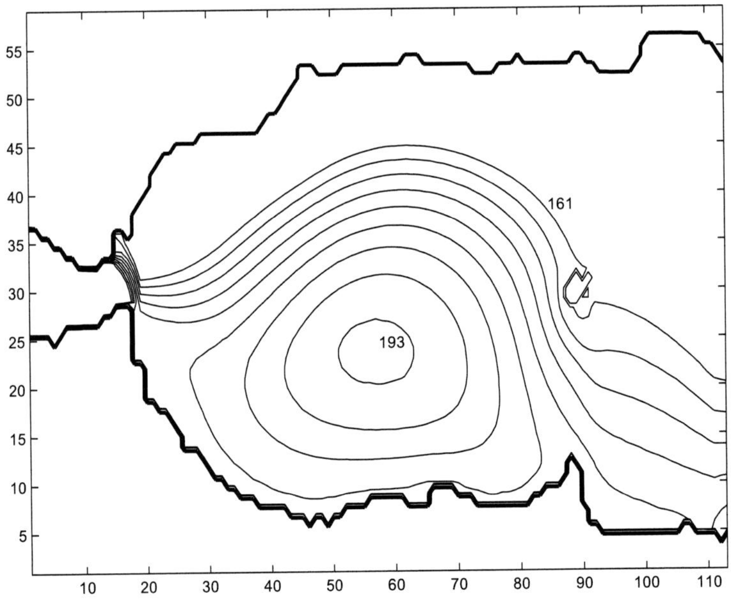

Figure 12. Thickness (m) of the upper water layer computed with the reduced gravity model. Distance between isolines is 4 m.

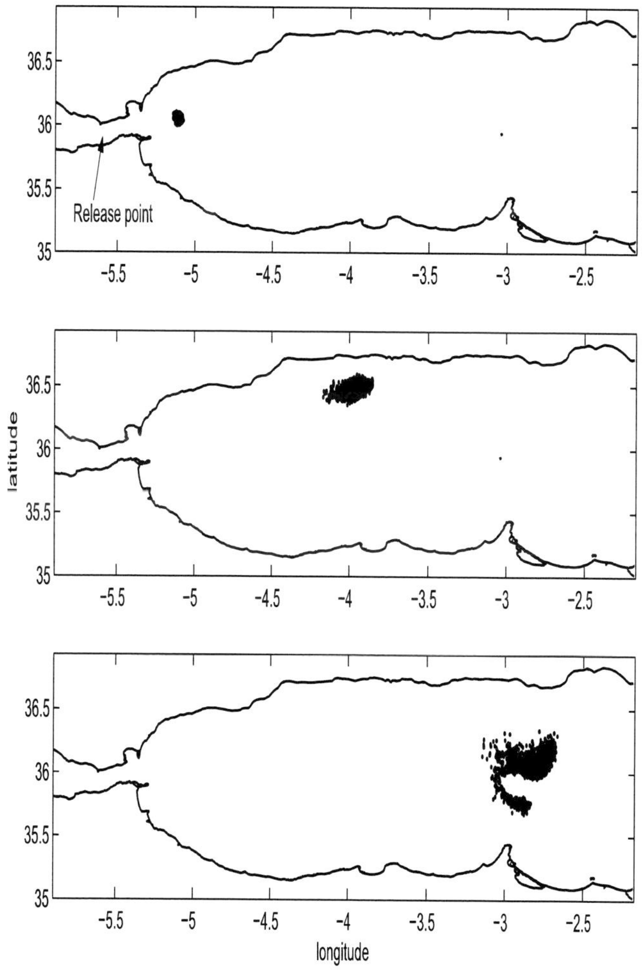

Figure 13. Snapshots showing the position of particles 2.5 (up), 12.5 (middle) and 27.5 (down) days after an instantaneous release in the central part of the Strait.

in longer simulations due to the longer time elapsed between snapshots.

A release of the same characteristics as before has been introduced in cell (20,28) at the same time and date, which would correspond to an accident occurring close to the harbour of Ceuta. The patch of course moves to the east, but in this case the complete patch crosses through the passage between Cape Tres Forcas and Alborán Island. The evolution of the patch may be seen in figure 14.

Another experiment consisted of simulating an accident in the central part of Alborán Sea [cell (55,30)]. Now all particles remain trapped in the gyre. Indeed, the time required to complete a loop is of the order of 60 days. Some results may be seen in figure 15, where the anticyclonic rotation of the patch can be observed. As an example, the map of concentrations, computed from the density of particles per water volume unit, is presented in figure 16.

Some examples of model results in the case of oil spills are presented now. Values

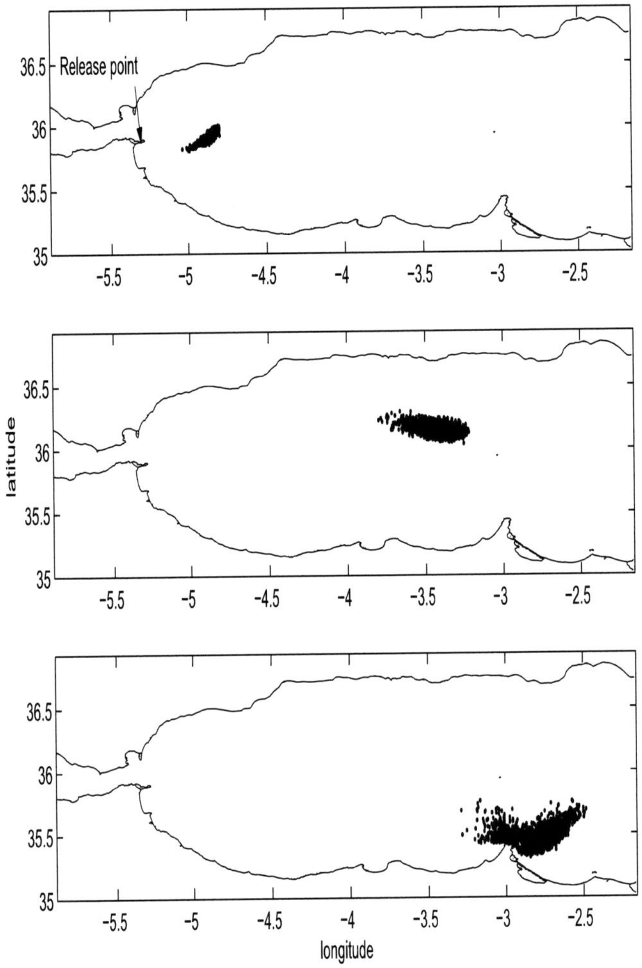

Figure 14. Snapshots showing the position of particles 4.2 (up), 20.8 (middle) and 37.5 (down) days after an instantaneous release occurring off the Ceuta harbour.

for the parameters required by the oil spill model must be set first. Oil density typically ranges from 660 kg/m^3 in the case of paraffin to 1200 kg/m^3 in the case of aromatic (policyclic) and naphtheno-aromatic oils [34]. A density equal to 900 kg/m^3 will be used in the simulations, that is between the value 870 kg/m^3 used in some calculations carried out for the Arabian Gulf spills of 1991 [6], and the value 950 kg/m^3 used in some hypothetical oil spill simulations for the Irish Sea [33].

The oil droplet size is assumed to be in the range $60 - 600$ μm, that are typical values [6, 33, 34]. The same values for the oil e-folding times as in [6] have been used: 25 h and 250 h for evaporation and decomposition respectively. It has finally been taken that evaporation occurs only if the age of the droplet is smaller than three times the evaporation e-folding time.

An example of results is presented in figure 17, that also illustrates the case of a continuous release. In this case, the accident originates an oil spill off the harbour of Gibraltar

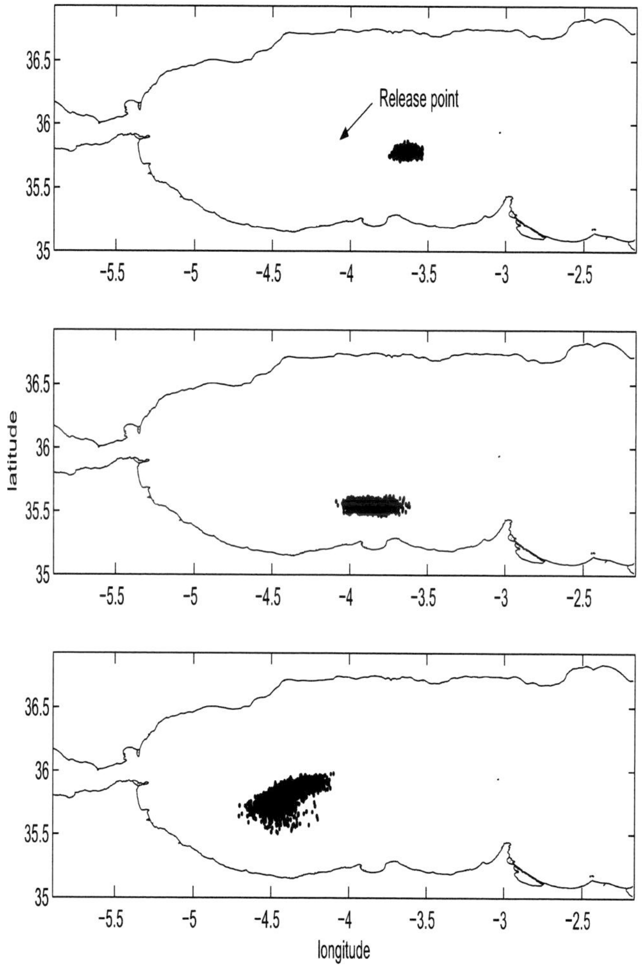

Figure 15. Snapshots showing the position of particles 15 (up), 30 (middle) and 55 (down) days after an instantaneous release in the central part of the Alborán Sea.

[cell (17,35)] again at 0:00 hours on January 1, 2005. The release lasts for 5 days and the movement of oil droplets is computed during 30 days. A very elongated patch, in the direction of the mean current, that travels along the Spanish coast is observed. As can be seen in the lowest panel of figure 17, the coast is contaminated from -5 to -4 degrees, in longitude, approximately. The area of Gibraltar is contaminated as well. During the 5 days of release the total number of particles that are introduced is 36000, at a rate of 25 new particles each time step. Only 215 remain in water and 520 are beached after one month. This implies that 1.4 % of the released oil has reached the coast and that 0.6 % is still being transported by water. The remaining oil has decayed or has been evaporated.

The effects of winds will be illustrated in what follows. The simulation shown in figure 13 (stable chemical pollutant release occurring at the central part of the Strait of Gibraltar) has been repeated but with a 10 m/s wind blowing from the east. Results are presented in figure 18. If this figure is compared with figure 13, it can be seen that the wind tends to

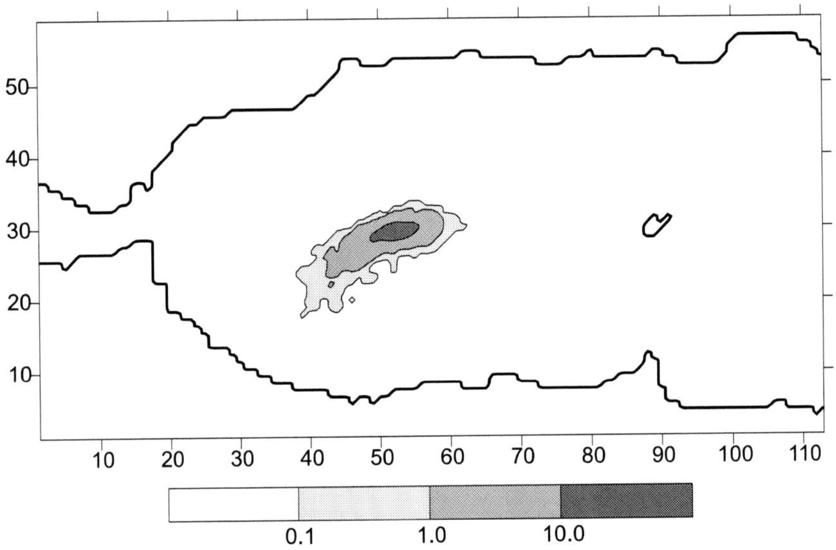

Figure 16. Concentration map (units/m^3) obtained 60 days after the release in figure 15. Numbers in the axes give the grid cell number.

retain the contamination inside the Alborán Sea for a longer time. Also, a tail of particles is produced along the Spanish coast and some contamination is retained in the cyclonic gyre existing north of Gibraltar. Thus, this region would remain contaminated for a longer time as a consequence of the wind. Finally, it can also be seen that more intense mixing of contamination along a south-north direction is produced than under calm conditions. West winds, in the same direction as the average currents, produce the opposed effect: a faster contaminant flushing off from the region.

As another example of the results that may be obtained with the model, the time evolution of the number of particles at compartment $(40, 45)$, that is located close to the town of Málaga, in the Spanish coast, is presented in figure 19 for the same experiment (east wind). It can be seen that contamination arrives 200 hours after the accident and that the maximum number of particles (thus maximum concentration) is obtained after 232 hours. The patch completely crosses after 278 hours. Since in an instantaneous release simulation 3000 particles are used and the total amount of contaminant released is 1×10^{12} units, each particle corresponds to 3.3×10^8 units. The maximum number of particles is 29, that implies a maximum concentration equal to 6.7 units/m^3.

If the accident occurred more in the south, for instance in the passage between Cape Tres Forcas and Alborán Island, the patch is divided in two branches by the east winds: one is flushed to the east by the residual current and the other is transported to the west and then trapped in the gyre.

It must be pointed out that it is not realistic to have these wind conditions persisting without change during such a long time. However, the aim of the experiment is simply to show the effect of winds on the behaviour of the contamination patch.

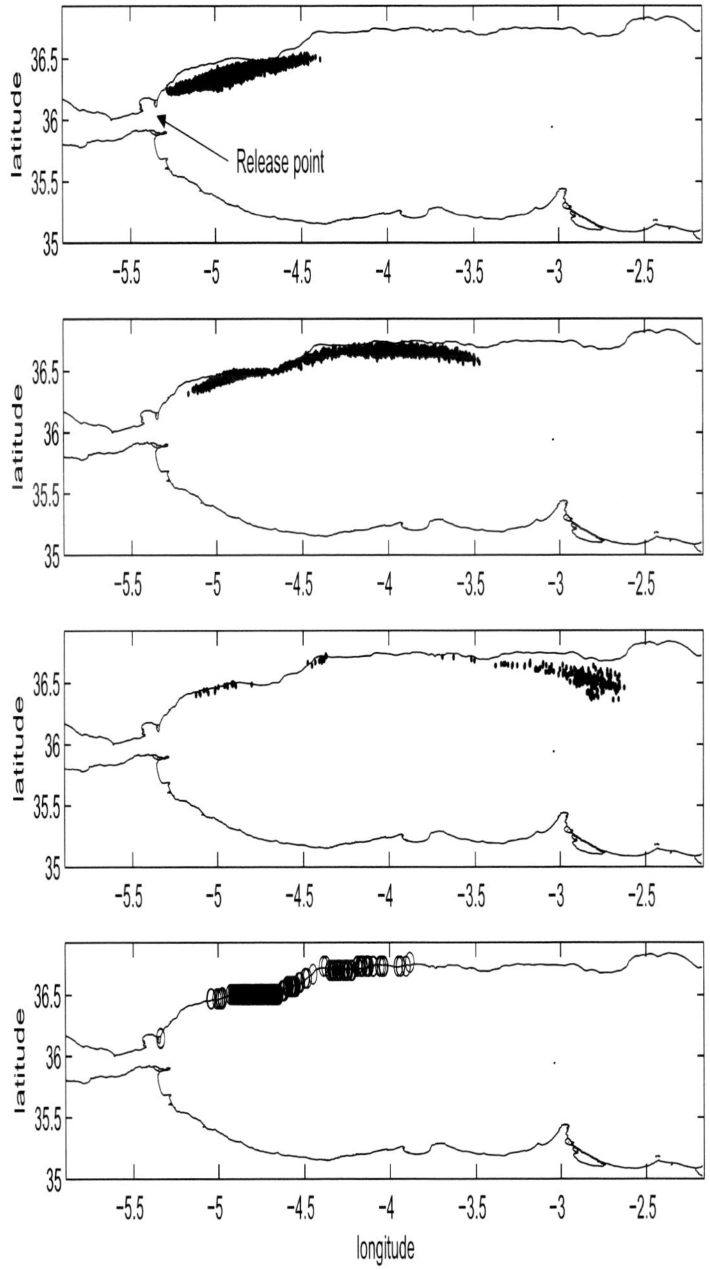

Figure 17. Snapshots showing the position of particles 7.5, 15 and 27.5 (from top to bottom) days after an oil release occurring off Gibraltar harbour. Circles in the lowest panel show particles that have been beached during the simulation.

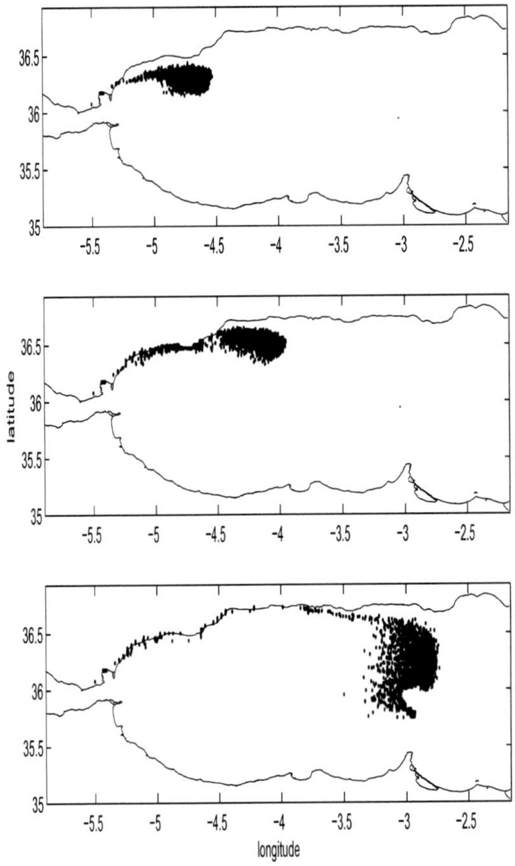

Figure 18. Snapshots showing the position of particles 2.5 (up), 12.5 (middle) and 27.5 (down) days after an instantaneous release of a stable chemical contaminant in the central part of the Strait. A 10 m/s wind blowing from the east is included.

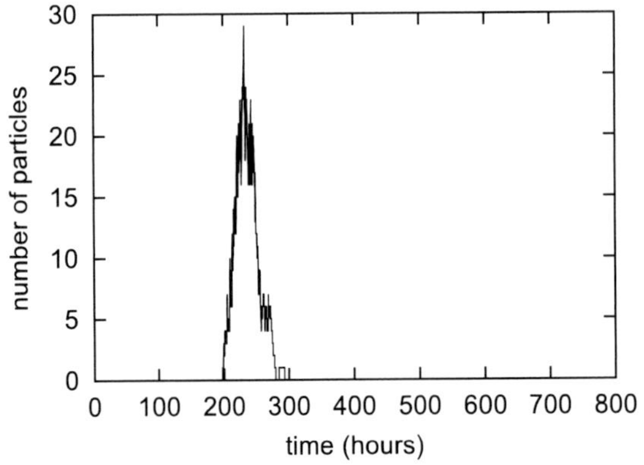

Figure 19. Time evolution of the number of particles at compartment (40,45) for the east wind experiment.

5. Conclusion

A particle-tracking model that simulates the dispersion of chemical, radioactive and oil spills in the Alborán Sea has been developed. The contamination release is simulated by a number of particles whose paths are computed. Diffusion and decay are simulated by a Monte Carlo method. Specific processes for oil spills must also be included: buoyancy, evaporation and beaching of oil droplets. Advection is calculated from a simple equation that presents the advantage over finite difference approaches of not introducing artificial numerical dispersion. The currents required to calculate the advective transport are obtained from appropriate hydrodynamic models. Contaminant concentrations are obtained from the density of particles per water volume unit.

A 2D barotropic model is used to calculate tidal currents over the domain. This model provides the tidal constants that are used by the dispersion model to reconstruct the tidal current at any time and position in the model domain. This model has been validated through the comparison of computed tidal elevations and phases with measurements in the area.

The residual circulation is obtained from a reduced gravity model. This is the simplest approach able to simulate the major features of circulation above the pycnocline. Essentially it is a two-layer model in which the lower layer is infinitely deep and at rest. It was not possible to carry out a quantitative validation of this model, but computed currents are, at least, in qualitative agreement with observations available in the region. In particular, the main circulation feature (the western Alborán gyre) is adequately reproduced by the model. Equations in both the barotropic and reduced gravity models are solved by standard explicit finite difference schemes.

Some experiments concerning hypothetical releases in the region have been carried out. The average speed of a contamination patch released in the Strait of Gibraltar as it travels towards the eastern boundary of the domain is of some 9 cm/s. If the accident occurs in the central part of the Alborán Sea, particles are trapped in the gyre. The time required to complete a rotation has been estimated to be of the order of two months.

The simulation of a continuous release has been illustrated with the case of an oil spill occurring off Gibraltar harbour. The model predicts that a significant portion of the Spanish coast is contaminated.

Winds have a significant effect on the dispersion patterns in the area. Thus, dominant east winds tend to retain contaminants inside the Alborán Sea, also enhancing mixing along the transverse direction (north-south). The Spanish coast is more affected by contamination than under calm conditions. On the other hand, west winds produce a faster cleaning of the sea.

As a final conclusion, it can be said that this kind of models is a very useful tool that may be used to support the decision making process in response to an emergency situation. But models can contribute to improve the knowledge of the physical oceanography of the studied area as well.

References

[1] Harms, I.H., Karcher, M.J. and Dethleff, D. *J. Mar. Syst.* **27**, 95 (2000).

[2] Gomez-Gesteira, M., Montero, P., Prego, R., Taboada, J.J., Leitao, P., Ruiz-Villareal, M., Neves, R. and Perez-Villar, V. *Oceanol. Acta* **22**, 167 (1999).

[3] Schonfeld, W. *J. Mar. Syst* **6**, 529 (1995).

[4] Periáñez, R. and Elliott, A.J. *J. Environ. Radioactivity* **58**, 13 (2002).

[5] Nakano, M. and Povinec P.P. *J. Environ. Radioactivity* **69**, 85 (2003).

[6] Proctor, R., Flather, R.A. and Elliott, A.J. *Cont. Shelf Res.* **14**, 531 (1994).

[7] Proctor R., Elliott, A.J. and Flather, R.A., *Mar. Poll. Bull.* **28**, 219 (1994).

[8] Elliott, A.J., Wilkins, B.T. and Mansfield, P. *Mar. Poll. Bull.* **42**, 927 (2001).

[9] Echevarría, F., García-Lafuente, J., Bruno, M., Gorsky, G., Goutx, M., González, N., García, C.M., Gómez, F., Vargas, J.M., Picheral, M., Striby, L., Varela, M., Alonso, J.J., Reul, A., Cozar, A., Prieto, L., Sarhan, T., Plaza, F. and Jiménez-Gómez, F. *Deep Sea Res. II* **49**, 4115 (2002).

[10] Tsimplis, M.N. and Bryden, H.L. *Deep Sea Res.* **47**, 2219 (2000).

[11] Perkins, H., Kinder, T., and La Violette, P. *J. Phys. Oceanog.* **20**, 242 (1990).

[12] Werner, F.E., Cantos-Figuerola, A. and Parrilla, G. *J. Phys. Oceanog.* **18**, 373 (1988).

[13] Preller, R.H. *Prog. Oceanog.* **16**, 113 (1986).

[14] Mañanes, R., Bruno, M., Alonso, J., Fraguela, B. and Tejedor, L. *Oceanol. Acta* **21**, 33 (1998).

[15] Tejedor, L., Izquierdo, A., Kagan, B.A. and Sein, D.V. *J. Geophys. Res.* **104**, 13541 (1999).

[16] Tsimplis, M.N., Proctor, R. and Flather, R.A. *J. Geosphys. Res.* **100**, 16223 (1995).

[17] Jensen, T.G. *J. Mar. Syst.* **16**, 297 (1998).

[18] Flather R.A. and Heaps N.S. *Geophys. J. Royal Astron. Soc.* **42**, 489 (1975).

[19] Elliott, A.J. and Clarke, S. *Hydrograp. J.* **87**, 19 (1998).

[20] Periáñez, R. Modelling the Dispersion of Radionuclides in the Marine Environment; Springer-Verlag: Heidelberg, 2005.

[21] Pugh, D.T. Tides, Surges and Mean Sea Level; Wiley: Chichester, 1987.

[22] Riddle, A.M. *Cont. Shelf Res.* **18**, 441 (1998).

[23] Elliott, A.J. *German Journal of Hydrography* **39**, 113 (1986).

[24] Dick, S. and Schonfeld, W. *German Journal of Hydrography* 48, 27 (1996).

[25] Elliott, A.J. *Spill Sci. Tech. Bull.* **5**, 39 (1999).

[26] León-Vintró, L., Mitchell, P.I., Condren, O.M., Downes, A.B., Papucci, C. and Delfanti, R. *Sci. Tot. Environ.* **237**, 77 (1999).

[27] Periáñez, R. *Mar. Poll. Bull.* **49**, 613 (2004).

[28] Cummins, P.F. and Lagerloef, G.S.E. *Deep Sea Res. I* **51**, 2105 (2004).

[29] Chern, C.S. and Wang, J. *J. Oceanography* **61**, 271 (2005).

[30] Morales-Maqueda, M.A., Willmott, A.J. and Darby, M.S. *Clim. Dynam.* **15**, 89 (1999).

[31] Arruda, W.Z., Nof, D. and O'Brien, J.J. *Deep Sea Res. I* **51**, 2073 (2004).

[32] Kowalick, Z. and Murty, T.S. Numerical Modelling of Ocean Dynamics; World Scientific: Singapore, 1993.

[33] Elliott, A.J. *Estuar. Coastal Shelf Sci.* **61**, 569 (2004).

[34] Korotenko, K.A., Mamedov, R.M., Kontar, A.E. and Korotenko, L.A. *J. Mar. Syst.* **48**, 159 (2004).

[35] Periáñez, R. *Estuar. Coastal Shelf Sci.*, **54** 809 (2002).

[36] Manzella, G.M.R. and Elliott, A.J. *Mar. Poll. Bull.* **22**, 553 (1991).

[37] Heburn, G.W. and La Violette, P.E. *J. Geophys. Res.* **95C2**, 1599 (1990).

[38] Bryden, H.L., Candela, J. and Kinder, T.H. *Prog. Oceanog.* **33**, 201 (1994).

[39] Beranguer, K., Mortier, L. and Crepon, M. *Prog. Oceanog.* **66**, 341 (2005).

[40] Vargas-Yáñez, M., Plaza, F., García-Lafuente, J., Sarhan, T., Vargas, J.M. and Vélez-Belchi, P. *J. Mar. Syst.* **35**, 229 (2002).

[41] Gascard, J.C. and Richez, C. *Prog. Oceanog.* **15**, 157 (1985).

[42] Arnone, R.A., Wiesenburg, D.A. and Saunders, K.D. *J. Geophys. Res.* **95C2**, 1587 (1990).

[43] Sannino, G., Bargagli, A. and Artale, V. *J. Geophys. Res.* **107C8**, 3094 (2002).

INDEX

A

D

E

F

G

H

I

J

K

L

M

Q

R

S

U

V

W

X

Y

Z